AF587354

POWDER METALLURGY RESEARCH TRENDS

POWDER METALLURGY RESEARCH TRENDS

LOTTE J. SMIT
AND
JULIA H. VAN DIJK
EDITORS

Nova Science Publishers, Inc.
New York

For permission to use material from this book please contact us:
Telephone 631-231-7269; Fax 631-231-8175
Web Site: http://www.novapublishers.com

LIBRARY OF CONGRESS CATALOGING-IN-PUBLICATION DATA

Powder metallurgy research trends / Lotte J. Smit and Julia H. van Dijk (editor).
p. cm.
ISBN 978-1-60456-852-3 (hardcover)
1. Powder metallurgy. I. Smit, Lotte J. II. Dijk, Julia H. van.
TN695.P674 2008
671.3'7--dc22

2008023079

Published by Nova Science Publishers, Inc. ✢ New York

CONTENTS

PREFACE

Powder metallurgy is a forming and fabrication technique consisting of three major processing stages. First, the primary material is physically powdered, divided into many small individual particles. Next, the powder is injected into a mold or passed through a die to produce a weakly cohesive structure (via cold welding) very near the dimensions of the object ultimately to be manufactured. Finally, the end part is formed by applying pressure, high temperature, long setting times (during which self-welding occurs), or any combination thereof. In powder metallurgy or ceramics it is possible to fabricate components which otherwise would decompose or disintegrate. All considerations of solid-liquid phase changes can be ignored, so powder processes are more flexible than casting, extrusion, or forging techniques. Controllable characteristics of products prepared using various powder technologies include mechanical, magnetic, and other unconventional properties of such materials as porous solids, aggregates, and intermetallic compounds. This book provides new research on this field from around the globe.

Chapter 1 - The corrosion behavior of finely dispersed powders based on Fe and Fe-Si in 0.85% NaCl solution at 37^0C has been studied. The powders were prepared by mechanical milling in organic media and in the presence of graphite. It has been demonstrated that together with the increase of dispersion the corrosion stability of the powders can grow in the process of mechanoactivation. First of all, it can be the case when a surface protective layer is formed when milling occurs in the presence of long-chained surfactants such as oleic acid. If iron is milled in the absence of long-chained surfactants, for example in acetic acid, isopropyl alcohol, and vinyltriethoxisilane, the surface protective characteristics are much worse despite the formation of thick oxide films and doping of the surface layer with silicon. Second, a considerable corrosion stability increase can result from the formation of amorphous phases (Fe-Si-C) in particle bulks in the process of mechanoactivation. At the same time, the increase of the carbide phase content, the encapsulation of the particles in graphite, and the change of grain size (5-100 nm) do not essentially influence the powder corrosion stability.

The second part of the paper is focused on the corrosion-electrochemical behavior of Fe-C powders prepared by dynamic compaction at 500^0C, which represented nanocomposites (grain size 40 nm) α-Fe+Fe_3C with the cementite content starting from 9-92 wt.%. It has been stated that the pitting corrosion stability in borate solutions with the admixture of NaCl increases stepwise with the increase of Fe_3C (~50 wt.%) content or with the formation of the ordered net-like cementite structure.

Chapter 2 - Fine and ultrafine powders in the form of metal, ceramic, pharmaceutical, and food ingredients have numerous applications in manufacturing and powder metallurgy, development of sensors, thermal barrier coatings, catalysts, pigments, drugs, and biotechnology, to name a few. There are several methods of powder production, such as wet chemistry, e.g., sol-gel and emulsion, mechanical methods, e.g., ball milling, gas phase methods, e.g., chemical and physical vapor deposition, liquid phase spray methods, e.g.,molten metal spray atomization, spray pyrolysis, and spray drying, and liquid/gas phase methods, e.g., flame spray pyrolysis.

Chapter 3 - Factors determining kinetics of physical and chemical conversion, structure of reaction product, and parameters of dynamically loaded reacting powder body surface radiation are investigated by the computer simulation method. The model takes into account structural parameters of the powder material capable of exothermal chemical reactions, an opportunity for phase transfers, formation of a product of chemical reaction, change of an aggregative state of the reacting medium, mechanical activation of the reacting components, the forced filtration of the molten component of a mixture, and change of parameters of a condition on each step of physical and chemical conversions. The algorithm of the numerical solutions of imitating simulation problems of physical and chemical behavior of reactive powder materials has been developed on the basis of the created model, providing the solution of the related problems of a macrokinetics of chemical transformations, mechanical modification of powder materials during dynamic loading, thermal balance and a filtration of a liquid phase. All parameters of the model are repeatedly specified during each step of time. The method for interpretation of experimental results of investigation of physical and chemical processes in the multicomponent reacting systems received by optical pyrometry methods for an estimation of a degree of realization of various stages of physical and chemical conversions, structure modification and other state parameters of reacting powder body, registration of changes of chemical reactions stages, both on the surfaces and in the bulk of the model compact has been developed. The governing factors of researched processes have been investigated during the computer simulation results analysis.

Chapter 4 - Powder metallurgy techniques have been extensively used to fabricate net shape components, and are popular in the current trends in materials processing. Laser Engineered Net Shaping (LENS®) is one of the fastest growing laser metal deposition processes, which has combined high power laser deposition and powder metallurgy technologies with advanced methodologies of rapid prototyping, to convert complex virtual objects (CAD solid models) into functional advanced structural components without the need for part-specific tooling. It is an additive versus a subtractive material forming process. With LENS® processing, the microstructure can be tailored by controlling both process parameters and composition. In addition to fabricating complex geometries of fully dense metals that require minimal finish machining, the mechanical properties have typically been found to be superior to those of components obtained by conventional processes. In this chapter, recent research and progress associated with the LENS® process are reviewed, such as laser materials processing and rapid manufacturing; metal alloys, composites, and cermets development with the LENS® process; the effect of process parameters on microstructure, properties, and build of part height; the application of *in-situ* molten pool sensor and Z-height control subsystems; thermal behavior measurement with thermal imaging and thermocouple methods, and numerical simulation; as well as a cost-benefit analysis of the LENS® process.

Special emphasis is placed on LENS® deposited materials and process control, along with trends and challenges of laser direct fabrication of bulk metallic glasses and nanocrystalline materials, and analysis of residual stress and porosity in deposited materials.

Chapter 5 - In order to further disclose the indirect SLS process, Forming mechanisms and behaviors of selective laser sintering of metal/polymer composite powders within one laser beam point, one laser scanning line and one laser scanning domain are mainly theoretically investigated. Some mathematical and physical models are built to explain these mechanisms and behaviors. It is indicated that all forming behaviors result from the same thermodynamic mechanism which is the decreasment of powders surface energy. The interaction between rigid powders and variable materials decides the energy states during a certain period. Mover, the variation of forming behaviors in laser beam spots, scanning lines and scanning areas are affected by energy distribution of laser beam spot, laser energy accumulation in the prolonged orientation of scanning line and powder material's effective energy density absorbed from laser respectively. The difference between the initial surface energy and final one is the indicate how stable the final mixed material system is. This work enlarges the theory of laser forming during indirect SLS and provides basis for indirect SLS process, which is a good contribution to the development of indirect SLS technology.

Chapter 6 - In order to evaluate distributions of mechanical properties and fracture toughness in ceramic-metal FGMs, mechanical properties and fracture toughness have been investigated on ceramic-metal non-graded composites (non-FGMs) and FGMs. The materials were fabricated by powder metallurgy using partially stabilized zirconia (PSZ) and austenitic stainless steel (SUS 304). Vickers hardness, Young's modulus and bending fracture strength are examined on smooth specimens of non-FGMs. The Vickers hardness continuously decreases with an increase in a volume fraction of SUS 304 metal phase, while the Young's modulus and fracture strength exhibit the low value in the non-FGMs with balanced composition of each phase. This suggests that the interfacial strength between the ceramic and metal phases is very low. The fracture toughness is determined by conventional tests for several non-FGMs with each material composition and by a method utilizing stable crack growth in FGMs. In contrast with the Young's modulus and fracture strength, the fracture toughness increases with an increase in a content of SUS 304 on both FGMs and non-FGMs and it is higher in the FGMs than in the non-FGMs. Under the assumption that the high fracture toughness in FGMs is attributed to the residual compressive stress in the ceramic-rich region created in the fabrication process, the residual stress distribution in the FGMs is estimated from the difference in fracture toughness between the FGMs and non-FGMs.

Chapter 7 - Conventionally, hardness is a primary parameter to determine the wear resistance of high volume ceramic reinforced metal matrix composites. However, this study reports that the friction coefficient and wear characteristics of these metal matrix composites are strongly controlled by their microstructural parameters and are independent of their hardness. Ratio of average grain size to the mean free path of binder phase was found to be a controlling microstructural parameter for wear performance. Metal matrix ceramic reinforced composites with higher ratio of grain size to mean free path of the binder phase can exhibit higher wear resistance and lower friction coefficient.

Chapter 8 - Reactive hot pressing (RHP) is widely recognized as an effective technique to prepare dense metal matrix composites (MMCs) reinforced with in-situ ceramic particulates/whiskers in one processing step. Ultrafine ceramic reinforcements are formed in-situ by exothermic reactions between elemental and/or compound powders during hot pressing. They are distributed uniformly within the metal matrix of MMCs. Moreover, the reinforcement-matrix interfaces are clean, resulting in a strong interfacial bonding. These enable effective load transfer from the metal matrix to ceramic reinforcements during mechanical loadings. Inherently, the in-situ MMCs exhibit improved mechanical strength, rendering them potential candidate materials for structural engineering applications. In the present review article, recent developments in the structure and mechanical properties of the RHP prepared aluminum, copper and titanium-based composites reinforced with in-situ ceramic particulates/whiskers are addressed and discussed. Particular attention is paid to the beneficial strengthening effects of in-situ TiB_2 particulates or TiB whiskers in the composites.

Chapter 9 - A simple route to produce high strength bulk nanostructured aluminium is presented in this work,. Firstly, a thermally stable nanocomposite Al powder is prepared by attrition milling of elemental Al in the presence of ammonia gas or urea. Both substances are dissociated during milling, enriching the nitrogen and carbon content of aluminium powder.

Then, the milled powder is consolidated to full density by a conventional press-and-sinter powder metallurgy (P/M) technique. Sinterability was improved by using copper powder, as additive, during milling. Shaped specimens possess a stable microstructure, with a grain size of about 150 nm, and their ultimate tensile strength (550MPa) surpass that of most commercial Al alloys, especially at high temperatures.

This new consolidation method may be of interest with a view to manufacturing large batches of small parts, a typical task for the automotive industry.

In: Powder Metallurgy Research Trends
Editors: L. J. Smit and J. H. Van Dijk

ISBN: 978-1-60456-852-3

Chapter 1

FORMATION OF CORROSION STABILITY OF FE-BASED POWDERED AND COMPACTED MATERIALS

A.V. Syugaev and S.F. Lomayeva
Physicotechnical Institute UrBr RAS, 132 Kirov st., Izhevsk, Russia

ABSTRACT

The corrosion behavior of finely dispersed powders based on Fe and Fe-Si in 0.85% NaCl solution at 37^0C has been studied. The powders were prepared by mechanical milling in organic media and in the presence of graphite. It has been demonstrated that together with the increase of dispersion the corrosion stability of the powders can grow in the process of mechanoactivation. First of all, it can be the case when a surface protective layer is formed when milling occurs in the presence of long-chained surfactants such as oleic acid. If iron is milled in the absence of long-chained surfactants, for example in acetic acid, isopropyl alcohol, and vinyltriethoxisilane, the surface protective characteristics are much worse despite the formation of thick oxide films and doping of the surface layer with silicon. Second, a considerable corrosion stability increase can result from the formation of amorphous phases (Fe-Si-C) in particle bulks in the process of mechanoactivation. At the same time, the increase of the carbide phase content, the encapsulation of the particles in graphite, and the change of grain size (5-100 nm) do not essentially influence the powder corrosion stability.

The second part of the paper is focused on the corrosion-electrochemical behavior of Fe-C powders prepared by dynamic compaction at 500^0C, which represented nanocomposites (grain size 40 nm) α-Fe+Fe_3C with the cementite content starting from 9-92 wt.%. It has been stated that the pitting corrosion stability in borate solutions with the admixture of NaCl increases stepwise with the increase of Fe_3C (~50 wt.%) content or with the formation of the ordered net-like cementite structure.

INTRODUCTION

In recent years, interest in nanocrystalline metal materials has been spurred by the discovery that a decrease in the crystallite size below a certain threshold value can dramatically change the properties of crystallites. Such effects arise when the average grain size does not exceed 100 nm and are most pronounced for sub-10-nm grains. Because of such small sizes, the specific area of the grain boundaries is very high and developed, and hence, nanocrystalline materials significantly differ in their properties from common polycrystals [1].

One of the simplest and most effective ways of producing nanocrystalline materials is mechanoactivation (mechanical milling and fusion) [2, 3]. A variety of processes that occur during its performance result in the production of nanocrystalline composite materials and make it possible to control their structural, phase, morphological, thermal and other properties. The Fe-based powders thus prepared can be applied as magnetic data carriers, magnetic drug carriers, in magnetic liquids, etc.

It is known that ductile metals are milled only in the presence of adsorbents [4-6]. Mechanoactivation in organic media both increases the dispersivity of powders and stabilizes them by forming a protective surface layer that can vary in composition and structure depending on a milling medium and time [7-11]. The powders prepared by mechanoactivation can be used as precursors for producing compacted nanocomposites.

The focus of the present study has been the influence of phase composition and surface layer structure on the corrosion behavior of nanocrystalline Fe and Fe-Si powders formed in the course of milling in organic media. Namely, the influence of the cementite content and its nanocrystalline state on the electrochemical characteristics of α-Fe+Fe_3C composites prepared by compacting powders mechanoactivated in carbon-containing media will be studied.

MATERIALS AND METHODS

The powders were prepared by mechanical milling of carbonyl iron, containing less than 0.03wt.% of carbon, or a Fe-Si alloy (20at.%Si) in a Pulverizette 7 planetary ball mill in the following organic media:

1. Fe or Fe-Si in heptane alone and heptane with oleic acid (0.3 wt %). The milling time (t_{mil}) varied from 1 to 99 h;
2. Fe in heptane with triethoxy(vinyl)silane (0.3 wt %), t_{mil}=1-99 h;
3. Fe in oleic acid and triethoxy(vinyl)silane (10wt.% solutions in heptane, 100%), in acetic acid and isopropyl alcohol (100%), t_{mil}=24 h. An excess of oleic acid taken in concentrations 10% and 100% was removed by a multiple washing of the powders in heptane;
4. Fe with graphite in the atomic ratio 85:15 in argon atmosphere; t_{mil}=16 h.

Vessels and balls were made of ShKhl5 steel containing C 1% and Cr 1.5% in order to minimize external contamination of the powders during pulverization. Forced air cooling was used to prevent the vessels from being heated above 60^0C during milling.

To prepare α-Fe+Fe_3C composites, powders obtained by milling Fe with graphite taken in proportion $Fe_{100-x}C_x$ (x=5, 10, 15, 20, 25 at.%) in Ar medium (t_{mil}= 16 h) as well as by milling Fe in 0.3 weight % vinyltriethoxisilane (t_{mil}= 48 h) were used. The powders were compacted by a magnetoimpulse compacting method [12], which makes it possible to prepare materials whose densities approach theoretical values and which preserve their nanocrystal state due to relatively low heating temperatures (500^0C). The compacting was attended with a continuous pump-down using a high-pressure impulse 3 GPa in amplitude and 300 μs in duration. Before compacting the samples had been subjected to vacuum degassing and held at the temperature of the following compaction (500^0C) for an hour. The samples obtained were disks with a diameter of 15 mm and thickness from 2 to 4 mm.

The powders were annealed under argon at 400, 500 and 800^0C for an hour. The annealing scheme involved heating to a given temperature at a rate of 60 K/min, holding at this temperature, and cooling at a rate of 100 K/s.

X-ray diffraction analysis was performed on a DRON-3M diffractometer (FeK_α radiation). When detecting structural and substructural parameters, the shapes of the $K_{\alpha 1}$ and $K_{\alpha 2}$ reflections were approximated by the Voigt function. Carbonyl iron powder annealed at 850^0C for 2 h was used as a standard. The mean grain size was determined by harmonic analysis [13].

Mössbauer resonance studies were carried out on a YaGRS-4M spectrometer (constant acceleration mode, ^{57}Co in the Cr matrix as a γ-radiation source). The distribution functions P(H) of hyperfine magnetic fields (HFMFs) were determined from the spectra by using the generalized regular algorithm [14].

The analysis of the chemical states of the elements in the surface layers of particles was carried out by X-ray photoelectron spectroscopy (XPS) on an ES-2401 spectrometer with a magnesium anode. The spectrometer chamber was evacuated to a residual pressure of 10^{-5} Pa; line positions were measured to within 0.2 eV. The spectrometer was calibrated against the $Au4f_{7/2}$ line (84.0 eV). The bonding energy E_b of the line of C1s electrons in the alkyl group was taken to be 285.0 eV.

IR spectra were recorded on an FSM-1202 FTIR spectrometer (resolution 4 cm^{-1}) in the 1000÷4000 cm^{-1} range. The spectra of the powders obtained with oleic acid were recorded for the particles forming a sol in heptane. The sol was applied to CaF_2 windows and dried.

The size distributions of the particles were determined on an Analysette 22 laser diffraction microanalyzer. The surface morphology of α-Fe+Fe_3C composites was examined by atomic force microscopy with a P47-SPM-MDT scanning probe microscope.

Electron microscopy studies were performed with a Jeol JEM2000 FXII microscope (accelerating voltage 200 kV) and a JAMP-10S Auger spectrometer.

CORROSION AND ELECTROCHEMICAL TESTS

Just like meaningful real-life corrosion processes, the corrosion of the majority of finely dispersed Fe-based powders runs in neutral media. For example, the powders can be affected by atmospheric corrosion; and in the case of magnetic biomaterials the powders represent water-based pulp with pH close to 7. Therefore, saline (0.85% NaCl at 37°C) was selected as a model medium for corrosion tests.

As opposed to acid media, where direct monitoring of the evolving hydrogen volumes is possible, no straightforward methods of detecting corrosion losses in Fe-based finely dispersed powders exist. Circumstantial methods based on measuring magnetic characteristics are used here to estimate the stability of finely dispersed materials [15, 16].

To determine metal corrosion losses in neutral and alkaline media a weight method has been widely used. After the test all surface and pitting products of metal corrosion must be removed by acid etching in the presence of an inhibitor or by cathode polarization, with weight losses in unoxidized metal being as low as 3-5% of the total corrosion losses. Such an approach was shown inapplicable in the case with finely dispersed iron (300-500 nm) [17], since a few seconds was enough for the powder to dissolve in the etching solutions recommended, even if the inhibitor content was increased an order of magnitude.

A technique for investigating corrosion stability of Fe powders in neutral media was suggested in [17], which was based on determining the difference in the volumes of hydrogen evolved during dissolution of the sample in acid before and after the corrosion test. Hydrogen volumes were equivalent to the content of unoxidized metal in samples, while their difference – to corrosion losses.

In kinetic studies this method requires multiple tests to be taken with different times of samples exposure in corrosion medium. Furthermore, it is invalid for the investigation of the powders with the increased stability to acid etching, for example the ones based on Fe-Si alloys.

The basic reaction during iron corrosion in neutral media is oxygen reduction. This is why we suggest direct monitoring of the content of oxygen absorbed in the course of the powder corrosion. This approach is appropriate for acid-stable powders and does not require multiple tests for corrosion kinetics study. In case a protective layer is present on the powder particle surface, an incubation period is observed during which oxygen absorption speed is relatively small. The destruction of this stabilizing layer results in a more intense oxygen absorption, which is reflected on the absorbed oxygen volume – exposure time curve as a distinct bend. There is a correlation between the incubation period duration and the surface layer protective characteristics.

The speed of the powder corrosion after the protective layer destruction can be determined with a tangent of a tilt angle of straight areas of exposure time - absorbed oxygen volume curves. This technique is however not very exact, since not only metal is oxidized by oxygen, but iron (II) compounds formed during corrosion as well.

Electrochemical investigations of compacted composites α-Fe+Fe_3C were conducted in 1) borate buffer solution 0.3M H_3BO_3 (in the presence of 5M NaOH to reach pH = 7.4) 2) the same borate buffer solution + 0.01 M NaCl. The investigations were performed with a potentiometer IPC-Pro at room temperature in conditions of natural aeration. Chloride-silver and platinum electrodes were taken as reference and auxiliary electrodes, respectively. All the potentials are given with reference to a standard hydrogen electrode, and currents are converted to the visible area of the sample surfaces. Epoxy resin was used to isolate the samples. Distilled water was used to prepare the samples. Before the electrochemical tests the sample surfaces were smoothed out with abrasive paper and polished with Al_2O_3 wetted in distilled water.

Corrosion potential was determined and potentiometric studies were held after the electrode had been held in the solution for no less than 1 h until a steady state was reached. Anodic polarization was conducted starting with corrosion potential with the scan velocity –

0.5 mV/s. Cathodic activation of the samples was conducted at –0.8 V for 10 minutes. Local activation potential was determined at current density of 20 μA/cm^2.

RESULT AND DISCUSSION

1. Corrosion of finely dispersed iron-based systems

1.1. Milling in Heptane and Heptane with Addition (0.3wt.%) of Oleic Acid

The micrographs of the powders (t_{mil}=24 h) are shown in figure 1. Upon milling in heptane, particles have a stonelike form, whereas those milled in oleic acid solution represent thin (to 1 μm) plates. Figure 2 demonstrates the particle size distributions (t_{mil}=24 h). The mean size of the particles prepared in heptane at t_{mil}= 24 h is equal to 18 μm; in oleic acid solution , to 8 μm; and at t_{mil}= 99 h, to 30 and 4 μm, respectively. Over long-term treatments, the fine milling in heptane reaches its limit and a reverse process (coagulation) becomes prevalent. The use of a surfactant (oleic acid) preventing aggregation allows us to obtain small particles with narrow particle size distribution.

The phase composition of powders prepared by milling in both media over 1÷47 h is identical and does not differ from the initial composition: powders contain only one bcc-phase, although the lines in diffractograms are broader (figure 3). On milling over 99 h, peaks corresponding to cementite Fe_3C appear in diffractograms. The grain sizes were calculated by broadening the X-ray lines (figure 4a). After just one hour of milling in heptane, the grain size falls to 11 nm; after milling in oleic acid solution, to 24 nm; i.e., t_{mil} = 1 h was enough to obtain α-Fe in a nanocrystalline state. An increase in treatment time leads to a decrease in the grain size to 4 nm.

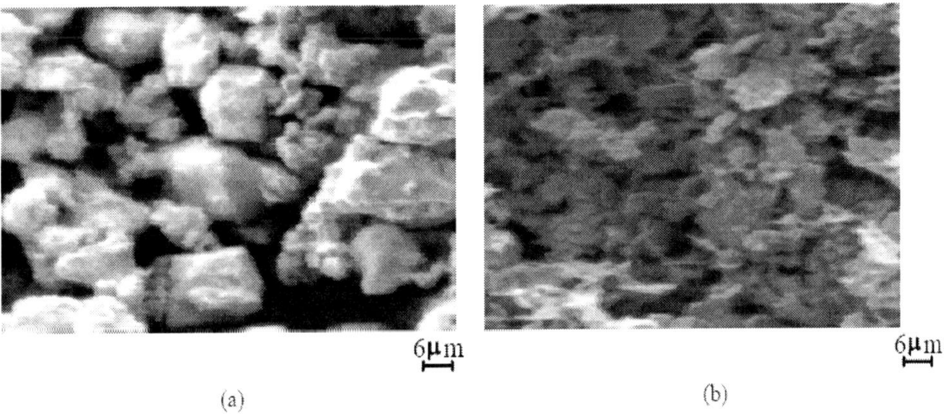

Figure 1. Micrographs of powders after 24 h milling in heptane alone (a) and heptane with oleic acid (b).

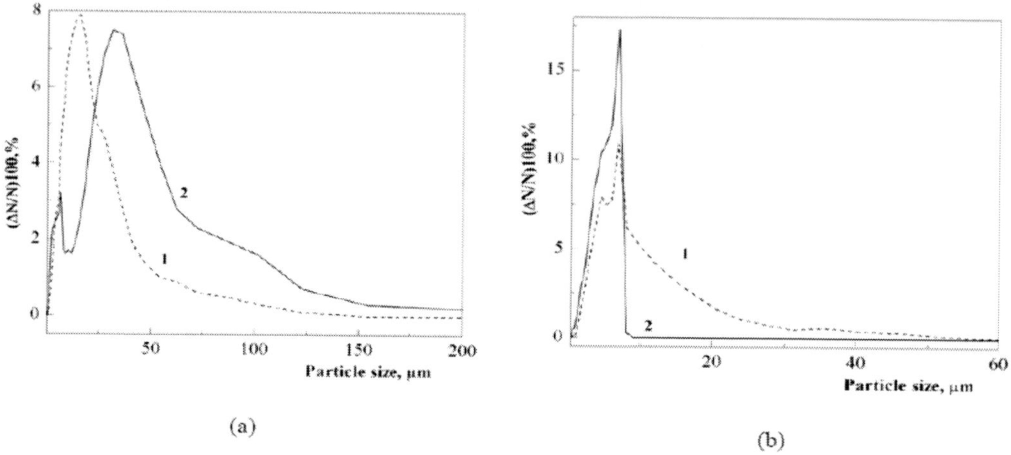

Figure 2. Size distributions of particles prepared by iron milling in heptane alone (a) and heptane with oleic acid (b) for (1) 24 h and (2) 99 h.

No changes (compared with the initial powder) are observed in Mössbauer spectra of the powder after milling over 1 h in heptane (figure 5). A new component with a lower value and wide distribution P(H) of HFMFs appears after milling for 3 h and more (Fig. 5b). Its appearance is clearly seen from the positions of the spectra with respect to the nonresonance level represented by horizontal solid lines (figure 5b). After the thermal treatment of the samples (heating up to 500^{0}C), a second sextet appears in the spectra of the powders milled over 3-47 h; the value of H=208 kOe obtained for this sextet agrees well with that for carbide Fe_3C [18, 19]. At t_{mil}=99 h, carbide Fe_3C begins to form directly in the course of milling.

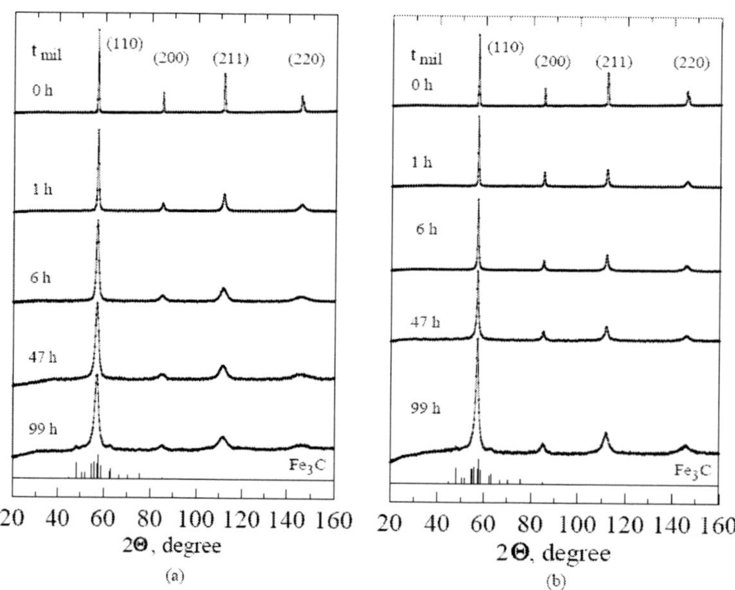

Figure 3. X-ray diffractograms of iron powder milled in heptane alone (a) and heptane with oleic acid (b) over various times. Indices of reflection of the bcc-structures are indicated at the top of the figure, positions of the most intense reflections of Fe_3C are shown at the bottom part of the figure.

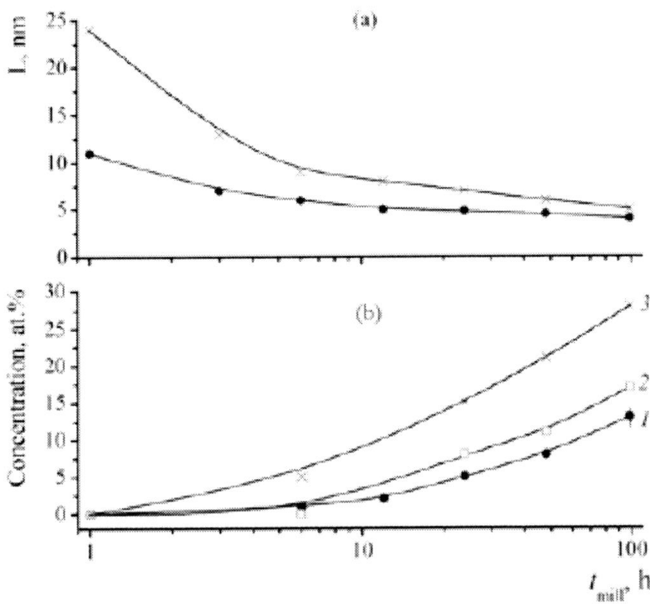

Figure 4. (a) The mean grain size ‹L› and (b) concentration of impurities (1 – carbon, 2 – oxygen, 3 – total concentration of carbon and oxygen) as function of the time of iron milling in heptane alone (solid points) and heptane with oleic acid (crosses, squares).

Mössbauer spectra of the powders prepared in a surfactant solution at t_{mil}=1-24 h (figure 6) are similar to those described above; however, the formation of Fe_3C begins after just 47 h of milling. After annealing of the powders milled over 12 h and more, sextets appear, for which H_1 and H_2 values are equal to 490 and 460 kOe, respectively, which corresponds to Fe_3O_4 [20].

The amounts of Fe_3C and Fe_3O_4 formed at each milling time were determined from the Mössbauer spectra of the milled and thermally treated powders (500^{0}C). Assuming that, after thermal treatment, all of the carbon is in the Fe_3C phase and all of the oxygen is in the Fe_3O_4 phase, we calculated the dependence of the total content of carbon and oxygen in the particle bulk on the time of milling (figure 4b). The amount of carbon at t_{mil}=99 h reaches 12 at. % in both media, and the amount of oxygen is equal to 17 at.% in oleic acid solution. Carbon and oxygen are the products of the decomposition of heptane and oleic acid during mechanical activation.

The analysis of Mössbauer spectra revealed that the maximum amount of carbon present in the form of a solid solution in iron is as high as 8 at. % (2 wt.%); further saturation with carbon leads to the precipitation of carbide Fe_3C. It follows from all the results obtained that milling of iron in heptane and oleic acid solution results in the diffusion of carbon and oxygen from the particle surface layers along the grain boundaries into the particle bulk with the formation of amorphous-like carbide and oxycarbide phases [21, 22].

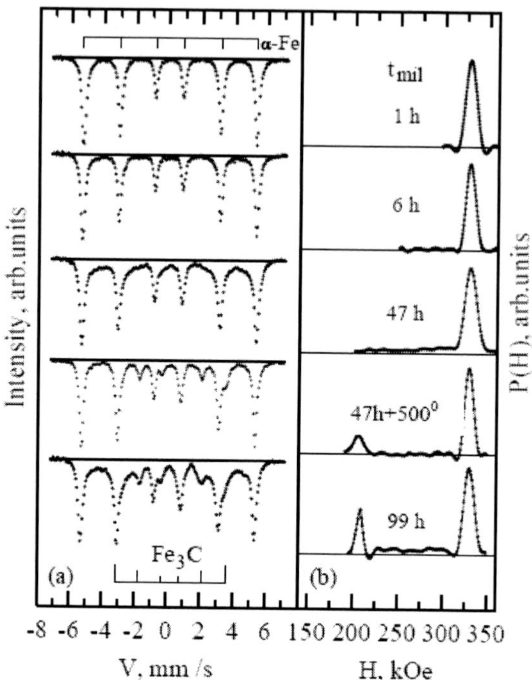

Figure 5. (a) Mössbauer spectra and (b) corresponding functions of hyperfine magnetic field distribution P(H) of iron powders milled in heptane over various times.

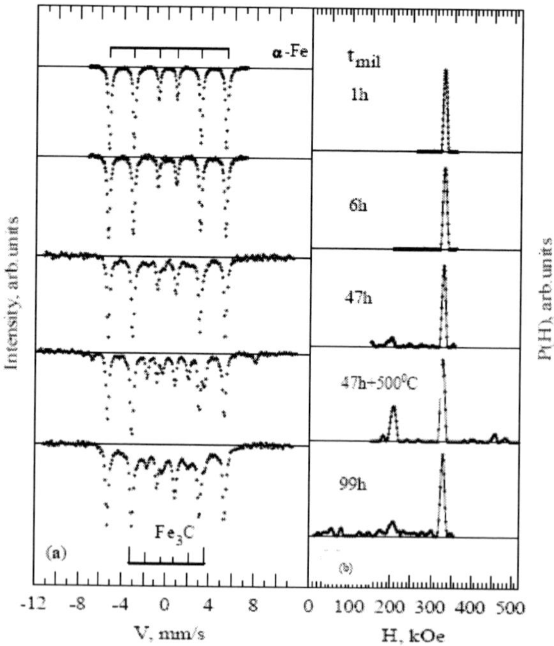

Figure 6. (a) Mössbauer spectra and (b) corresponding functions of hyperfine magnetic field distribution P(H) of iron powders milled in heptane with oleic acid over various times.

Plots of the volume of the absorbed oxygen vs. the exposure time in a corrosive medium are shown in figure 7. In these tests, powders milled in heptane (figure 7a) exhibit equal corrosion resistances, and their incubation period is short (about 10 min for all powders). Therefore, the protective effect of the surface layer formed by milling in heptane alone is insignificant. These powders corrode much more intensely than the initial carbonyl iron, which correlates with their higher dispersivity.

The corrosion resistance of the powders obtained in oleic acid solution (figure 7b) grows with the increase of t_{mil}: the incubation period lengthens from 10 min at t_{mil}=6 h to 480 min at t_{mil}=99 h. Thus, the powder particles become covered with a protective layer during milling in the presence of oleic acid.

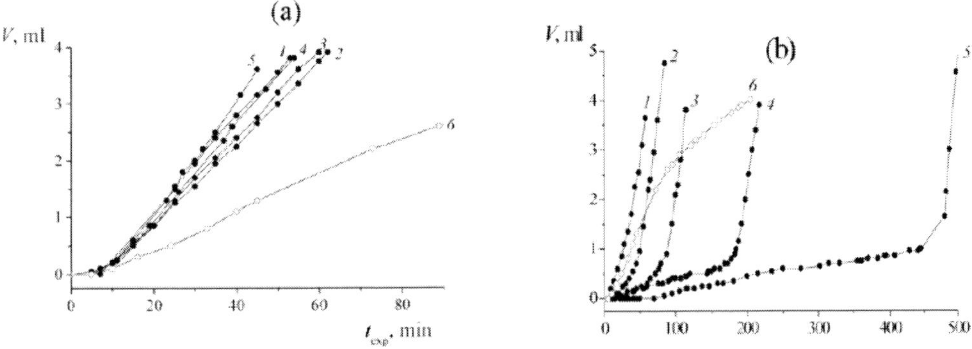

Figure 7. Plots of the volume of absorbed oxygen vs. the exposure time for powders milled in heptane alone (a) and heptane with oleic acid (b) in a corrosive medium; t_{mil}=(1) 6, (2) 12, (3) 24, (4) 47 and (5) 99 h. Curves 6 refer to the initial carbonyl iron. From hereon the batch weight for corrosion tests made up 100 mg.

The powder surface was examined by XPS; the C1s and Fe3p spectra of the powders (t_{mil}=24 h) are shown in figure 8. The powders placed in the spectrometer chamber were (I) not exposed to air (under dodecane); (II) exposed to air for 2 h, (III) one year, and (IV) three years; powder V was obtained from powder IV by treating it with heptane plasma. The surface layer is composed of C (60 at. %), O (30 at. %), and Fe (10 at. %), irrespective of milling medium and time.

In the C1s spectra, a line with E_b = 284.0 eV corresponds to graphite, a line with E_b=285.0 eV, to carbon atoms of alkyl groups, and lines with E_b > 286.0 eV, to carbon atoms of oxygen-containing organic groups [23, 24].

In the Fe3p spectra, a line with E_b≈53.0 eV corresponds to unoxidized iron, a line with E_b≈55.7 eV, to Fe_2O_3, and a line with E_b ≈58.5 eV, to FeOOH [23, 25, 26]. The presence of an unoxidized Fe line suggests that the thickness of the protective surface layer does not exceed 10 nm [27]. Hence, the protective layer contains graphite-like structures, oxides, and organic compounds.

After a 2-h exposure of the powder milled in heptane to air, the intensity of the oxide line in its Fe3p spectra increases. This suggests that the oxide component mainly forms in air rather than in a milling medium, because oxygen dissolved in heptane is not enough to produce such an amount of oxides. For the powder prepared with oleic acid and exposed to air for 2 h, the intensity of the oxide line in its Fe3p spectrum does not increase since the oxide has already formed during milling in the presence of oleic acid which serves as an oxygen source when it decomposes at mechanoactivation. During the third year of the powder exposure to air, the oxide layer continues to grow. The organic component of the surface also oxidizes, which is manifested by somewhat more intensive lines of oxygen-containing organic groups in the C1s spectra.

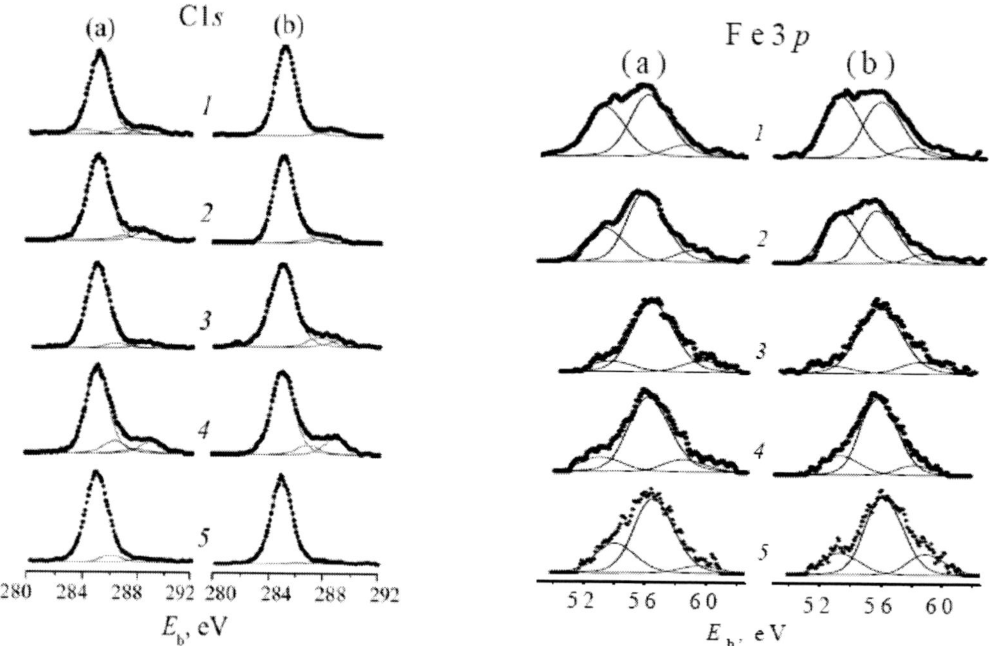

Figure 8. C1s and Fe3p spectra of powders (t_{mil}=24h) obtained in heptane alone (a) and heptane with oleic acid; the exposure to air for (1) 0, (2) 2h, and (3) 1 and (4) 3 years, respectively. Samples 5 was obtained from powders 4 by plasma treatment.

To clear out the role of the oxide and hydrocarbon layers in the protective characteristics of the surface layer the powders were treated by heptane low-temperature plasma. Such a treatment is used to apply protective hydrocarbon films to metal surfaces. Thus, plasma treatment allows indirect estimation of the contributions from the organic and oxide components to the protective properties of the surface layer. Plasma treatment was carried out under conditions [28] providing the coverage of metal plates with the best protective film. The powders were fixed on a magnetic support.

The hydrocarbon plasma treatment equally affected the surface composition of the powders. The total carbon content increased (to 80%) and a line with E_b=285.0 eV in the C1s spectrum became more intense (figure 8-5), suggesting the growth of a hydrocarbon film. At the same time, the amount of the surface oxides reduced: the total oxygen content diminished (to 15%) and the intensities of an unoxidized iron peak in the Fe3p spectrum increased.

The corrosion behavior of the powders prepared in heptane was not affected by plasma treatment (figure 9), while the incubation period for the powders prepared in the presence of oleic acid decreased four times.

Thus, despite the hydrocarbon film thickness increased, the protective characteristics of the surface layer of the powder prepared in the presence of oleic acid became worse due to the oxide reduction in the course of plasma treatment. It therefore follows that an oxide component plays an important role in the surface layer protective characteristics. In more detail the role of the different components of the surface layer in the formation of corrosion stability was analyzed when the influence of milling medium on Fe powder corrosion behavior was studied.

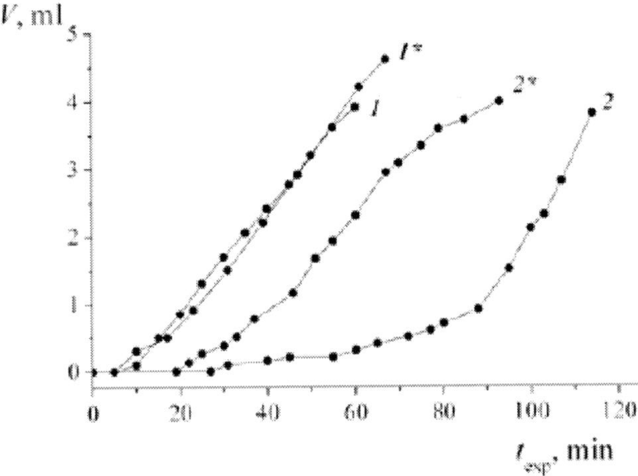

Figure 9. Plots of the volume of absorbed oxygen vs. the exposure time for powders milled (t_{mil}=24 h) in heptane alone (1) and heptane with oleic acid (2) in a corrosive medium. Curves 1* and 2* refer to the respective plasma-treatment powders.

1.2. Influence of Milling Medium Chemical Nature on Corrosion Stability

The composition and properties of the powders were studied depending on 1) the concentration of oleic acid – 10% solutions in heptane and 100%; 2) the length of a hydrocarbon tail in a carboxylic acid molecule – oleic and acetic acids; 3) functional group – acetic acid and isopropyl alcohol.

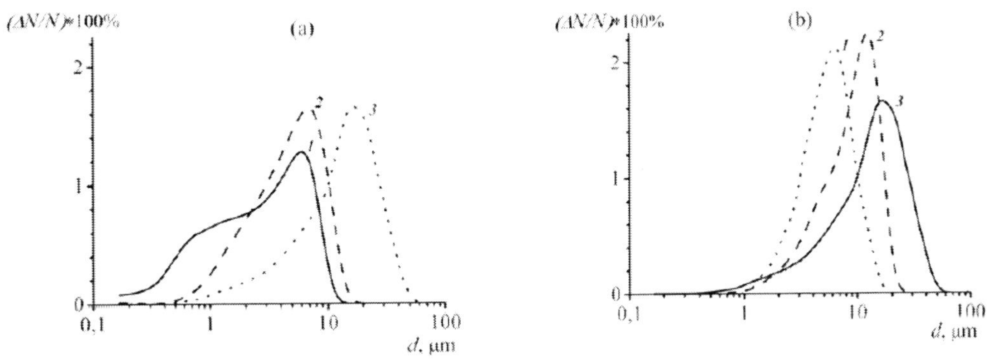

Figure 10. Size distributions of particles prepared by iron milling in (a): (1) acetic acid, (2) isopropyl alcohol and (3) oleic acid; (b): 10% solution of oleic acid (1) without air exposure and (2) with three-month air exposure, (3) – 100% oleic acid.

The particle size distributions are shown in figure 10. According to their dispersity the powders form a series: 100% oleic acid, 10% oleic acid, isopropyl alcohol, acetic acid. The particle size increase with the increase of oleic acid content is presumably associated with the agglomeration of the particles due to the co-polymerization of oleic acid molecules with oxygen (formation of oxygen “bridges”) [29]. For the same reason the exposure of the powders to air (10% oleic acid) results in the agglomerate size increase (figure 10-b).

The X-ray diffraction patterns of the powders prepared at different concentrations of oleic acid and the respective Mössbauer spectra with the functions P(H) are shown in figures. 11a and 11b. The diffraction patterns contain broadened bcc reflections; their positions suggest the constant lattice parameter, which equals to that of pure α-Fe. The increased intensity at the base of the reflection (110) in the range of carbide phases reminds of the halo of the amorphous phase (AP). The intensity of the amorphous phase halo decreases with the increase in the oleic acid concentration. The diffraction patterns were used to calculate the α-Fe grain size, which varies from 6 to 14 nm (see table 1).

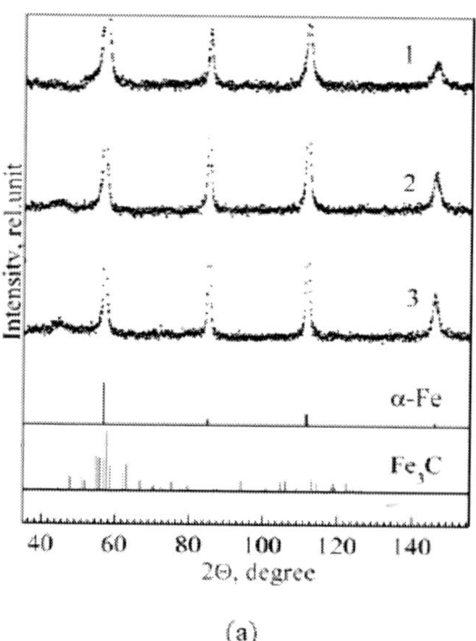

(a)

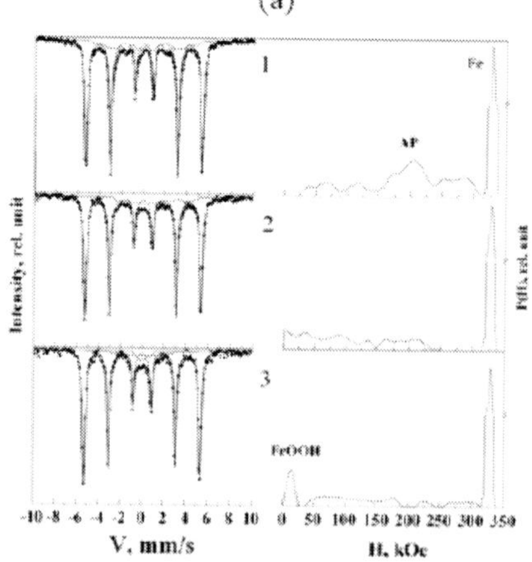

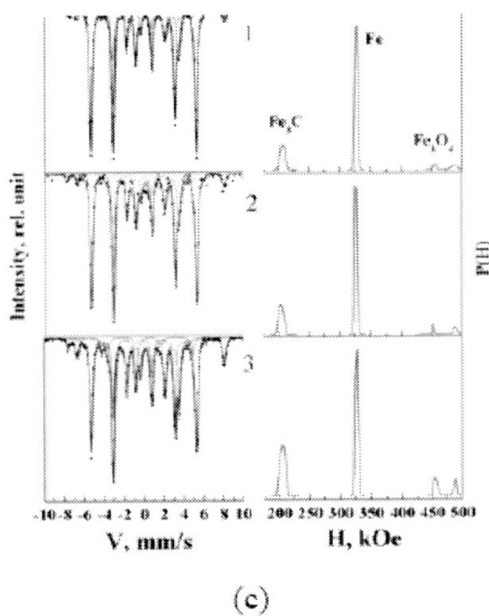

(c)

Figure 11. (a) X-ray diffraction patterns and Mössbauer spectra of (b) original and (c) annealed powders obtained in (1) 0.3 and (2) 10% oleic acid in heptane and (3) oleic acid alone.

Since precise phase identification from our powder diffraction data is complicated because of the presence of a big amount of amorphous phase, we used Mössbauer spectroscopy for this purpose. The spectra and the functions P(H) show, apart from α-Fe, a component with a wide HFMF distribution; its intensity also decreases with an increase of the oleic acid concentration. This component indicates the presence of a set of clusters with a long-range order corresponding to carbide and oxycarbide phases [30-32]. The peak at P(H)≈209 kOe is due to cementite [18], while the P(H) range from 0 to 50 kOe is contributed by FeOOH. Annealing of the powders resulted in the formation of the Fe_3C and Fe_3O_4 phases from the amorphous phase and FeOOH (figure 11c). In addition, the annealing resulted in the formation of carbides and oxides from the oleic acid adsorbed on particles. This explains the increase of the number of these phases with a decrease in the amorphous phase content of the initial samples as the oleic acid concentration increases.

Phase analysis of the powders (see the table 1) showed that an increase in the oleic acid concentration leads to larger grain sizes. Fine milling is prevented by the plasticizing effect of the surfactants [33], which becomes stronger with an increase in the oleic acid concentration. It is well-known [30-32] that during mechanoactivation the effective saturation of nanocrystalline metals with impurities and alloying elements does not begin until the grain size reaches a limiting low value (5 to 7 nm). This is responsible for the smaller amount of the amorphous phase at the higher concentration of oleic acid.

X-ray diffraction patterns and Mössbauer spectra with the HFMF distribution functions P(H) of the powders prepared in isopropyl alcohol, acetic and oleic acids are shown in figure 12. The smallest grain size (< 3 nm) and, accordingly, the most dramatic changes in the phase composition (table 1) were reached in isopropanol. Carbides constitute the major phase when milled in alcohol (figures 12b, 12c). The oxide phase quantity is smaller than in the case of carboxylic acids. This suggests that an alcohol OH group is less reactive than a carboxy group in oxide formation during mechanoactivation.

Table 1. Phase composition, the corrosion rate tan α, the average size d_{av} of powder particles and the average grain size ‹L› in powder particles

Medium		Phase composition (number of Fe atoms in the phase, %) ±2%						Characteristics of the original powders		
		Original			annealed			tan α*, ml/ (g min)	d_{av}, μm	<L>, nm
		α-Fe	AP	FeO-OH	α-Fe	Fe_3C	Fe_3O_4			
Oleic acid	0.3%	80	20		72	23	5	1.2	8	6
	10%	88	12		68	26	7		9	10
	100%	86	9	5	51	37	12		15	14
Acetic acid		64	22	14	60	23	17	3.2	4	5
Isopropanol		51	49		48	49	3	1.1	5	<3
Heptane		80	20		80	20	0	0.8	18	4

α* is the slope of the linear segment in the plot of the absorbed oxygen vs. the exposure time in the corrosive medium after the protective layer was broken.

It has been shown above, that the corrosion of the powders prepared by milling in heptane and in 0,3% oleic acid is preceded by an incubation period, during which the corrosion rate is low because of the formation of a protective surface layer. The powders prepared in 10 and 100% oleic acid do not corrode at all (for the exposure times studied) and are more resistant to corrosion than the powder prepared in 0.3% oleic acid solution (figure 13). The corrosion of the powders obtained in isopropanol and acetic acid shows no incubation period.

The corrosion rates of the powders increase in the order: heptane<isopropyl alcohol<0.3% oleic acid (after the destruction of the protective layer)<acetic acid, which correlates well both with the number of oxide phases in powders (table 1) and dispersivity.

XPS studies showed that the surface layer on all the powders includes oxides and organic compounds (figure 14). The Fe3p spectra of the powders obtained in isopropyl alcohol and acetic acid are not contributed by unoxidized iron, which suggests the formation of thicker oxide films on the particle surface. In the presence of oleic acid, the oxide layer is substantially thinner (< 10 nm), but exhibits better protective properties.

The C1s spectrum of the powders prepared in oleic acid shows a line with $E_b \approx 288.7$ eV, which can be assigned to carboxylate complexes of iron [23]. The characteristic feature of the Fe2p spectra is a distinct shake-up satellite with $E_b \approx 718.0$ eV (figure 14b). It indicates the presence of Fe(III) in the surface layer, probably as its complexes with a carboxy group of oleic acid [23, 34].

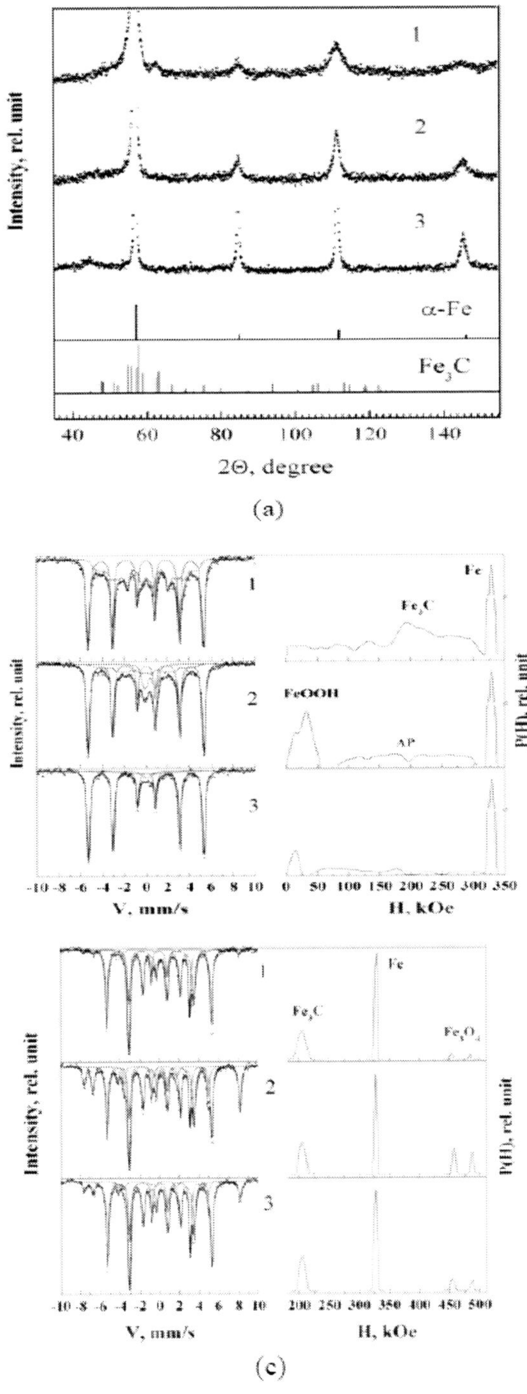

Figure 12. (a) X-ray diffraction patterns and Mössbauer spectra of (b) original and (c) annealed powders obtained in (1) isopropyl alcohol, (2) acetic acid and (3) oleic acid.

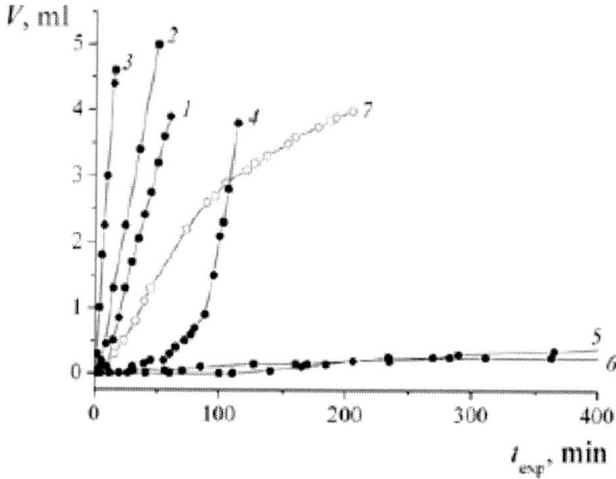

Figure 13. Plots of the volume of absorbed oxygen vs. the exposure time in a corrosive medium for powders milled in (1) heptane, (2) isopropyl alcohol, (3) acetic acid, (4) 0.3% oleic acid in heptane, (5) 10% oleic acid in heptane, (6) 100% oleic acid. Curves 6 refer to the initial carbonyl iron.

The presence of such complexes was confirmed by IR spectroscopy. The absorption band at 1710 cm^{-1} in the IR spectra of oleic acid and iron particles obtained by milling in its presence relates to the stretching vibrations of carbonyl group (figure 15). This band is much less intense for iron particles, but their spectrum contains absorption bands at 1440 and 1590 cm^{-1} due to a carboxylate complex COO^- [34]. Thus, the IR spectra suggest the adsorption of oleic acid molecules as carboxylate complexes that form a protective layer on the particle surface.

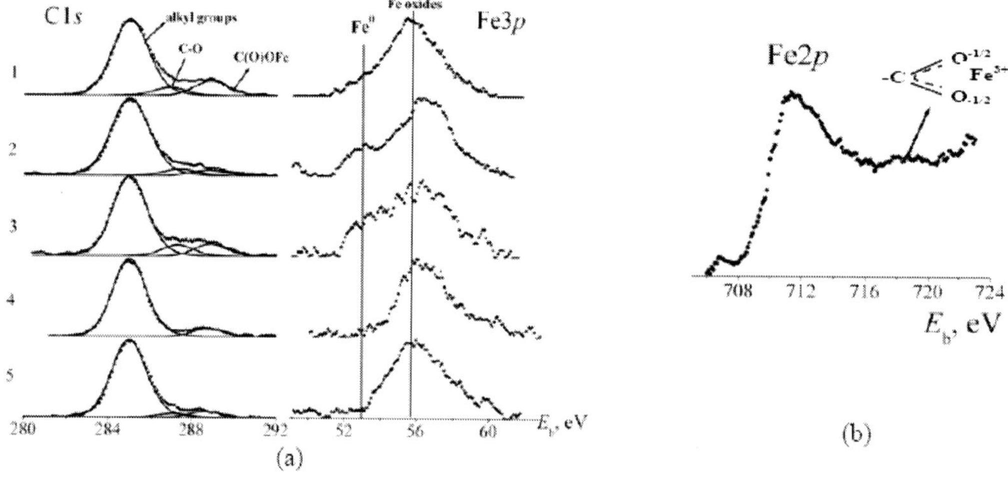

Figure 14. C1s and Fe3p spectra of powders obtained by milling in different organic media: (1) 0.3% oleic acid in heptane, (2) 10% oleic acid in heptane, (3) oleic acid, (4) acetic acid, and (5) isopropyl alcohol; (b) Fe2p spectrum of the powder obtained in 0.3% oleic acid.

Hence, the surface of nanocrystalline powders, prepared by milling iron with oleic acid, is mainly protected from corrosion by the oxide layer and the barrier layer of chemisorbed oleic acid or products of its partial degradation.

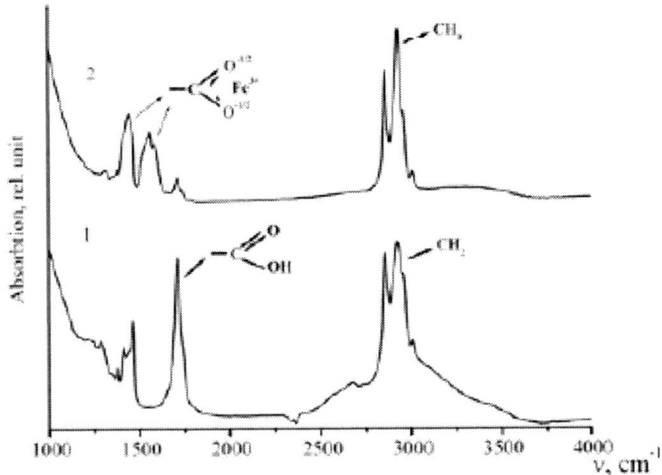

Figure 15. IR spectra of (1) oleic acid and (2) the powder obtained in 0.3% oleic acid.

1.3. Effect of Structural-Phase Composition on Corrosion Stability

It is known, that the uniform corrosion of iron and most of its alloys in neutral media is a diffusion-controlled process [35]. Its rate is limited by oxygen delivery to the metal, being weakly dependent on the metal purity grade and the presence of inclusions such as, e.g., carbides [35, 36]. Unlike traditional materials, Fe-C systems prepared by mechanoactivation are unbalanced to a higher degree: they are highly dispersed and can be enriched in metastable phases (e.g., amorphous Fe-C phases (up to 65%) [32] or cementite (up to 95%) [37]). Literature data on the effect of considerable contents of transient phases on the corrosion behavior of iron—carbon alloys in neutral media are absent.

Here we consider the systems representing different ratios of phase mixtures, prepared by iron milling in graphite or heptane: (1) nanocrystalline α-Fe + amorphous Fe-C phase, (2) α-Fe +Fe_3C, and (3) α -Fe +carbon.

Mössbauer spectra showed that the intergranular disordered Fe-C phase is synthesized during milling after the annealing carbide is formed (figure 16). The quantitative phase composition determined from the Mössbauer spectra and the grain size calculated by broadening the X-ray lines are given in table 2. The content of the amorphous Fe-C phase upon the pulverization in graphite is significantly higher than in heptane. The reason is that during milling in heptane, both carbon and hydrogen diffuse to the bulk metal. Being more diffusible, hydrogen occupies part of defects and thus decreases the carbon solubility and promotes the formation of carbide phases. To attain specific structural-phase characteristics, powders were subjected to thermal treatment. The highest amount of cementite was obtained at 500^0C (1 h). At 800^0C, the powders are identical in phase composition (cementite, 20 at %;

α -Fe, 80 at %). Annealing the powder milled with graphite at 700^{0}C (10 h) gave a mixture of α-Fe with the highest graphite content (table 2).

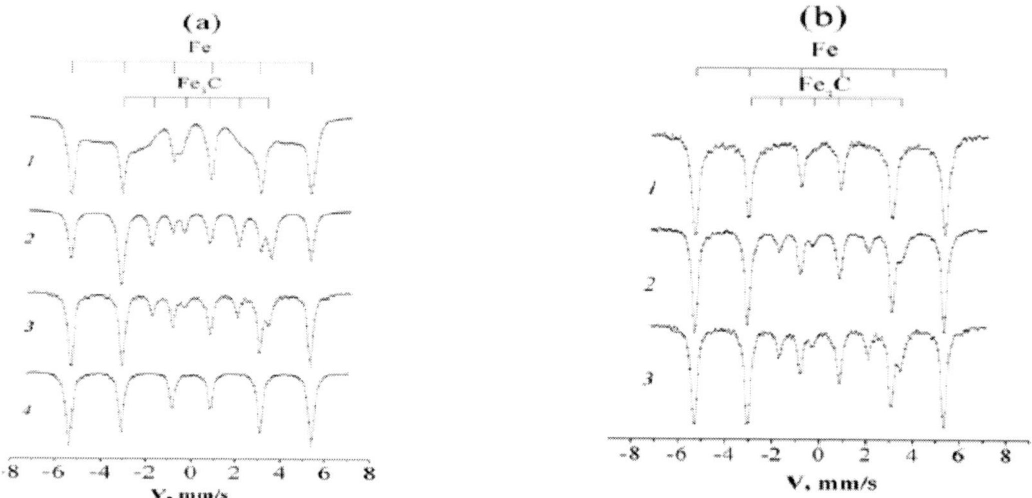

Figure 16. Mössbauer spectra of the powders milled in (a) graphite and (b) heptane: (1) without annealing, (2-4) powders annealed at 500 (1h), 800 (1h), and 700^{0}C (10h), respectively.

Table2. Phase composition (in atomic fractions of the phases) and the grain size ‹L› in α-Fe

Medium	t_{mil}, h	T_{ann}, °C	T_{ann}, h	AP	Fe_3C	α-Fe	<L>, nm
Graphite	16	-	-	65	0	35	4
		500	1	0	60	40	80
		800	1	0	20	80	200
		700	10	0	0	100	>200
heptane	48	-	-	20	0	80	4
		500	1	0	20	80	80
		800	1	0	20	80	200

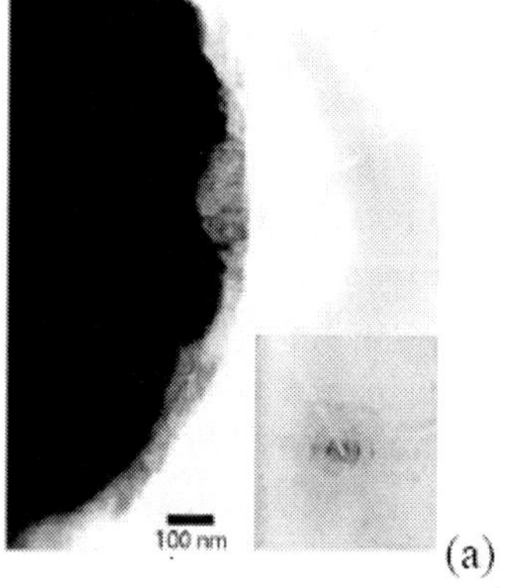

(a)

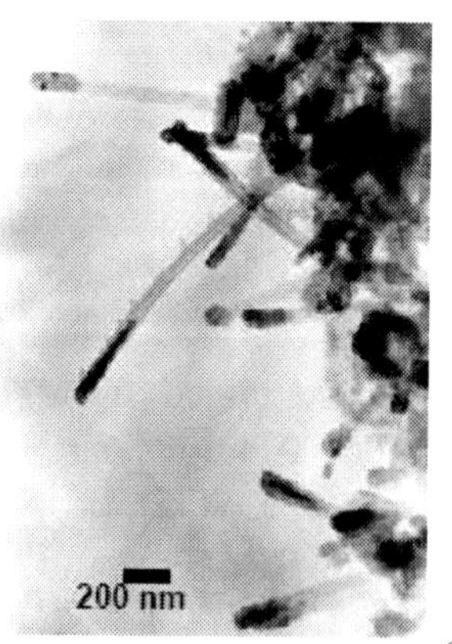

(b)

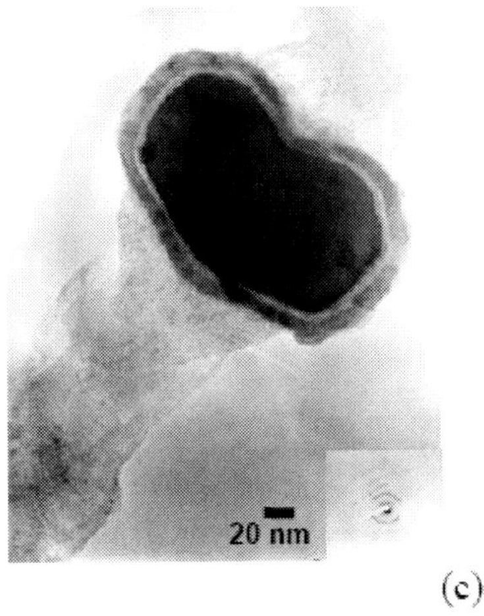

Figure 17. Electron microscopy images of the particles milled in (a) graphite and annealed at 700^{0}C (10 h) and (b,c) heptane and annealed at 800^{0}C (1h).

To elucidate the carbon state upon annealing, we performed electron microscopy studies. The particle edge of the powder milled in graphite and annealed at 700^{0}C (10 h) shows a surface layer 50 to 70 nm thick, which contrasts with the bulk of the particle (figure 17a). The microdiffraction pattern (figure 17a, inset) of this area corresponds to graphite carbon. The structures formed in the annealed powder (800^{0}C, 1 h) milled in heptane are shown in figure 17b. Carbon is mainly present in an amorphous form and as nanofibers 20 to 40 nm thick with cementite nanoparticles inside. The image of the particle is shown in figure17c. The particle is covered with a two-layer shell, the upper layer (~10nm thick) being made up of iron oxides (XPS data).

The XPS spectra are shown in figures 18 and 19; the quantitative analysis data for the powder surface are given in table 3. In the Fe2p spectra, Fe^0 is responsible for the line at 707.0 eV; FeO, for the line at 709.5 eV; Fe_3O_4, for the lines at 708.3 and 710.6 eV; Fe_2O_3, for the line at 711.0 eV; and FeOOH, for the line at 711.9 eV [25, 26].

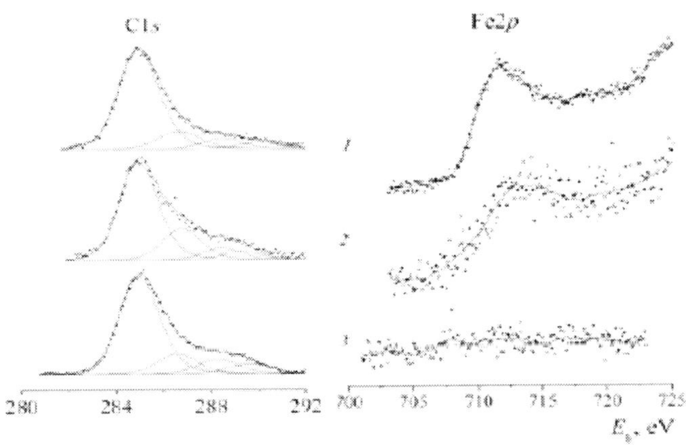

Figure 18. C1s and Fe2p spectra of powders (1) milled in graphite and annealed at (2) 500^{0}C (1 h) and (3) 700^{0}C (10h).

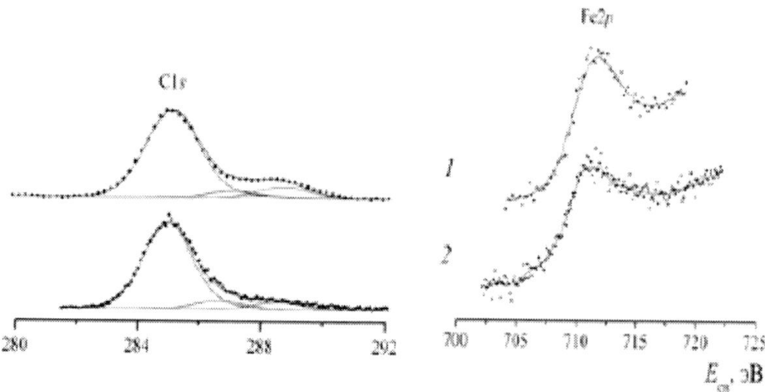

Figure 19. C1s and Fe2p spectra of powders (1) milled in heptane and (2) annealed at 800^{0}C (1 h).

Table 3. Composition of the surface layer (at %)

Medium	t_{mil}, h	T_{ann}, ^oC	t_{ann}, h	C	O	Fe
Graphite	16	-	-	50	34	16
		500	1	70	25	5
		800	1	75	25	0
		700	10	73	27	0
Heptane	48	-	-	54	33	13
		800	1	61	27	12

At the surface of the powder annealed at 500^{0}C (1h) and pulverized in graphite, the carbon content is substantially increased, which is due to the decomposition of Fe_3C in the particle bulk and diffusion of carbon to the surface. Upon annealing at 800 (1 h) or 700^{0}C (10 h), the signal from iron disappeared from the spectra, which corresponds to the formation of thick graphite layers at the surface of the powder particles (figure 18-3). In the powder milled in heptane (800^{0}C, 1 h), the carbon content in the surface layer increases only slightly because cementite does not decompose. In this case, the carbon layers are significantly thinner than those on the powders pulverized in graphite.

The Cls spectra of the annealed powders milled in graphite contain no graphite line with E_b=284.0 eV, although the electron microscopy studies revealed its presence at the surface. According to [38], amorphous carbon and disordered graphite are characterized by $E_b \approx 285.0$ eV, for which reason the upper layer of the graphite film can be regarded as disordered. Apparently, this is the reason for the strong surface oxidation of this layer: in the Cls spectra, the C-O groups with E_b>286.0 eV are responsible for 30% the line intensity.

According to the Fe2p spectra of the milled powders, their surface is oxidized to FeOOH. Annealing of the powder pulverized in graphite not only increases the carbon content of the surface layer, but also results in the oxidation of both graphite (Cls spectrum shows more lines corresponding to the oxidized carbon groups) and iron (at 500^{0}C, the peak of Fe2p spectrum is substantially shifted to the higher E_b values).

The plots of the volume of the absorbed oxygen vs. the exposure time in a corrosive medium, which characterizes the corrosion kinetics, are displayed in figure 20. For the powder milled with graphite, the incubation period decreases from 25 to 5 min with an increase in annealing temperature. For the powder milled in heptane, the incubation period lasts for 10 min even upon thermal treatment. Oxide films on amorphous alloys are known to exhibit enhanced protective properties [36]. The high content of the disordered Fe-C phase in the powder milled with graphite favors the formation of a more protective surface oxide film. It should be noted that the encapsulation of Fe particles into a graphite film (up to 70 nm) did not enhance the protective properties of the surface.

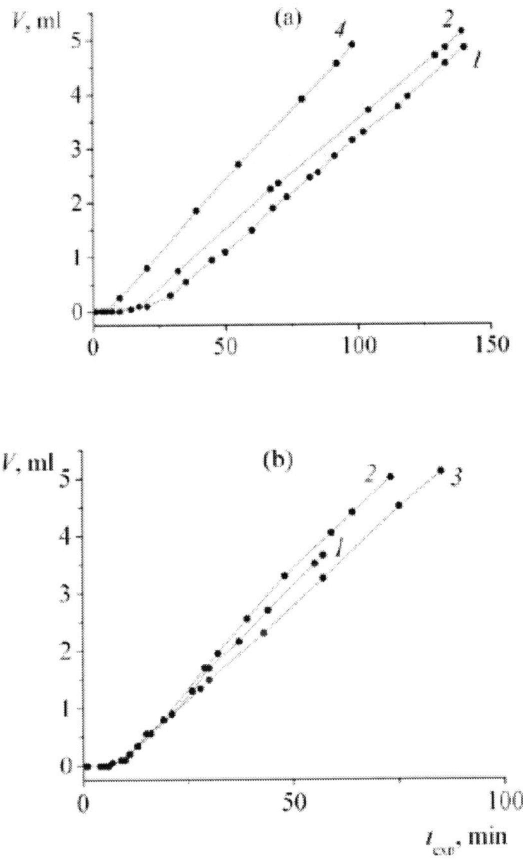

Figure 20. Plots of the volume of absorbed oxygen vs. the exposure time in a corrosive medium. The powders were milled in (a) graphite and (b) heptane: (1) without annealing and (2-4) powders annealed at 500°C (1 h), 800°C (1 h), and 700°C (10h), respectively.

Hence, the high content of the disordered Fe-C phase at the surface of the powders milled in carbon-containing media favors the formation of a more protective oxide film. Other alterations in the structural-phase state (presence of cementite in different amounts, change in grain size, formation of surface carbon films) do not affect the corrosion behavior of highly dispersed Fe-C powders in neutral media.

1.4. Milling in Heptane and Heptane with an Organosilicon Additives

The use of a silicone additive was expected to result in the increase of dispersity and corrosion stability, because silicon introduced into the particles bulk makes Fe more fragile and stable. SiO_2 could be the other product of vyniltriethoxysylane (TEVS) destruction during mechanoactivation and thermal treatment. The creation of SiO_2 - based layers on finely dispersed iron surfaces is used for their stabilization [39, 40].

Particle size distributions are presented in figure 21. The mean particle sizes were 4.6 and 1.8 μm at milling times t_{mil}= 24 and 99 h, respectively. A fairly large amount of particles with a size of approximately 0.5 μm was present in the powders. The dispersity of the powders prepared in the presence of TEVS was higher than that of the powders prepared in the presence of oleic acid. The prepared particles have the shape of thin scales (figure 22) because of plastic deformation during the mechanical activation of the particles.

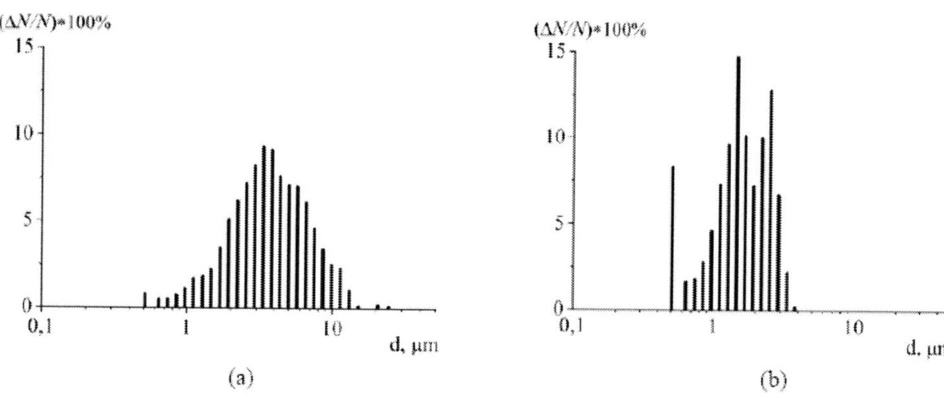

Figure 21. Size distributions of particles prepared by iron milling in heptane with TEVS for (a) 24 h and (b) 99 h.

Figure 22. Micrographs of powder particles after (a) 24 h and (b) 48 h milling in heptane with TEVS.

X-ray diffraction patterns of the powders prepared at t_{mil}= 24, 48, and 99 h are represented in figures 23 (curves 1-3). When the milling time is no longer than 24 h, diffraction patterns (figure 23, curve 1) demonstrate broadened reflections of a bcc lattice whose positions indicate an unchangeable lattice parameter equal to that of pure α-Fe. An increase in the intensity resembling a halo of an amorphous phase is observed at the base of the reflection (110). When t_{mil} increases to 48 h (figure 23, curve 2), additional reflections of cementite Fe_3C arise, which are also markedly broadened. A further rise in milling time leads to an increase in the content of both cementite and the amorphous-like phase.

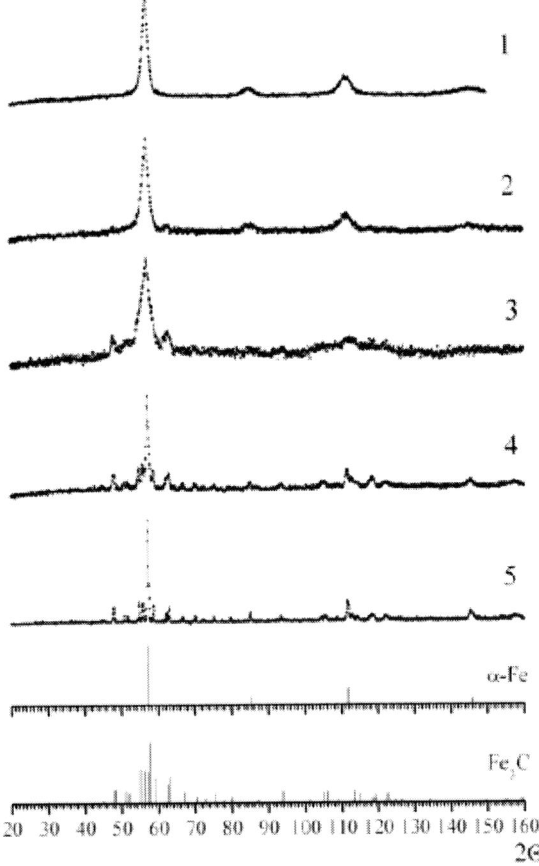

Figure 23. X-ray diffractograms of powders milled in heptane with TEVS over various times: (1) 24 h, (2) 48, and (3-5) 99 h; (1-3) initial powders, (4-5) powders subjected to thermal treatment for 1 h at (4) 500 and (5) 800^{0}C.

A component with a wide and smooth HFMFs distribution is present in Mössbauer spectra and P(H) functions (figure 24), with the component intensity increasing with a rise in milling time. The presence of this component is indicative of the formation of a disordered Fe-C phase in the intergrain regions [30, 32, 41-43]. At t_{mil}=48 h, a new component with H≈208 kOe appears in the Mössbauer spectrum, thus indicating the formation of carbide Fe_3C.

Note that the formation of carbide during the mechanical activation proceeds more intensely in the presence of TEVS than in pure heptane. This can be explained by the fact that, when the mechanical activation is performed in the presence of organosilicon additives, the grain sizes are substantially smaller (<2nm) than those milled in pure heptane (4 nm).

Hence, the surface area of the intergrain boundaries, along which carbon atoms can penetrate into the particle bulk, is enlarged. A more drastic decrease in the grain size is associated with the diffusion of silicon atoms into the particle bulk and the accumulation of microdistortions.

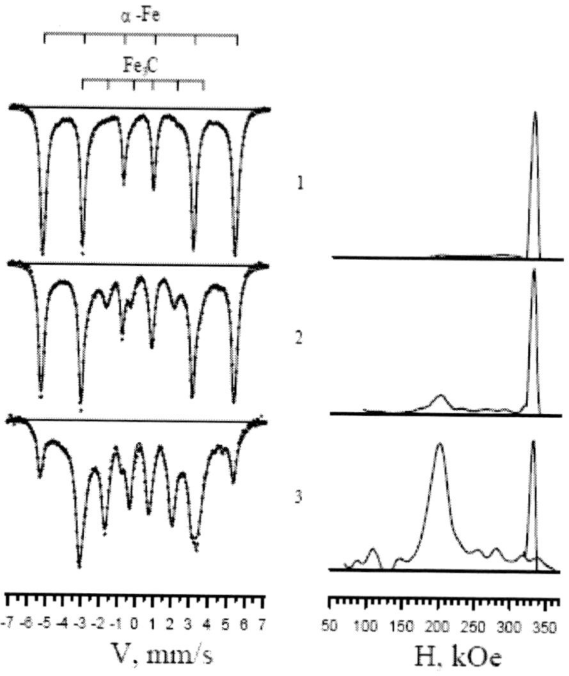

Figure 24. Mössbauer spectra and corresponding distribution functions P(H) of powders milled over various times: (1) 24, (2) 48, and (3) 99 h.

Let us consider the effect of thermal treatment on the phase composition of the prepared powders. A noticeable narrowing of the lines is observed in the X-ray diffraction patterns (figure 23, curves 4, 5), while the contribution of the amorphous halo disappears. In general, it can be stated that, when the annealing temperature increases, cementite is formed from the intergrain disordered phase. Cementite atomic fractions in the powders calculated from the Mössbauer spectra are listed in table 4 as functions of annealing temperature for the systems prepared by the mechanical activation of iron in the presence of TEVS and the mechanical activation of dry Fe-graphite mixtures in argon atmosphere (at C content in the initial mixture of 5-20 at. %). As is seen from table 4, the thermal treatment (800°C) of the powders prepared by the mechanical activation of iron and graphite results in the transitions proceeding as shown in the diagram of stable states of the iron-carbon system; i.e., cementite is dissolved to form a-Fe and graphite. Despite the fact that, at t_{mill}= 99 h, the content of carbon in the powders prepared by mechanical activation in the TEVS solution in heptane is on the order of 20 at. %, the amount of cementite decreases insignificantly and is close to its initial content at all annealing temperatures, thus indicating its thermal stability.

Table 4. The fraction of Fe3C in powders prepared by the dry milling of iron-graphite mixtures and the milling of iron in the presence of TEVS and annealed at different temperatures (by the data of Mössbauer spectroscopy)

Fe_3C, atom fraction	Fe(100-x)C(x)			Fe+TEVS	
	x=5	*x*=10	*x*=20	48 h	99 h
500^{0}C	0,24	0,42	0,75	0,44	0,75
800^{0}C	0,16	0,22	0,27	0,39	0,66

The stability of cementite can be explained by the presence (in the intergrain regions) of nanointerlayers of a third phase separating α-Fe and Fe_3C and preventing carbon from diffusion out of cementite into γ-Fe at high temperatures. Such a phase can be silicate structures or silica. Because the interaction of the material with the organic milling medium and its destruction products occurs across the particle surface and so does the diffusion of foreign atoms into the particle interior, information on the composition of intergrain boundaries can be obtained by studying the particle outer surface.

XPS studies were performed on the powders prepared at t_{mil}= 99 h and under different exposure conditions: in air or under a layer of dodecane preventing contact with air; exposure time 1 h and one year (figure 25). Quantitative analysis data for the powder surface are given in table 5.

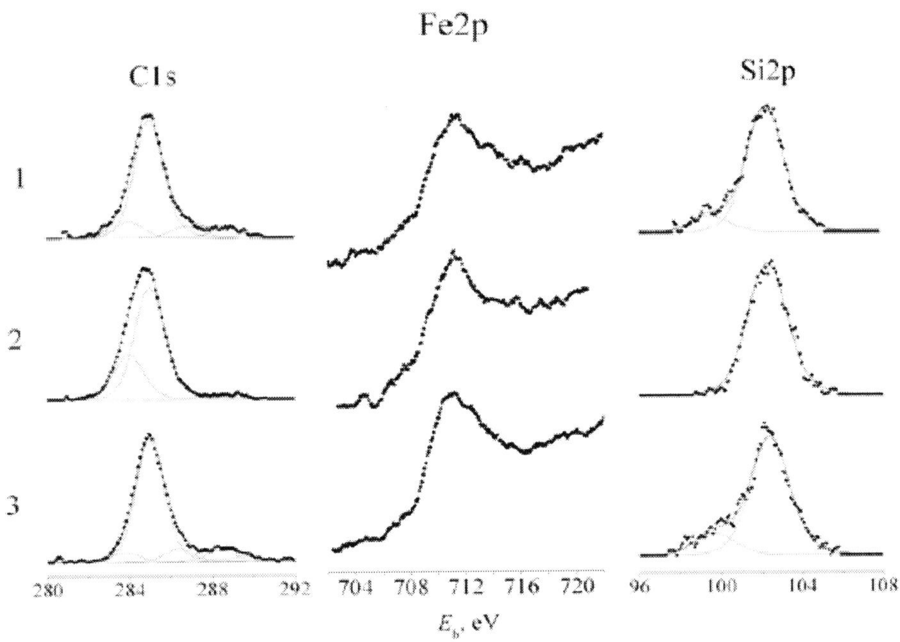

Figure 25. C1s, Fe2p and Si2p spectra of the powders (0.3% TEVS) exposed to air for (1) 0, (2) 1h, and (3) 1 year; t_{mil}=99 h.

Table 5. Composition of the surface layer (at %) of powders particles prepared by the milling of iron in the presence of TEVS; t_{mil}=99 h

Exposure to air	C	O	Fe	Si	Fe/Si
0	43	38	13	6	2,2
1 h	61	26	8	5	1,6
1 year	44	37	15	4	3,7
1 year + annealing at 800^0C	61	29	7	3	2,3

The Si2p spectra contain the lines at E_b≈99.5 (unoxidized Si or Fe-Si structures) and E_b≈104.0 eV (SiO_2); the line at E_b ~ 102.2 eV can be assigned to both silicate and organosilicon structures [23, 24, 44].

Hence, the surface layer of iron powders contains iron oxides, Fe-Si structures, organic (including organosilicon) compounds, silicate structures, and graphite. The presence of unoxidized Si in the surface layer suggests the cleavage of such strong bonds in TEVS as Si-O.

Compare the spectra of air-exposed powders. First, the Si2p spectra of a powder placed under dodecane in the spectrometer chamber and of a powder exposed to air for one year contain a line of unoxidized Si, which is absent from the spectra of powders exposed to air for 1h. Second, for both zero and one-year exposures, the Fe/Si ratio is increased (table 5); i.e., the surface layer is depleted of silicon. Apparently, these changes are due to partial dissolution of organosilicon compounds in dodecane and their removal from the surface upon exposure to air.

The Fe2p and Si2p spectra of a thermally treated powder are shown in figure 26. The reduced Fe/Si ratio (table 5) indicates that the surface layer becomes enriched with Si. The Fe2p spectrum contains no shoulder of unoxidized iron at E_b=707.0 eV. The line of unoxidized silicon disappears from, and a line of SiO_2 appears in the Si2p spectrum. Thus, the surface layer of the annealed powder is oxidized to form a structure characteristic of Fe-Si alloys (iron silicates and SiO_2) [45]. It can be assumed that the same phases can be formed in the intergrain regions due to thermal treatment. These phases prevent carbon from diffusion out of carbide phases, thus increasing the thermal stability of the latter.

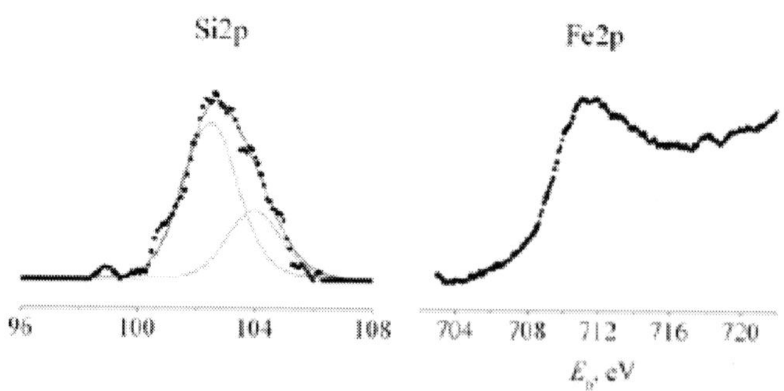

Figure 26. Si2p and Fe2p spectra of the annealed powder (0.3% TEVS); t_{mil}=99 h.

The plots of the absorbed oxygen volume vs. the exposure time in the corrosive electrolyte for powders obtained at t_{mil}=99 h in heptane, heptane + oleic acid, and heptane +

TEVS are shown in figure 27. The addition of TEVS virtually does not affect the corrosion behavior of the powders. The protective effect of the resulting surface layer is substantially weaker than for the powders milled with oleic acid. To find out whether this is due to the insufficient concentration of TEVS in solution, we milled iron powders in 10% TEVS in heptane and pure TEVS. In both cases, the corrosion behavior of the powders (including annealed ones) remained the same, regardless of the concentration of TEVS (table 6).

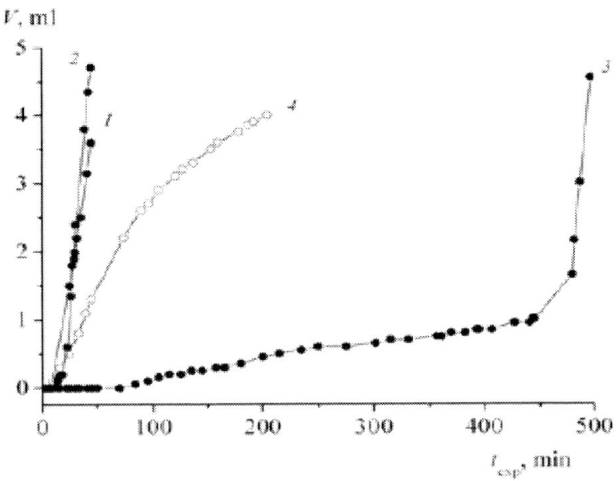

Figure 27. Plots of the volume of absorbed oxygen vs. the exposure time in a corrosive medium for the powders obtained at t_{mil}=99 h in (1) heptane, (2) heptane+0.3% TEVS, and (3) heptane+0.3% oleic acid; curve 4 refers to the original carbonyl iron.

Table 6. Incubation period and the corrosion rate tan α (after the protective layer was broken) of powders obtained at different concentrations of TEVS; t_{mil}=24 h

Parameters	TEVS concentrations, wt%				
	0,3	10	10*	100	100*
Incubation period, min	11	19	5	19	5
corrosion rate tan α, ml/(g min)	1,0	1,0	0,7	1,5	0,7

* additionally annealed at 800^{0}C.

Table 7. Incubation period and the corrosion rate tan α (after the protective layer was broken) for different t_{mil} values and air exposure times

Параметры	t_{mil}, h; exposure to air						
	0*	1	24	48	99		
		1 year			1 h	1 year	1 year**
Incubation period, min	5	5	11	18	31	22	5
corrosion rate tan α, ml/(g min)	0,3	0,8	1,0	1,5	2,0	2,0	1,1

* Original carbonyl iron

** additionally annealed at 800^{0}C.

As noted above, the surface of the air-exposed powders is depleted of organosilicon compounds and the incubation period shortens (table 7). Therefore, organosilicon compounds can enhance the protective properties of the surface, though only slightly. It should be noted that the surface layer of annealed powders does not inhibit their corrosion, though it is enriched with silicon and contains SiO_2.

2. Corrosion of Finely Dispersed Fe-Si Systems

Fe-Si alloys are known to excel in fragility and corrosion resistance. This is why we took this alloy for the same series of investigations that was performed with carbonyl iron. We expected to obtain more dispersed and corrosion-stable powders. At silicon content 20 at.% Fe-Si still preserves high magnetic characteristics.

The images of the powders are shown in figure 28, the particle size distributions – in figure 29. The mean size of particles (t_{mil}=99 h) prepared in heptane is equal to 22 μm; in oleic acid solution, to 2 μm. Fe-Si powders have dispersity higher than Fe powders prepared under the same conditions.

(a)

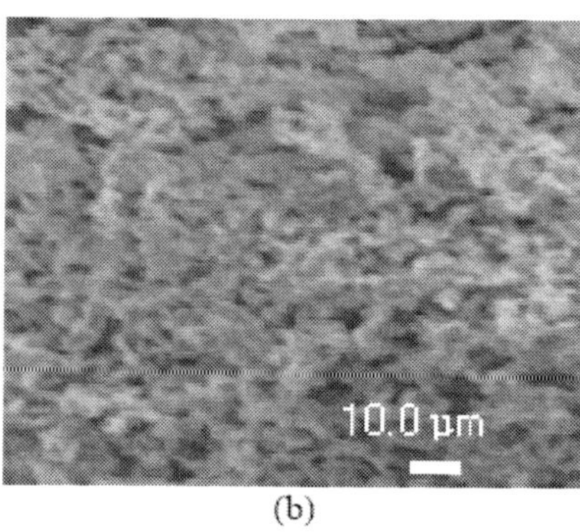

(b)

Figure 28. Micrographs of Fe-Si particles after 99 h milling in heptane alone (a) and heptane with oleic acid (b).

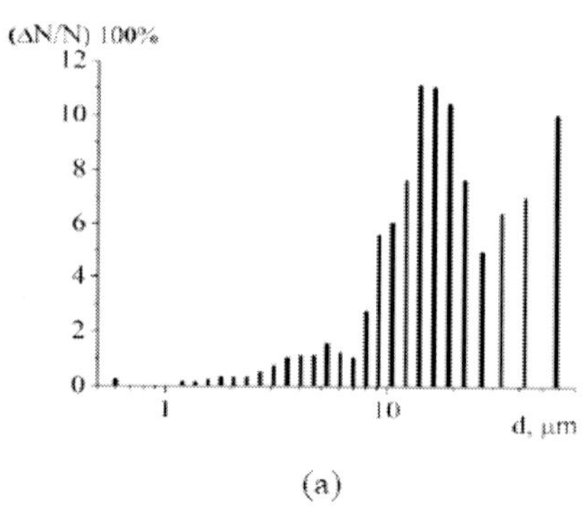

(a)

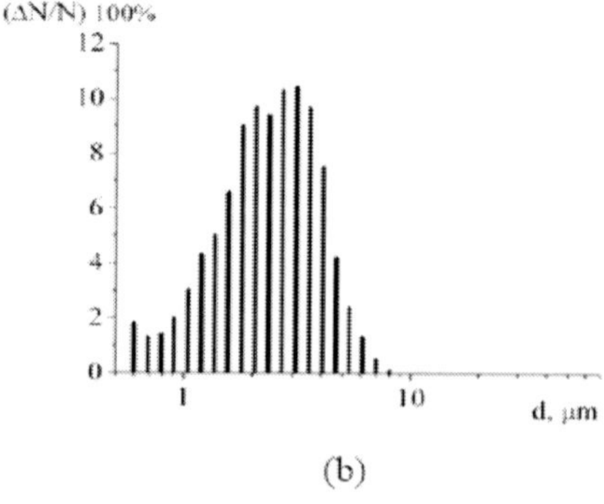

(b)

Figure 29. Size distributions of Fe-Si particles after 99 h milling in heptane alone (a) and heptane with oleic acid (b).

In figure 30 we present the powders diffraction patterns. The general scheme of the phase transformations during milling and further annealing is presented in figure 31. The halo near the base of the line (100) for the powders milled in heptane is associated with the formation of a Fe-Si-C amorphous phase whose amount increases with an increase in t_{mil}. In the powders obtained with the presence of oleic acid, in addition to a small amount of amorphous phase, iron silicate-carbide Fe_8Si_2C forms at t_{mil}=48 and 99 h [46, 47]. The process of iron silicate-carbide formation in Fe--Si powders is similar to that of cementite formation in Fe powders. In particular, the products of the milling medium destruction (C, O, H) diffuse along the intergrain boundaries into the particles; solid solutions and amorphous-like phases form, which is followed by the formation of metastable chemical compounds. Note, that in the presence of oleic acid the formation of carbide phases begins at lower t_{mil}.

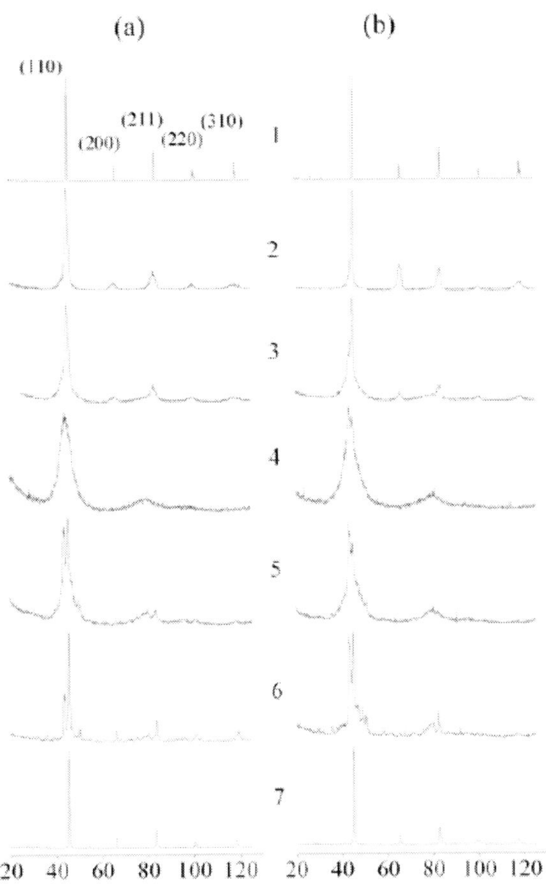

Figure 30. X-ray diffractograms of Fe-Si powders milled in heptane alone (a) and heptane with oleic acid (b) over various times: (1) 0, (2) 24 h, (3) 48 h, (4) 99 h; (5-7) the sample (4) upon annealing at 400, 500 and 800^{0}C, respectively; star (*) denotes the most intense reflection for the phase Fe_8Si_2C.

At 400 and 500^{0}C, the amorphous phase undergoes crystallization, thus yielding iron silicate-carbide and Fe_3Si in the powders prepared in heptane, or iron silicatecarbide and Fe_3C in the powders milled with oleic acid. At 800^{0}C, the powder obtained in heptane transforms back to the state that corresponds to a Fe80Si20 alloy. In the powder obtained with oleic acid, a Fe87Si13 alloy forms; it is depleted in Si compared with the initial alloy. The depletion is caused by SiO_2 formation in the particle bulk; this process consumes oxygen released from oleic acid during milling process.

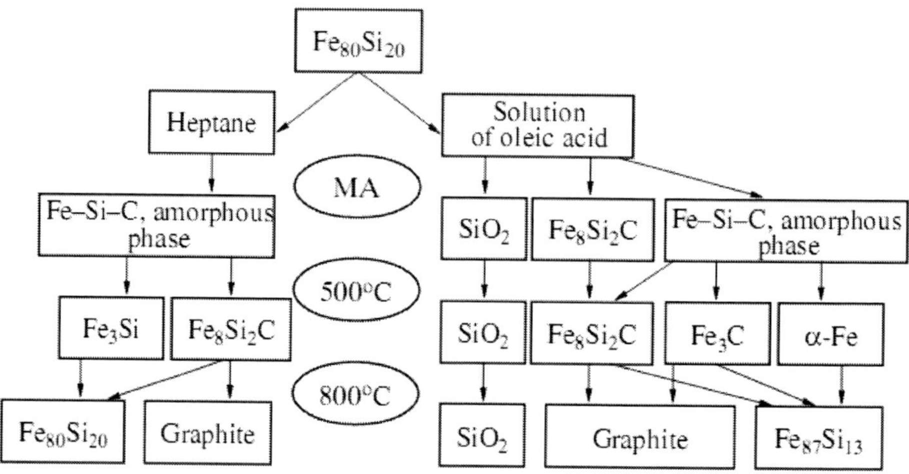

Figure 31. Scheme of phase transformation observed during the mechanical activation (MA) and subsequent annealing.

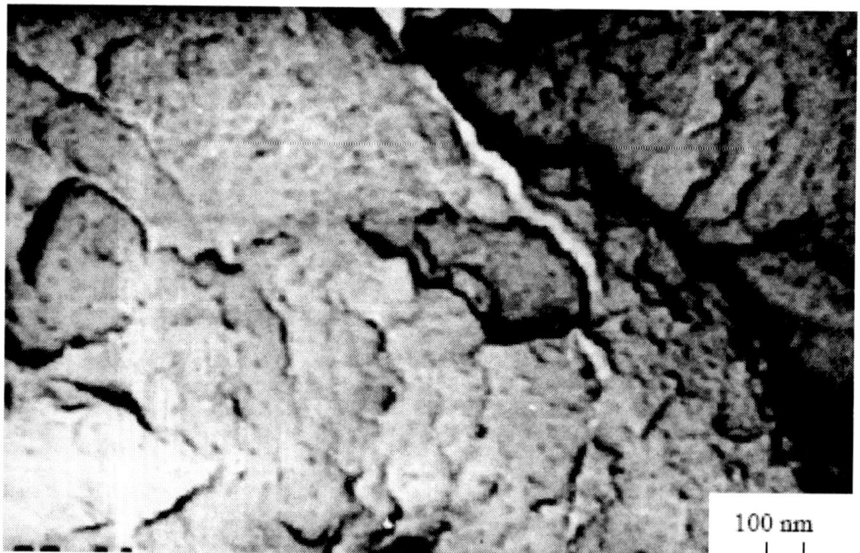

Figure 32. Electron microscopy image of the powder milled in heptane (t_{mil}=99 h) upon annealing at 800^0C (1h).

The destruction of the carbide phases at 800^0C does not result in the appearance of graphite-phase reflexes in x-ray diffraction patterns (figure 30). To reveal the state of carbon, we performed electron microscopy studies of the powders milled in heptane (figure 32). In the microphotographs we see high-dispersed inclusions, sized ~20 nm, in the alloy matrix. According to [48], such high-dispersed graphite inclusions have been formed directly in the microscope column during the annealing (at 300^{0}C for 3 h) of a Fe-Si-C alloy prepared by quick quenching. Thus, upon annealing, carbon precipitates in the form of graphite inclusions in ferrite matrix, which agrees with the XPS data (see below).

The surface layer structure was studied as a function of (1) milling medium and duration, (2) exposure to air, and (3) annealing temperature. At any t_{mil}, similar changes in the spectra are produced by exposure to air and annealing; therefore, we shall restrict our consideration to the spectra of the powders obtained at t_{mil}=99 h (figure 33).

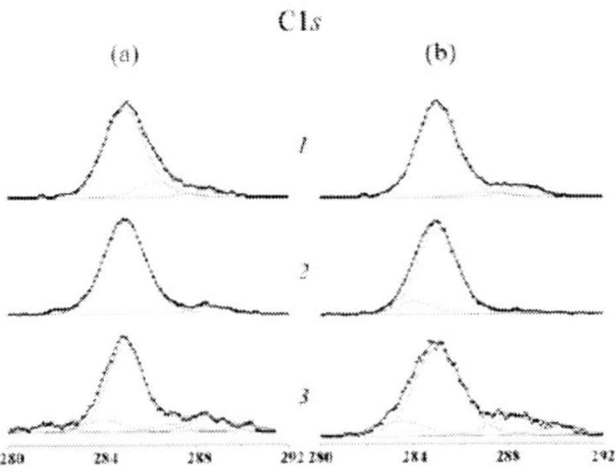

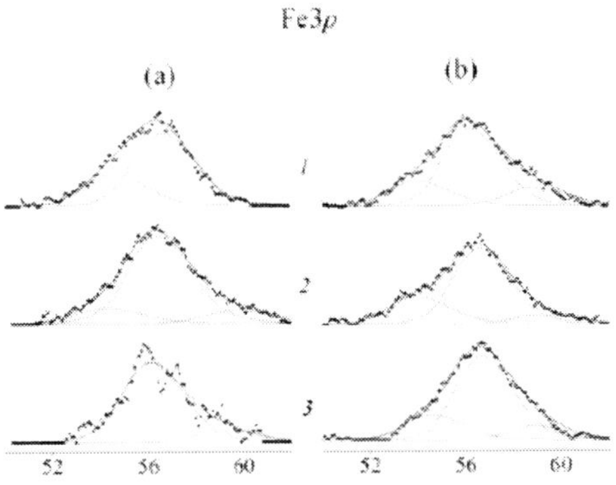

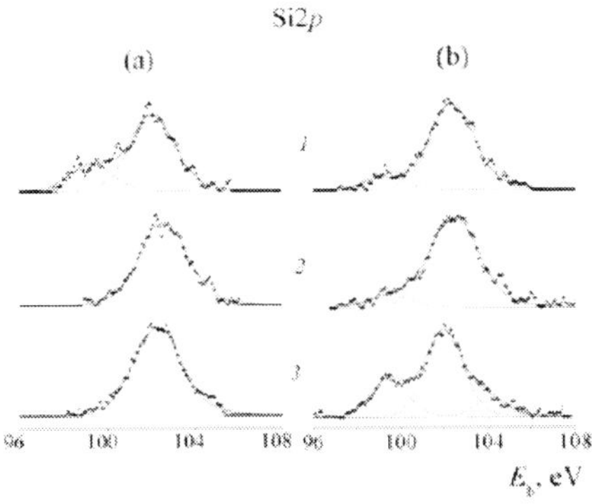

Figure 33. C1s, Fe3p and Si2p spectra of powders milled (t_{mil}=99h) in heptane alone (a) and heptane with oleic acid (b): (1) without exposure to air and upon exposure to air for (2) 3 month and (3) 1 year.

From the analyses of the spectra we concluded that the surface layer of the Fe-Si powders contains carbides, graphite, oxides, and organic compounds. The composition of the powders surface layers is summarized in table 8. With an increase in t_{mil}, no significant changes in the surface layer occurred, with the only exception that the content of Si at the surface of the powders prepared with oleic acid increased, due to the SiO_2 formation during milling.

We now turn to the changes in the surface layer of the powders during their exposure to air. In the C1s spectra, the contribution from oxygen-containing organic groups increases; the lines of graphite and carbides appear. All this is associated with partial destruction of the organic film. In the Fe3p and Si2p spectra of the powders milled in heptane, the line of the Fe-Si alloy disappears *per se*, which points to the powder surface oxidation in air, although in the case of the powder milled in the presence of oleic acid the surface is not oxidized in air. Thus, as in the case of Fe-based systems, the formation of oxide films at the Fe-Si powder particles depends on their preparation medium.

Table 8. Dependence of surface layer composition (at %) of Fe-Si powders particles* on the milling time t_{mil} (h) and the annealing temperature T_{ann} (^{0}C)

Medium	t_{mil}, h	T_{ann}, ^{0}C	C	O	Fe	Si	Fe/Si
Heptane	24		53	32	10	5	2.0
	48		54	30	10	6	1.7
	99		49	37	10	4	2.3
	99	400	58	30	4	8	0.5
	99	500	61	27	2	10	0.2
	99	800	56	31	0	13	0
Heptane with olcic acid	24		50	35	10	5	2.0
	48		49	36	9	6	1.5
	99		47	39	7	7	1.0
	99	400	52	33	5	10	0.5
	99	500	52	38	4	6	0.6
	99	800	73	21	0	6	0

During annealing, the powders surface layer transforms as follows. With an increase in the annealing temperature, the surface of the powder milled in heptane becomes enriched in silicon, while the silicon content in the surface of the powder milled in oleic acid solution practically does not change (table 8).

Let us discuss the effects of a milling medium and duration on the powders corrosion behavior (figure 34). The common feature of all the powders studied is the decrease in the corrosion rate, observed after the system has reached a certain level of corrosion losses. Such behavior is characteristic of any Fe-Si alloys; it is due to the iron selective dissolution, the surface enrichment with silicon, and the SiO_2 passive film formation [35, 36].

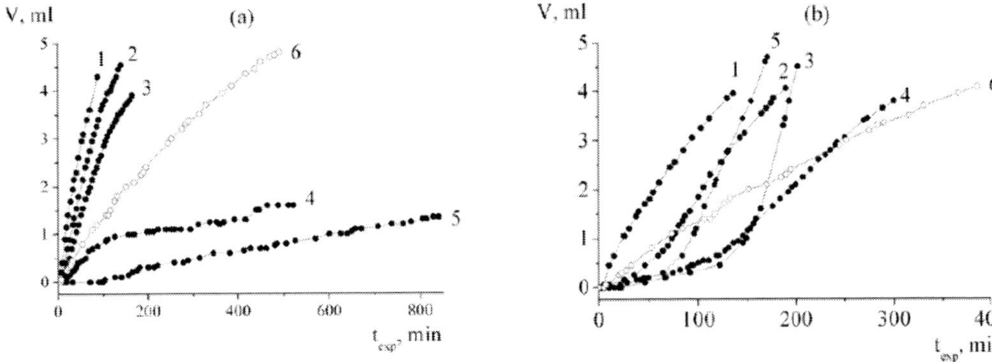

Figure 34. Plots of the volume of absorbed oxygen vs. the exposure time for Fe-Si powders milled in heptane alone (a) and heptane with oleic acid (b) in a corrosive medium; t_{mil}=(1) 1, (2) 12, (3) 24, (4) 48 and (5) 99 h; curves 6 refer to the initial alloy Fe80Si20.

With an increase in t_{mil}, the mean corrosion losses of the powder prepared in heptane decreases (figure 34a); this means that some deceleration appears, that is, the powder passivation is facilitated. At t_{mil}=99 h, the powder is initially passive; it corrodes with constant (small) rate. The enhancement of the corrosion resistance of the powder milled in heptane agrees with the accumulation of the amorphous phase. According to [36], the passive films form easier at homogeneous and uniform surfaces of amorphous alloys; they also have better protective ability.

The powders milled in the presence of oleic acid show some incubation period that lasts 5 to 120 min, depending on t_{mil}. The existence of an incubation period is caused by the formation of a protective film at the powder surface, the destruction of which makes corrosion accelerate. It is noteworthy that the initial protective film at the powders milled with oleic acid differs in its nature from the film which is capable of bringing about the stable passive state of the powders milled in heptane (t_{mil}=99 h).

As in the case of Fe-based powders, the protective properties of the surfaces of the powder milled with oleic acid are due to an oxide film containing chemisorbed products of incomplete mechanochemical destruction of oleic acid, which form directly during milling. The decrease in the incubation period length at t_{mil}=99 h is due to an increase in defects concentration in the surface layer, caused by the silicate-carbide phase segregation and the alloy transfer to a two-phase state (figure 30).

When the incubation period comes to its end for the powders milled with oleic acid, the protective film destructs, and the corrosion rate increases with an increase in t_{mil}. Such behavior can be explained as follows: unlike the powders ground in heptane, no amorphous phase accumulates in the bulk of the powder prepared with oleic acid particles during milling. In addition, the alloy becomes significantly depleted of silicon.

Let us discuss the annealing effect on the corrosion behavior of the powders (figure 35). After annealing at 400^0C the corrosion resistance of the powder milled in heptane decreases because of the amorphous phase crystallization. The increase in the corrosion resistance after annealing at 500 and 800^0C is caused by a substantial enrichment of the surface with Si (table 8), leading to a SiO_2 film formation.

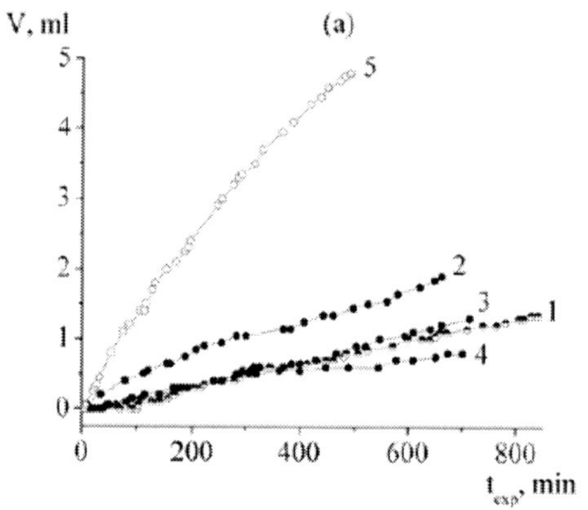

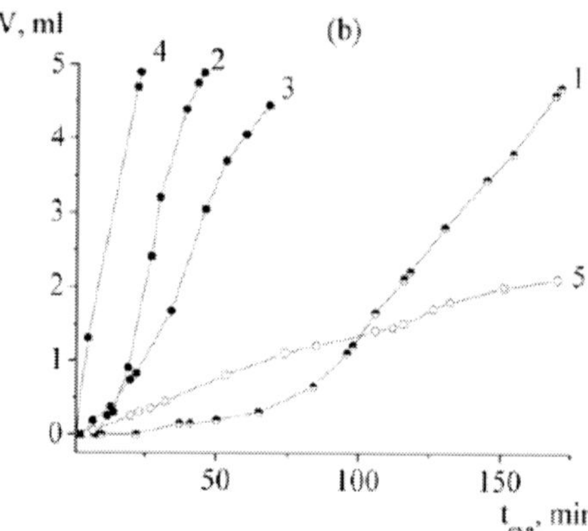

Figure 35. Plots of the volume of absorbed oxygen vs. the exposure time for Fe-Si powders milled in heptane alone (a) and heptane with oleic acid (b) in a corrosive medium; t_{mil}: (1) 99 h; (2-4) sample (1) upon annealing at 400, 500 and 800^{0}C, respectively; curves 6 refer to the initial alloy Fe80Si20.

As a result of annealing, the corrosion resistance of the powder obtained with oleic acid decreases significantly, because the amorphous phase crystallizes and the protective oxide-organic film undergoes destruction. The SiO_2 formed at the surface of the powder milled with oleic acid does not protect against corrosion. Some improving of the corrosion resistance (500^{0}C) can be due to the formation of the maximum amount of the iron silicate-carbide phase in the powders. Although the powder obtained with oleic acid upon annealing at 800^{0}C corresponds to a Fe87Si13 alloy, its corrosion resistance appears to be much lower than that of the Fe-based powders prepared in all the media studied. This may be associated with the presence of a significant amount of SiO_2 phase inclusions in the bulk of material.

3. Pitting Stability of Compacted Nanocrystalline Systems Fe+Fe$_3$C in Neutral Media

For practical application of compacted nanocrystalline materials and better understanding of their fundamental characteristics it is important to study their electrochemical behavior. Besides, mechanoactivation followed by compacting makes it possible to synthesize model nanocrystalline objects, the study of which will widen the concept of the corrosion behavior of important traditional materials such as construction steels.

The corrosion-electrochemical behavior of Fe-C alloys is known to depend on the nature of the evolving excess phases, their quantity and microstructure. Most papers are devoted to the influence of widespread metallurgic impurities (sulfides, oxides, etc.), the quantity and type of the cementite evolved [49-53]. The systematic study of the cementite influence has only been conducted as regards relatively low concentrations – up to 1.5 weight % [49-52].

The present part of the paper deals with the investigation of the laws of anodic dissolution and pitting formation of compacted nanocrystalline systems (NS) α-Fe+Fe$_3$C depending on cementite content (9 to 92 wt.%) in neutral chloride-containing media. To compare the behavior of the α-Fe+Fe$_3$C system in nanocrystalline and polycrystalline states cast steel (grain size 40 μm) with the same carbon content (1.3 wt.%) as in Fe95C5 has also been studied.

In figure 36 demonstrated are X-ray diffractograms of NS α-Fe+Fe_3C. Phase content, grain size and lattice parameters found from the diffractograms are given in the table 9.

The corrosion potential of all the investigated α-Fe+Fe_3C systems in borate buffer solution (pH=7.4) is in the passivation area (50÷150 mV), which corresponds to the formation on a metal surface of a protective oxide film including, according to [54], Fe_3O_4/Fe_2O_3. Anode curves taken from corrosion potential are also typical of metal in a passive state: active dissolution areas are absent and the metal dissolves with one and the same relatively slow speed within all the range of potentials (see figure 37-1). Current increase close to 1250 mV is connected with the beginning of oxygen emission at the passive surface of NS.

Upon cathodic polarization a passive layer is removed from the surface, and activation areas of the dissolution of ferrite ($Fe+2H_2O \rightarrow Fe(OH)_2+2H^++2e$; $E_{max} \approx -300$ mV) and cementite ($Fe_3C + 6H_2O \rightarrow 3Fe(OH)_2+C+6H^++6e$; $E_{max} \approx -100$ mV) components appear in anodic curves (figure 37-2). The maximum dissolution current values for α-Fe and Fe_3C phases (see figure 38) correlate with the content of these phases in NS α-Fe+Fe_3C (see table 8). The increase of the cementite content causes a detrimental effect on the α-Fe + Fe_3C passivation, which manifests itself in the increase of the currents of dissolution from a passive state (from 2 μA/cm^2 for Fe95C5 to 40μA/cm^2 for Fe75C25). It is noteworthy that the anodic curves measured at the nanocomposite Fe95C5 and cast polycrystalline steel practically coincide (figure 38-1, 2). Therefore, a nanocrystalline state (grain size 40 nm) has no connection with anodic processes in neutral media (in the absence of depassivating ions).

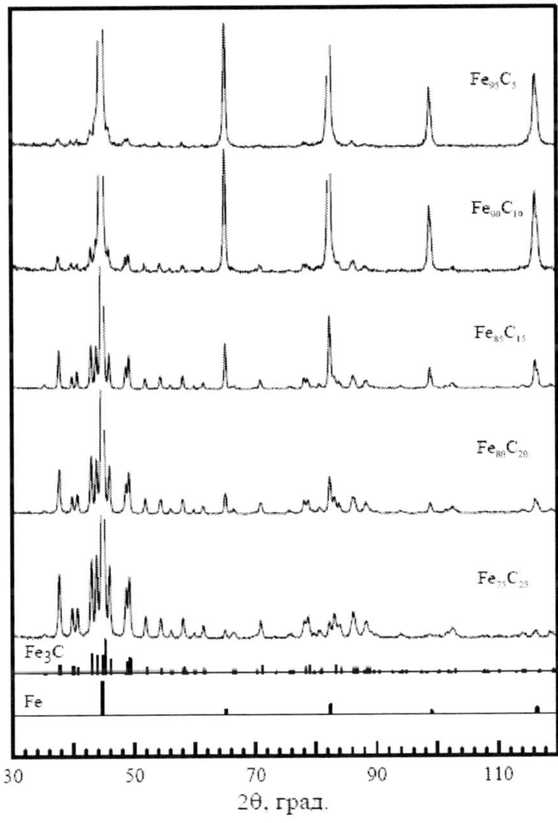

Figure 36. X-ray diffractograms of NS α-Fe+Fe_3C.

Table 9. Phase composition and grain size <L> for NS α-Fe+Fe_3C

Sample	Phase composition, wt.% (±3%)		<L>, нм (±1 нм)	
	α-Fe	Fe_3C	α-Fe	Fe_3C
Fe95C5	91	9	38	48
Fe90C10	84	16	40	49
Fe85C15	45	55	49	47
Fe80C20	24	76	39	32
Fe75C25	8	92	42	29
Fe+TEVS	73	27	15	20

For a passive state metal it is highly probable that pitting corrosion can arise in media that contain depassivating ions such as Cl^-. The corrosion process here localizes in defective areas of a passive film [36, 54]. The activating action of the ions is only manifested at potentials which exceed local activation potential (E_{la}). It is these values that we used for the comparative analysis of the pitting stability of NS α-Fe + Fe_3C.

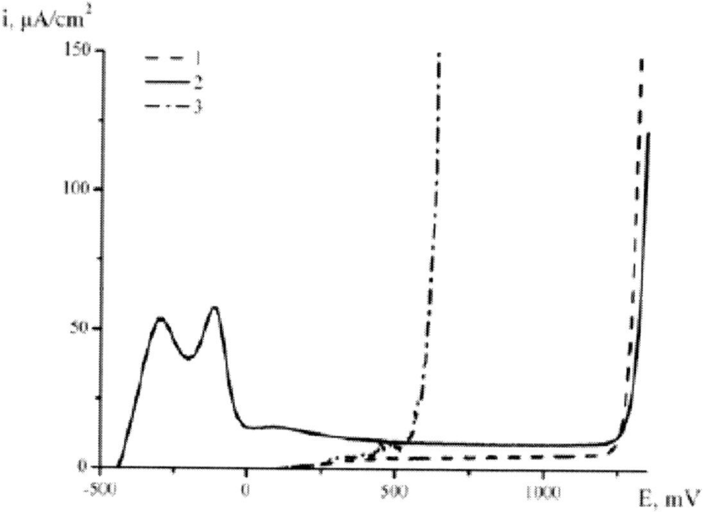

Figure 37. Anodic curves taken for NS Fe85C15: (1,2) – in borate buffer solution (pH 7.4); (3) - in borate buffer solution (pH 7.4) with added NaCl 10^{-2} mol/l. (2) – cathodic prepolarization 10 min at -0.8 V.

To investigate the influence of the cementite content on the NS α-Fe + Fe_3C pitting stability a borate buffer solution, pH=7.4, with the admixture of 0.01 M NaCl was taken. In the figure 37-3 you can see the influence of the added Cl^- ions on the form of the anodic curve for NS Fe85C15. Current increase at E_{la}≈500 mV is associated with pitting formation on the sample surface. The dependence of E_{la} on cementite content (see figure 39) demonstrates a clear stepwise behavior of the pitting stability increase with the increase of Fe_3C content. This can be due to the change of the phase which serves as a base of material (matrix). Ferrite serves a base in low-carbon samples (Fe95C5, Fe90C10, cast steel), while cementite is a base in high-carbon samples (Fe85C15, Fe80C20, Fe75C25).

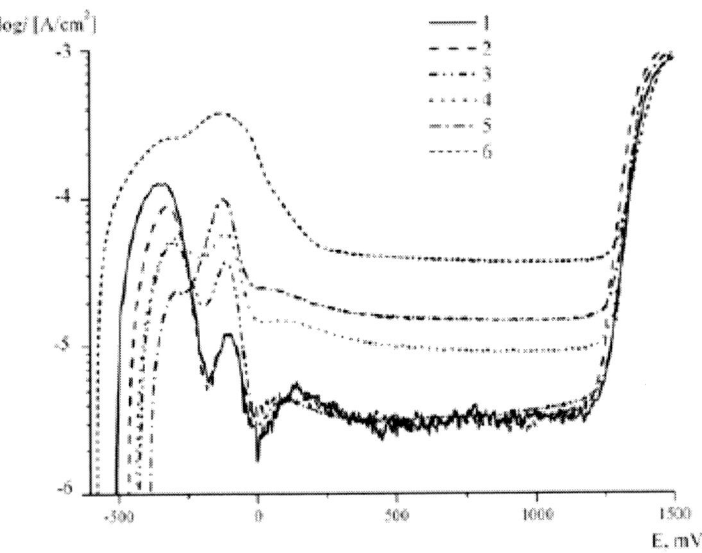

Figure 38. Anodic curves taken for cast steel and NS α-Fe+Fe_3C in borate buffer solution (pH 7.4): (1) steel; (2) Fe95C5; (3) Fe+TEVS; (4) Fe85C15; (5) Fe80C20; (6) Fe75C25. Cathodic prepolarization: 10 min at -0.8 V.

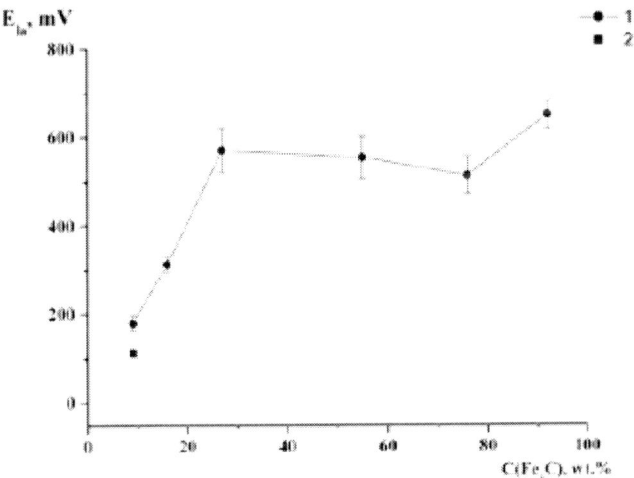

Figure 39. Dependence of the local activation potential (E_{la}) on the Fe3C content for NS α-Fe+Fe_3C (1) and cast steel (2).

The fact of such a transition is proved by the atomic force microscopy investigation of a NS surface after the selective etching of ferrite in a 3% HNO_3 alcohol solution. For a ferrite matrix samples surface etching was mainly observed along the boundaries of the particles from which NS were compacted (figure 40-a). For a cementite matrix (figure 40-b,c) dark areas correspond to the etched ferrite and light areas belong to cementite. Special attention should be drawn to NS Fe+VTES, which contains 27 weight % Fe_3C. This quantity seems not enough for matrix formation. However, cementite inclusions in NS Fe+VTES form an ordered netlike structure at the expense of the encapsulation of Fe_3C grains with silicate structures. The formation of such cementite structure not only results in the increase of pitting stability, but also lends high microhardness to H=6.0±0.5 GPa in comparison with the other NS, for which H=14.0±0.5 GPa [55].

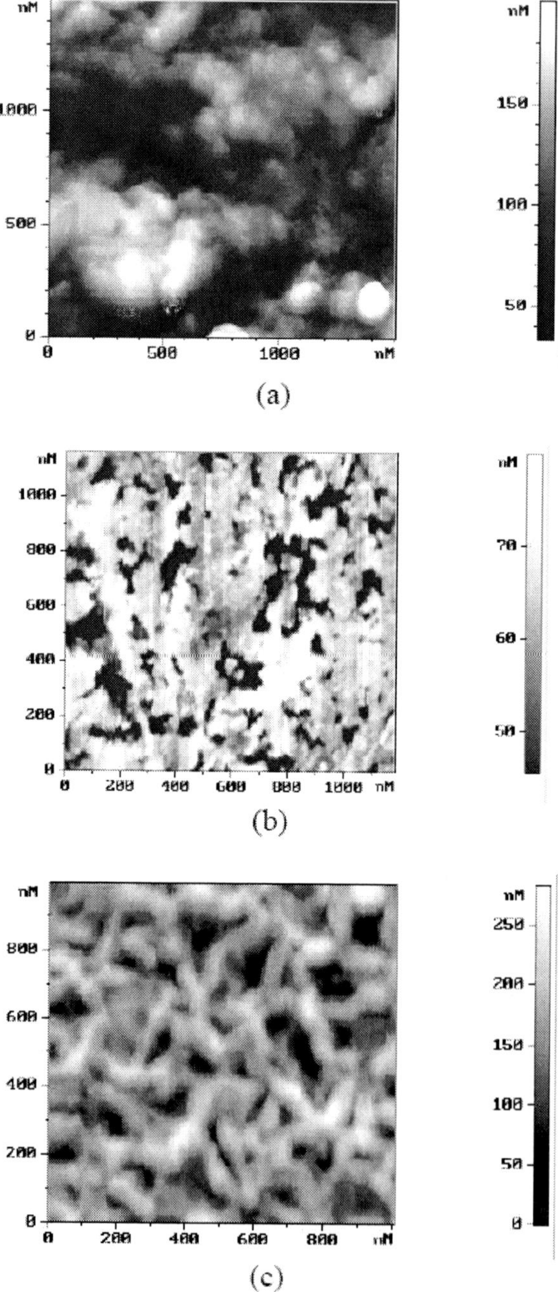

Figure 40. Atomic force microscopy images of NS Fe95C5 (a), NS Fe85C15 (b) and NS Fe+TEVS (c) after the selective etching of ferrite in a 3% HNO_3 alcohol solution.

It is well known that metal ionization in pitting runs at potentials which correspond to active dissolution [54, 56]. It was shown above that cementite dissolves at higher potentials in comparison with ferrite. In the case with the cementite matrix the pitting formed represents a plane whose walls are completely screened by cementite, and pitting does not develop without its electrochemical dissolution. Since there are not enough cementite inclusions in ferrite, and keeping in mind that it is ferrite that mainly dissolves, they lose their adhesion capacity with the material bulk, hence, the cementite dissolution is not compulsory for the pitting development.

CONCLUSIOS

The formation of a protective layer with high corrosion stability to neutral chlorine-containing media on Fe and Fe-Si powders runs directly in the process of milling in the presence of long-chained surfactants such as oleic acid. In the absence of a surfactant (acetic acid, isopropyl alcohol) the surface layer of the milled Fe-powder does not have anticorrosion characteristics, even despite the formation of relatively thick oxide films. In silicon media the enrichment of the surface layer in silicon and the formation of SiO_2 does not result in a considerable increase of the corrosion stability of the powders.

The corrosion stability increases when amorphous phases (Fe-Si-C) accumulate in the powders bulk during milling. The decrease of grain size, the formation of carbide, graphite or amorphous carbon layers on the powder particle surfaces do not considerably affect the corrosion behavior.

We can thus conclude that mechanical milling not only results in dispersity increase, but also leads to the corrosion stability increase of the powders. The corrosion stability can rise both at the expense of the surface protective layer creation and at the expense of the formation of a certain phase-structural state.

In the course of study of compact activated nanocomposites α-Fe+Fe_3C a stepwise increase of pitting stability was stated with the increase of Fe_3C content or the formation of ordered net-like cementite structure. The pitting stability can increase due to the necessity of electrochemical dissolution of a more stable cementite upon pitting formation and growth.

REFERENCES

[1] Cleiter, H. *Nanostruct. Mater.* 1995, vol .6. pp. 3-13.

[2] Fecht, H.J. *Nanostruc. Mater.* 1996, vol. 6, pp. 33-42.

[3] Suryanarayana, C. *Proc. Mater. Sci.* 2001. vol. 46, pp. 1-184.

[4] Gorokhovskii, G.A., *Poverkhnostnoe dispergirovanie dinamicheski kontaktiruyushchikh polimerov i metallov* (Surface Dispersion of Polymers and Metals in Dynamic Contact); Naukova Dumka: Kiev, 1972; 152 p.

[5] Rebinder, P.A., *Izbrannye trudy. Poverkhnostnoe yavleniya v dispersnykh sistemakh. Fiziko-khimicheskaya mekhanika* (Selected Works. Surface Phenomena in Dispersed Systems. Physicochemical mechanics); Nauka: Moscow, 1979; 381 p.

[6] Baranov, A.A.; Baranov, D.A. *Phys. Met. Metallogr.* 2003, vol. 96, pp. 400-413.

[7] Syugaev, A.V.; Lomayeva, S.F.; Reshetnikov, S.M. *Protection of Met.* 2004, vol. 40, pp. 226-231.

[8] Syugaev, A.V.; Lomayeva, S.F.; Reshetnikov, S.M. *Protection of Met.* 2005, vol. 41, pp. 263-268.

[9] Lomayeva, S.F.; Bokhonov, B.B.; Syugaev, A.V. et al. *Protection of Met.* 2005, vol. 41, pp. 465-471.

[10] Syugaev, A.V.; Lomayeva, S.F.; and Reshetnikov, S.M. *Protection of Met.*, 2006, vol. 42, pp. 315-322.

[11] Lomayeva, S.F.; Syugaev, A.V.; Reshetnikov, S.M. et al., *Protection of Met.* 2007, vol. 43, pp. 194-202.

[12] Popov., V.A.; Lesuer, D.R.; Kotov, I.A. et al. Proc. of Symp "Ultrafine Grained Materials II", 2002, Washington. pp. 289-295.

[13] Warren, B.E.; Averbach, J. *J. Appl. Phys.* 1950, vol. 21, pp. 595-605.

[14] Voronina, E.V.; Ershov, N.V.; Ageev, A.L.; Babanov, Yu.A. *Phys. Status Solidi B.* 1990, vol. 160, pp. 625-634.

[15] Chen, Z.; Li, F. *Hyperfine Interact.* 1998, vol. 112, pp. 101-106.

[16] Zhou, W.L.; Carpenter, E.; Lin, J. et al. *Europ. Phys. J. D.* 2001, vol. 16, pp. 289-299.

[17] Pletnev, M.A.; Dorfman, A.M.; Lyakhovich, A.M. et al. *Protection of Met.* 1999, vol. 35, pp. 32-35.

[18] Huffman, G.P.; Errington, P.R.; and Fisher, P.M. *Phys. Status Solidi.* 1967, vol. 22, pp. 473-481.

[19] Le Caër, G.; Matteazzi, P. *Hyperfine Interact.* 1991, vol. 66., pp. 309-318.

[20] Pople, J.A. *Mol. Phys.* 1958, vol. 1, pp. 168-175.

[21] Vasil'ev, L.S.; Lomayeva, S.F. *J. Mater. Scie.* 2004, vol. 3, pp. 5411-5415.

[22] Vasil'ev, L.S.; Lomayeva, S.F. *Colloid J.* 2003, vol. 65, pp. 639-647.

[23] Nefedov, V.I., *Rentgenoelektronnaya spectrockopiya khimicheskikh soedinenii* (X-ray Photoelectron Spectroscopy of Chemical Compounds); Khimiya: Moscow, 1984; 256 p.

[24] Beamson, G.; Briggs, D. *High Resolution XPS of Organic Polymer.* The Scienta ESCA300 Database; Wiley: Chichester, 1992; 582 p.

[25] Asami, K.; Hashimoto, K. *Corros. Sci.* 1977, vol. 17, pp. 559-570.

[26] McIntyre, N.S.; Zetaruk, D.G. *Anal. Chem.* 1977, vol. 49, pp. 1521-1528.

[27] Practical Surface Analysis by Auger and X-ray Photoelectron Spectroscopy, Briggs, D. and Seach, M.P.; Ed. Wiley: Chichester, 1983, 500 p.

[28] Lyakhovich, A.M.; Dorfman, A.M.; Povstugar, V.I. *Izv. Akad. Nauk, Ser. Fiz.* 2002, vol. 66, pp. 1054-1058.

[29] Mikhailik, O.M.; Povstugar, V.I.; Mikhailova, S.S. et al. *Colloid Surf.* 1991, vol. 52, pp. 325-328.

[30] Lomayeva, S.F.; Yelsukov, E.P.; Konygin, G.N. et al. *Colloid J.* 2000, vol. 62, pp. 579-587.

[31] Lomayeva, S.F.; Ivanon, N.V.; Yelsukov, E.P. *Colloid J.* 2004, vol. 66, pp. 186-191.

[32] Yelsukov, E.P.; Dorofeev, G.A.; Fomin. V.M. et al. *Phys. Met. Metallogr.* 2002, vol. 94, pp. 356-366.

[33] Likhtman, V.V.; Shchukin, E.D.; Rebinder, P.A. *Fiziko-khimicheskaya mekhanika metallov* (Physicochemical Mechanics of Metals); Khimiya: Moscow, 1984, 303 p.
[34] Kataby, G.; Cojocaru, M.; Prozorov, R.; Gedaken, A. *Langmuir.* 1999, vol. 15, pp. 1703-1708.
[35] Uhlig, H.H.; Revie, R.W. *Corrosion and Corrosion Control*; Wiley: New York, 1985, 450 p.
[36] Tomashov, N.D.; Chernova G.P. *Teoriya korrozii I korrozionno-stoikie splavy* (Theory of Corrosion and Corrosion-Resistant Structural Alloys); Metallurgiya: Moscow, 1986, 359 p.
[37] Yelsukov, E.P.; Dorofeev, G.A.; Konygin, G.N. et al. *Phys. Met. Metallogr.* 2002, vol. 93, pp. 278-288.
[38] Schlogs, R.; Boehm, H.P. *Carbon.* 1983, vol. 21, pp. 345-358.
[39] Wang, G.; Harrison, A. *J. Colloid Interface Sci.* 1999, vol. 217, pp. 203-207.
[40] Atarashi, T.; Kim, Y.S.; Fujita, T.; Nakatsuka, K. *J. Magn. Magn. Mater.* 1999, vol. 201, pp. 7-10.
[41] Yelsukov, E.P.; Lomayeva, S.F.; Konygin, G.N. et al. *Phys. Met. Metallogr.* 1999, vol. 87, pp. 114-119.
[42] Lomayeva, S.F.; Yelsukov, E.P.; Konygin, G.N. et al. *Nanostruct. Mater.* 1999, vol. 12, pp. 483-486.
[43] Lomayeva, S.F.; Yelsukov, E.P.; Konygin, G.N. et al. *Colloids Surf. A.* 1999, vol. 162, pp. 279-284.
[44] Laoharojanaphand, P.; Lin, T.J.; Stoffer, J.O. *J. Appl. Polymer Sci.* 1990, vol. 40, pp. 369-374.
[45] Lee, Y.P.; Bevolo, A.J.; Lynch, D.W. *Surf. Sci.* 1987, vol. 188, pp. 267-277.
[46] Yelsukov, E.P.; Maratkanova, A.N.; Lomayeva S.F. et al., *J. Alloys and Compouds.* 2006, vol. 407, pp. 98-105.
[47] Lomayeva, S.F.; Yelsukov, E.P.; Maratkanova, A.N. et al. *Phys. Met. Metallogr.* 2005, vol. 99, pp. 590-594.
[48] Chen, Y.C.; Chen, C.M.; Su, K.C. *Mater. Sci. Eng.* 1991, vol. A133, pp. 596-600.
[49] Reformatskaya, I.I.; Sulizhenko, A.N. *Protection of Met.* 1998, vol. 34, pp. 447-450.
[50] Reformatskaya, I.I.; Rodionova, I.G.; Beilin, Yu.A. et al. *Protection of Met.* 2004, vol. 40, p. 447-452.
[51] Khaldeev, G.V.; Kamelin, V.V.; Pevneva, A.V.; Zazhigina, T.V. *Zashch. Met.* 1984, vol. 20, pp. 218-223.
[52] Zazhigina, T.V.; Pevneva, A.V.; Khaldeev, G.V.; Kuznetsov, V.V. *Zashch. Met.* 1984, vol. 20, pp. 279-281.
[53] Kasparova, O.V.; Plaskeev, A.V.; Kolotyrkin, Ya.M. et al. *Zashch. Met.* 1985, vol. 21, pp. 339-345.
[54] Sukhotin, A.M. *Fizicheskaya khimiya passiviruyushchikh plenok na zheleze* (Physical Chemistry of Passivating Films on Iron); Khimiya: Leningrad, 1989; 320 p.
[55] Yelsukov, E.P.; Ivanov, V.V.; Lomayeva, S.F. et al. *Perspektivnie Materiali.* 2006, no. 6, pp. 59-63.

[56] Kaesche, H. *Korrosion Der Metalle;* Springer-Verlag: Berlin, Heidelberg, New York, 1979; 400 p.

In: Powder Metallurgy Research Trends
Editors: L. J. Smit and J. H. Van Dijk
ISBN: 978-1-60456-852-3

Chapter 2

POWDER PRODUCTION VIA SPRAY ROUTE

Morteza Eslamian and Nasser Ashgriz
Department of Mechanical and Industrial Engineering, University of Toronto,
5 King's College Road, Toronto, ON, M5S 3G8, CANADA

1. OVERVIEW

Fine and ultrafine powders in the form of metal, ceramic, pharmaceutical, and food ingredients have numerous applications in manufacturing and powder metallurgy, development of sensors, thermal barrier coatings, catalysts, pigments, drugs, and biotechnology, to name a few. There are several methods of powder production, such as wet chemistry, e.g., sol-gel and emulsion, mechanical methods, e.g., ball milling, gas phase methods, e.g., chemical and physical vapor deposition, liquid phase spray methods, e.g.,molten metal spray atomization, spray pyrolysis, and spray drying, and liquid/gas phase methods, e.g., flame spray pyrolysis.

Spray methods are simple one step methods suitable for production of a broad range of powders with controlled properties for specialty applications. These methods include but not limited to spray drying (SD) for production of pharmaceuticals and food powders, spray pyrolysis (SP) for preparation of non-agglomerated monodispersed fine and unltrafine ceramic powders, flame spray pyrolysis (FSP) for synthesis of ceramic and complex nanoparticles, and melt atomization (MA) mostly for production of metal powders. The other spray dependent methods of powder production include spray freeze drying (SFD) and emulsion combustion method (ECM).

MA is used as the first step in powder metallurgy (PM) to powder the primary material. The melt is atomized into small droplets, which solidify and form particles as they fly in the surrounding medium. MA is also the first step of spray forming (SF), even though the final product of SF is not in the powder form. In SF, the molten droplets hit a surface before they solidify.

SP relies on direct conversion of single solution droplets, generated by atomization of a precursor, to single particles in a one step process. In SP, a precursor solution is sprayed into a hot-wall reactor, where the solvent evaporates and the solute precipitates within the droplets to form the particles. SP is accompanied by a chemical decomposition of the precursor to the final product, which is usually a ceramic and a gaseous byproduct. In SD, the solution droplets directly dry and form the powder without any chemical reaction taking place. One of the applications of this method is in production of pharmaceutical and bio-powders.

In contrast to SD and SP, in FSP, the liquid precursor is entirely evaporated in a high temperature spray flame, formed by combustion of the atomized liquid precursor. In FSP nanoparticles are made in the gas phase. FSP is an emerging method of production of a broad spectrum of new materials, with tailored properties.

In this chapter, fundamentals and recent advances regarding the above mentioned spray methods of producing powders are considered. The theory and the fundamental governing equations for each method are introduced and discussed. The available models to simulate these processes are presented. These models are excellent tools to design the reactors and explore and predict the effect of reactor conditions on the morphology of prepared particles. Also, the experimental results on the effect of processing conditions on the morphology and other characteristics of the powders are reviewed. Some of the powders that have been successfully produced by each method are introduced and their applications are discussed.

2. Melt Atomization (MA)

2.1. Introduction

Melt atomization is used for the production of metal powders as in powder metallurgy (PM), in spray deposition for coating or in spray forming (SF) of bulk structures such as billets or tubes. Spray deposition and forming is not the subject of this chapter and will not be considered further.

Powder metallurgy (PM) is a fabrication technique consisting of three processing stages. First, the primary material is physically powdered. The powder is then injected into a mold or passed through a die to produce a weakly cohesive structure near the dimensions of the object, ultimately to be manufactured. Finally, the end part is formed by applying pressure, high temperature, long setting times, or any combination thereof.

In PM, it is possible to fabricate components, which otherwise would decompose or disintegrate. All considerations of solid-liquid phase changes can be ignored, so powder processes are more flexible than casting, extrusion forming, or forging techniques. Controllable characteristics of products prepared using various powder technologies include mechanical, magnetic, and other unconventional properties of such materials as porous solids, aggregates, and intermetallic compounds. PM commodities are today used in a wide range of industries, from automotive and aerospace applications to power tools and household appliances. Any fusible material can be atomized. Powders are produced by MA in amounts varying from a few kilograms for expensive dental materials to one million tons for steel or iron shot per year for a single plant. Particle sizes depend on the atomization conditions and their application, and vary from a few microns to several millimeters. For many applications it is important to achieve narrow size distributions and shapes. In addition to PM, powders may be prepared by comminuting, grinding, chemical reactions, or electrolytic deposition.

2.2. Atomization

The mechanisms of melt breakup and atomization have been extensively researched, showing that atomization typically consists of 3 steps: (1) primary breakup of the melt stream; (2) secondary breakup of molten droplets and ligaments; and (3) cooling and solidification of particles. The theoretical analyses of the atomization process to predict droplet sizes have only been able to provide results that are in moderate agreement with the experimental data.

There are several different techniques for the atomization of molten metals, many of which are derived from the powder metallurgy industry. The major atomization techniques used in MA are gas atomization, centrifugal atomization, and water atomization techniques.

2.2.1. Gas Atomization

In gas atomization technique, molten metal is forced to atomize using the kinetic energy of a gas. There are several different methods of combining a gas flow with the molten metal. One method is to force the molten metal through through an orifice at moderate pressures and inject a high speed onto the molten metal stream just after it leaves the nozzle. Once the liquid is atomized, the powder is segregated from the gas using gravity or cyclonic separation. In the cyclone the coarse and the fine powder fractions are separated from the atomization gas. The metal powder is collected in sealed containers which are located directly below the cyclones.

The usual performance index in gas atomization is the Reynolds number (Re) based on gas velocity, gas properties, and nozzle diameter. At low Re, the gas does not have enough energy to atomize the liquid jet, and the jet breaks into large droplets. As the gas Re is increased, the atomization efficiency improves, and droplet sizes become smaller. The atomization process produces a wide spectrum of particle sizes, necessitating downstream classification by screening and remelting a significant fraction of the grain boundary. Many other techniques, such as nozzle vibration, nozzle asymmetry, and multiple impinging streams, are developed to increase atomization efficiency, produce finer grains, and to narrow the particle size distribution. One of the main limitations in reducing the droplet sizes in the reduction of the melt nozzle diameter. It is difficult to eject molten metals through orifices smaller than a few millimeters in diameter. This limits the minimum size of powder grains produced by this method to approximately 10 μm.

The atomizing gas mass flow rate to molten metal mass flow rate ratio is another key parameter in controlling the droplet diameter and hence the cooling rate, billet temperature and resulting solid particle nucleant density. The gas-metal ratio (GMR) is typically in the range of 1.5 to 5.5, with yield decreasing and cooling rates in the spray increasing with increasing GMR. Atomizing gas is generally either N_2 and can be either protective or reactive depending on the alloy system, or Ar which is generally entirely inert but more expensive than N_2.

2.2.2. Centrifugal Atomization

Centrifugal atomization involves pouring molten metal at relatively low flow rates onto a spinning plate or disc. Provided the rotation speed is sufficient to create high centrifugal forces at the periphery to overcome surface tension and viscous forces, the melt is broken up into droplets. Droplet diameters produced by centrifugal atomization are dependent primarily on the rotation speed, (up to 20,000 rpm) and are typically in the range 20 to 1000 μm.

Centrifugal atomization is generally conducted under an inert atmosphere of Ar or N_2 to prevent oxidation of the fine droplets or can be operated under vacuum.

2.2.3. Water Atomization

Water atomization technique involves a thin jet of liquid metal impacted by high-speed streams of atomized water, which break the jet into drops and cool the powder before it reaches the bottom of the atomization chamber. In subsequent operations, the powder is dried. The advantage of water atomization is that metal solidifies faster with water than by gas. The solidification rate is inversely proportional to the particle size. Also, smaller particles are produced because the total atomization energy of water is much higher than that of gas. Therefore, one can produce smaller particles by water atomization. The smaller the particles, the more homogeneous the microstructure will be. Particles will have a more irregular shape and the particle size distribution will be wider. Also, some surface contamination may occur due to the oxidation and skin formation.

2.3. Applications

2.3.1. Metal and Non-Metal Powders

The main application of PM is in the production of metal powders for the manufacturing of mechanical components raging from welding rods to aircraft components. The way the powders are post-processed is important and depends on the properties of the metal. The post processing methods are explained by Schaefer and Kushner (1990).

The following list provides a few applications of metal powders. Many new applications are developed rapidly: aluminum powder with high reactivity used as solid rocket fuel and explosives; cobalt powder used to make strips for diamond synthesis; AgSnCu dental amalgams; zirconium powder used in batteries; Iron powder used as food additive; light bulb filaments; metal glasses for high-strength films and ribbons; electrical contacts for handling large current flows; and magnets (Yule and Dunkley 1994).

The atomization of non-metal melts is also practiced. For instance, by atomization of glassy materials, commodities such as ceramic fibers, glass wool, etc., are produced. Some special products of PM are as follows: Al_2O_3 whiskers coated with very thin oxide layers for improved refractories; iron compacts with Al_2O_3 coatings for improved high-temperature creep strength; linings for friction brakes; heat shields for spacecraft reentry into Earth's atmosphere; and filters for gases (Yule and Dunkley 1994).

2.3.2. Metal Matrix Composites

Metal matrix composites (MMC) are composite materials with at least two constituent parts, one being a metal. The other material may be a different metal or another material, such as a ceramic or organic compound. MMCs are made by dispersing a reinforcing material into a metal matrix. The reinforcement surface can be coated to prevent a chemical reaction with the matrix. The reinforcement material is embedded into the matrix. The reinforcement does not always serve a purely structural task (reinforcing the compound), but is also used to change physical properties such as wear resistance, friction coefficient, or thermal conductivity.

Several methods are currently being used to produce MMCs such as Al/SiC composite. These include stir casting, preform infiltration, powder metallurgy and spray forming or melt atomization (Ortiz 1996, Rohatgi, and Asthana, 1991, Stefanescu et al. 1990). The various PM-related processes currently in use in the fabrication of MMCs, are reviewed by Liu et al (1994), outlining the common problems encountered in each of these fabrication processes. New developments in fabrication techniques made in recent years have been enumerated and discussed in detail.

The current limitations to wider industrial applications of MMCs are inadequate toughness and high cost of these materials. The low ductility and fracture toughness of MMCs are mostly caused by the premature failure of the interfacial bonds, as a result of poor wetting and interracial reactions between metal powders and reinforcement particles. These problems are expected to be solved by improvements in the low-temperature and high-pressure processes. Some new techniques developed to fabricate high-performance MMCs have so far increased the cost of these final products because of the inclusion of expensive equipment and operations, such as complex blending processes, powder handling and consolidation. One of the most important directions for future research into the fabrication of PM MMCs is, therefore, to develop ways to lower the processing cost without compromising the mechanical properties.

The conventional method of the production of MMCs using MA approach is the spray atomization of molten aluminum and injection of SiC particles into the stream of molten aluminum particles (Figure 1*a*). In this process segregation due to gravity is avoided. Interfacial reactions between SiC and aluminum are minimized due to the short time the SiC particles are in contact with the molten aluminum. However, the mechanical properties of composites produced by this method have not been impressive (Wu and Lavernia 1991). This is due to different cooling rates and improper collision angles causing that many of the particles to not be engulfed by the molten aluminum during their flight. The result is that, particles are not strongly bonded to the matrix (Wu and Lavernia 1992). A method of the production of MMCs based on the centrifugal atomization of the molten matrix while injecting SiC particles just prior to atomization (Figure 1*b*) has been developed by Eslamian et al (2008). A stream of molten aluminum is forced to impinge onto a rotating disk (rotating at 24000 RPM). The molten metal is spreads on the rotating disk forming a thin sheet of metal, which later breaks to form small droplets. SiC particles are injected onto to the molten metal sheet prior to atomization. SiC particles penetrate into the melt and later form Al/SiC particles. This method has two advantages: One is that it reduces particle agglomeration by creating a pre-combined powder metallurgy MMC with evenly dispersed particles, which does not require mixing. The second is that it reduces the SiC particles contact time with the molten aluminum, therefore, the reaction times are minimized. The Al/SiC composite powder produced contained 18 vol.% of SiC particles and 1.2 vol.% of voids. For both aluminum and Al/SiC powders almost all of the large particles (>200 μm) had irregular shapes. They observed that the variation of ASTM grain size with particle size for Al/SiC powders produced using this method was not substantial. This is desirable for the synthesis of high quality metal matrix composites. The void volume fraction in Al/SiC composite particles prepared by this method is low. Low amount of void present in the matrix is also desirable for production of high quality strong metal matrix composites.

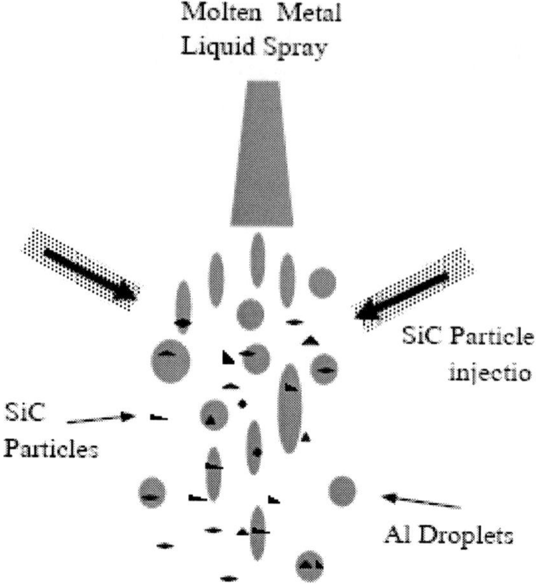

(a) Conventional spray method

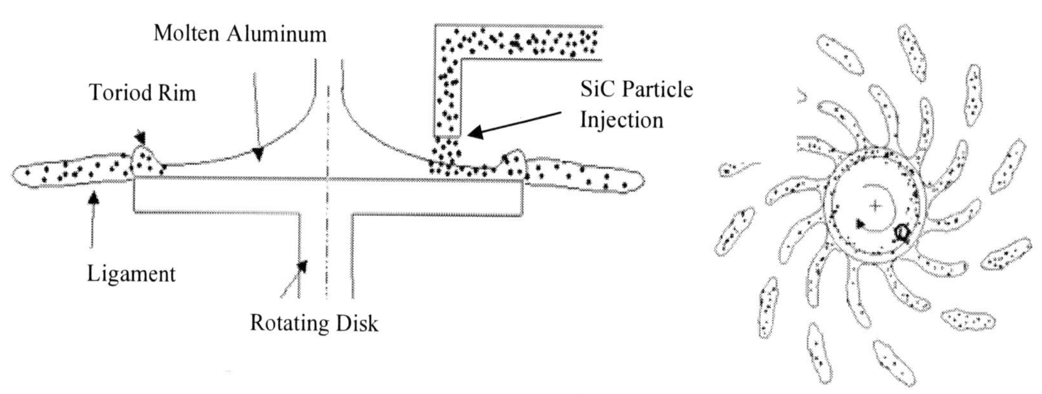

(b) Centrifugal atomization

Figure 1. (*a*) Schematic of conventional melt atomization method to produce MMC powders and (*b*) schematic of centrifugal atomization method for production of Al/SiC MMC (Eslamian et al. 2008).

2.4. Effect of Operating Parameters

Uslan et al. (1999) experimentally studied the effect of operating conditions on the characteristics of aluminum powder produced by gas atomization of melts. They found that: the geometry and dimensions of the nozzle are important in maximizing the delivery of the gas energy to the liquid metal stream; nitrogen gas atomization produces more spherical powders than air; and cooling rate is sensitive to particle size.

Grant et al. (1993) developed a mathematical model to describe the in-flight dynamic and thermal histories of gas atomized melt droplets. They used the model to investigate the effect of dynamic and thermal behavior of individual gas atomized droplets and cooling and solidification behavior of the overall spray. The main findings of their model are as follows:

2.4.1. Effect of Droplet Size

Droplet dynamic and thermal behavior is strongly dependent upon droplet diameter. Droplets smaller than 50 μm in diameter are accelerated and then decelerated rapidly by the gas to approach seeded flow. On the other hand, droplets greater than 50 μm in diameter are accelerated and decelerated relatively slowly by the gas. Droplet heat transfer coefficients fall with decreasing gas/droplet slip velocity during droplet acceleration. Five different droplet cooling regimes are identified: cooling in the liquid state; undercooling and subsequent recalescence; segregated solidification; eutectic solidification; and cooling in the solid state. Solid fraction evolution within droplets is quickest during recalescence and then slows with the onset of segregated and then eutectic solidification.

2.4.2. Effect of Initial Gas Velocity

Increasing the initial axial gas velocity has three effects on individual droplet dynamic and thermal behavior. There is a reduction in flight times, an increase in mean heat transfer coefficients and an increase in mean cooling rates. Although droplet cooling rates increase with increasing initial axial gas velocity, the reduction in droplet flight times is the dominant effect on thermal behavior, so that individual droplet solid fractions reduce at each axial distance.

2.4.3. Effect of Melt Flow Rate

Increasing the melt mass flow rate has no effect on individual droplet dynamic and thermal behavior, but affects spray cooling and solidification because of a shift to progressively larger mean droplet diameters. Droplet cooling rate is inversely proportional to droplet diameter so that the increase to larger mean droplet sizes and standard deviation produces an approximately inverse reduction in the spray solid fraction with increasing melt mass flow rate.

2.4.4. Effect of Melt Superheat

Increasing the melt superheat had no effect on individual droplet dynamic behavior. Superheated droplets travel to greater axial distances to lose their superheat and cool to the same temperature. The melt superheat has no effect on droplet size distribution so that the spray solid fraction reduces linearly with increasing melt superheat. However, a strong variation in melt viscosity and surface tension with melt superheat would affect the droplet size distribution produced by atomization, and therefore the spray solid fraction.

2.5. Recent Advances and Future Trends

Orban (2004) reviewed the new research directions in PM. Processes such as sintering and compaction are more or less common to all methods of PM. What differentiates between PM methods is the way the powder is produced. In MA method of powder production, the governing process is the atomization of melt and solidification of droplets. Therefore, no wonder that many researches are focused on the improvement of atomization technology to improve powder purity and compressibility through a better particle size distribution, a higher cleanliness, smaller particle median size (e.g. for aluminum particles), etc.

Anderson and Terpstra (2002) studied the performance of a high pressure gas atomizer, which is a method of producing fine and spherical powders, or spherical powders of a narrow particle size class. In another study, Antipas (2006) studied the breakup mechanism in gas atomization of liquid metals. He developed a model for the atomization of a liquid column perturbed by a flowing gas phase. The model is based on the concept of the formation of sinusoidal waves travelling along the surface of the liquid. According to Antipas, his model covers both primary and secondary atomization and can in principle describe the breakup of well-defined liquid shapes. Antipas's model was validated against the experimental data for the atomization of metals.

Kearns (2003) discussed the effect of the particle size on reactivity of aluminum particles and the market opportunities for nano-aluminum and the potential method of producing aluminum nanopowder, based on the conventional gas atomization technique. According to his results, even at 2 μm median size, there is evidence of exceptional reactivity. This size is at the limit of what can be produced today by gas atomization. Therefore, development of low-cost methods or nozzles to produce sub-micron powders is a potential research area in the field.

Another potential research area in MA is the production of alloys with special properties. Huttunen-Saarivirta (2004) reviewed the microstructure, fabrication methods and properties of quasicrystalline Al–Cu–Fe alloys. The major method of the production of such alloys is the centrifugal and gas atomization. Quasicrystalline materials constitute a new materials group with certain crystalline structural characteristics, such as the generation of Bragg peaks in the X-ray data and points in the electron diffraction pattern, but translational symmetry is forbidden for crystalline materials. Besides being theoretically interesting due to their complicated atomic structure, the unique properties of quasicrystalline materials such as low electrical and thermal conductivity, unusual optical properties, low surface energy and coefficient of friction, oxidation resistance, biocompatibility and high hardness also make them interesting for many practical purposes.

Huang et al. (2005) produced spherical powders and fibers with different ratios of bioglass and different diameters in the hot gas atomization process. Bioglass is a kind of glass material with a special composition that induces specific biophysical and biochemical reactions in the implant-tissue interface, forming bioactive bonding between them. By adapting process parameters, such as gas temperature or pressure, they concluded that gas atomization is a good process for different morphology requirements, though more detailed work is needed to improve the useful yield of powders or fibers. They observed that with regard to fiber production, gas pressure plays an important role in determining the final mean fiber diameter. For spherical powder production, hot gas is needed to prolong the spheroidization time in the spray. With proper process control and final classification of the product, as-atomized materials can be directly used within biomedical applications.

Melt atomization has been also used for rapid prototyping (Yingxue et al. 2004). In this method, however, instead of atomization of a stream of molten metal, the molten metal jet from a nozzle is broken at regular intervals by imposed vibrations from a piezo-electric transducer to form a stream of uniform droplets. Because of the precisely controlled dimension, trajectory and solidification procedure of the droplets, this technology is suitable for the rapid deposition forming of metal parts. According to Yingxue et al. (2004), in contrast to the usual manufacturing technology, molten metal uniform droplet spraying technology has several advantages. It is cheap and environmentally safe. The deposited components have a refined micro-structure including fine grains, low micro-segregation, absence of macro-segregation, modified primary or secondary phases and limited oxidation. Because of the rapid solidification of the droplets, they retain an amorphous instead of crystalline structure, giving the prototype enhanced mechanical properties.

3. Spray Drying and Pyrolysis

3.1. Introduction

In spray pyrolysis (SP), a solution is sprayed into a carrier gas forming small droplets; the solvent of the droplets is vaporized and the solute is precipitated, forming the initial solid particles; finally, chemical decomposition takes place, forming the final powder particles. A schematic diagram of the spray pyrolysis process is shown in Figure 2. Spray drying (SD) is similar to spray pyrolysis, except that there is no chemical decomposition in SD and usually the process temperature is lower (Eslamian et al. 2006). SP and SD techniques may produce fully filled or hollow particles depending on the operating conditions. Hollow particles are formed if at the onset of solute precipitation on the droplet surface, the solute concentration at the droplet centre is lower than the equilibrium saturation (Jayanthi et al. 1993). Chau et al. (2008) showed that Jayanthi's model is not accplicable for the formation of NaCl partciles.

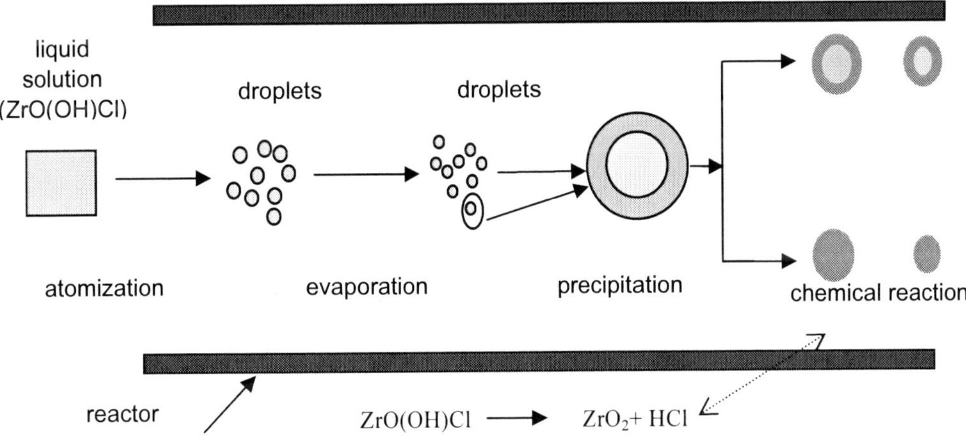

Figure 2. Basic steps of spray pyrolysis method for zirconium hydroxychloride (ZrO(OH)Cl). Once the (ZrO(OH)Cl) powders are relatively dried, they decompose to ZrO_2 and HCl gas. ZrO_2 powders remain in the collector and HCl gas goes to the carrier gas (Eslamian et al. 2006).

3.2. Theory and Mathematical Modeling

Several mathematical models for the conversion of solution droplets to particles have been developed, each of which focusing on one or few specific aspects of the process. Xiong and Kodas (1993) theoretically studied the SP process in a hot-wall tubular reactor, taking into account the time variation of vapor concentration within the reactor. However, they ignored most of the intra-particle phenomena, such as solute precipitation and particle evolution. Jayanthi et al. (1993), based on the solution concentration profile within a single droplet, proposed a model that for the majority of materials can fairly predict whether the particles are hollow or fully-filled. However, they did not consider the effect of multiple droplet evaporation and time variation of vapor concentration in the reactor. Lenggoro et al. (2000) combined the previous two models of Jayanthi et. al (1993) and Xiong and Kodas (1993), and investigated the effect of various parameters such as temperature, reactor flow rate, droplet number density, and initial droplet size on the morphology of powders produced by SP. Eslamian et al. (2006) modeled evaporation of isolated nano-sized solution droplets and particle evolution and evaporation of micro-sized solution droplets in reduced pressures and combined their model with the model of Jayanthi et al. (1993). It was shown that as the droplet size decreases to submicron or at reduced pressures the droplet evaporation rate and temperature history changes due to the non-continuum effects.

In all above mentioned models (Jayanthi et al., 1993, Xiong and Kodas, 1993, Lenggoro et al., 2000, and Eslamian et al. 2006), the droplet evaporation and particle evolution is modeled until the onset of solute precipitation on the droplet surface. In Jayanthi's model later modified by Lenggoro et al. (2000) and Eslamian et al. (2006), the droplet diameter at this point is considered to be the same as the final particle size. However, by comparing the experimental and simulation data, Eslamian et al. (2006) found that the final particle size is smaller than the droplet size at the onset of precipitation. To explain this, they introduced the induction period concept, during which the droplet shrinkage continues while the solute is precipitating on the droplet surface. Following a different approach, Farid (2003), modeled drying of single solution droplets focusing on spatial temperature variations within the droplet which is the case for drying of large droplets (~1 mm), such as milk. His model does not consider the solute concentration gradient and as a result cannot predict whether the particles will be hollow or fully-filled.

The model which is discussed here (Eslamian et al. 2009) provides a fundamental understanding of the basic phenomena involved in evaporation of solution droplets and evolution of particles as in SD and SP processes. It should be noted that, especially for industrial SD process, commercial packages are available to simulate the process but from a macroscopic point of view. These packages do not consider the detail of the molecular phenomena involved and just provide general information of the overall process.

3.2.1. Approach

In SD and SP, a precursor solution is atomized and carried into a hot-wall reactor, where the volatile species of the droplets (solvents) evaporate and mix with the carrier gas. During this process, droplet temperature and moisture content change, when evaporation of a large number of droplets is considered. This interaction between the droplets results in a continuous change in evaporation rate of each droplet. Once the solute concentration on the droplet surface reaches the Critical Supersaturation (CSS), the solute starts to precipitate on the droplet surface, while the solvent evaporation continues, until the particle is entirely dried. Here the droplets are considered as a discrete phase flowing in the carrier gas, which is considered as a continuum phase. These two phases continuously exchange momentum, heat and mass.

In a tubular reactor as it is typically used for SD and SP processes, the flow geometry is simple, and there is no need to numerically solve the transport equations. Therefore, in the continuum phase, the solution droplets are carried by the carrier gas at the average velocity of the flow. The carrier gas can be either laminar or turbulent. Analytical solutions are available for heat transfer from the reactor walls to the continuum phase, assuming constant heat flux or surface temperature. For the turbulent flows, empirical correlations are used to obtain the Nuselt number.

The solution droplets and the carrier gas have zero relative velocity. For small droplets (< 4 micron), the relative velocity between the droplets and the gas is small and so is the droplet breakup (Heine and Pratsinis, 2005). Therefore, droplet breakup, collision, and coalescence processes are negligible. The presence of a relative velocity between the droplet and the gas could cause internal motion within the droplet and asymmetric concentration equation within the droplet, for which very few drying models are available (Eslamian and Ashgriz 2006*a*). With no relative velocity between the droplets and the carrier gas, the droplets and particles move with the flow streamlines.

Evaporation of the discrete phase (droplets), which occurs symmetrically, alters the conditions of the continuum phase, such as the carrier gas moisture content and the bulk temperature, which in turn affect the droplet evaporation rate.

3.2.2. Model Description

The entire process of SD and SP, starting from the evaporation of the solution droplets to the final stage of particle formation may be divided into three main steps: shrinkage period, transition from shrinkage to constant-diameter period, called the induction period, and the constant-diameter period. The physical phenomena and as a result the governing equations during each period remain unchanged while some or all equations change as the process is switched from one step to another.

With solvent evaporation, the solute surface concentration increases until it reaches a critical value, called critical supersaturation (CSS). This is the end of the shrinkage period. At this point, there are two possibilities. If at the onset of precipitation, which will most likely occur at the droplet surface, the solute concentration everywhere within the droplet is equal to or is above the equilibrium saturation (ES), the solute starts to precipitate everywhere within the droplet, a phenomenon known as volume precipitation. Therefore, the whole droplet undergoes nucleation, crystallization and growth of the solute nuclei. In this case, the final particle will be fully-filled provided that the amount of solute available can fill the volume of the droplet, a condition that can be examined by the percolation theory (Jayanthi et al. 1993). For low initial solution concentrations or high reactor temperatures, it is likely that at the onset of precipitation on the droplet surface, solute concentration in a portion of the droplet, including droplet center, be less than the ES. If this is the case, solute precipitation occurs merely at locations where the local solute concentration is higher than ES. This results in the formation of a thin layer of solute on the droplet surface. As the droplet continues to evaporate and shrink, this layer thickens, until the thickness reaches a critical value. At this time, it is postulated that a rigid, hollow particle or a shell is established and droplet evaporation is switched to the third stage, called the constant-diameter period. This has been postulated for the vaporization of slurry droplets by Lee and Law (1991) and may be generalized to solution droplets.

Shrinkage Period

During the shrinkage or the regressing-diameter period, the solvent evaporates from the solution droplet leaving higher concentration of solute behind, while the droplet remains in the liquid phase. The droplet temperature depends on the ambient air temperature, humidity, and the rate of heat transfer from the hot surrounding gases to the droplet. For low evaporation rates, droplet temperature reaches the wet bulb temperature and it remains constant afterwards, provided the humidity does not change. In rapid evaporation, such as fuel vaporization and combustion in a high temperature environment, it is usually assumed that the droplet temperature reaches and remains at the fuel boiling point. During the shrinkage period, heat is continuously transferred from the surrounding gas to the droplet. The main portion of the heat is used for the phase change and the rest is stored in the droplet. The heat or energy balance on the droplet surface, assuming that the heat transfer from the surrounding air to the droplet is only by conduction, leads to the following equation:

$$mc_p \frac{dT_l}{dt} - 4\pi Rk(T_b - T_l) - L\frac{dm}{dt} = 0 \tag{1}$$

In equation (1), m is mass of the droplet, c_p is the droplet heat capacity, T_l is droplet temperature, R is droplet diameter, k is the droplet thermal conductivity, T_b is the reactor bulk temperature, L is the liquid latent heat of vaporization, and t is time. Considering uniform temperature droplet evaporation with no internal circulation, the equation for the spherically symmetric mass conservation of the solvent inside the droplet is as follows:

$$\frac{\partial W_l}{\partial t} - \frac{1}{\rho_l r^2}\frac{\partial}{\partial r}\left(r^2 \rho_l D_l \frac{\partial W_l}{\partial r} \right) = 0 \tag{2}$$

In equation (2), W_l denotes the solvent mass fraction, r is the radial coordinate, ρ_l is the droplet density and D_l is the mass diffusion coefficient of solute within the droplet. Prior to solving equations (1) and (2) along with the initial and boundary conditions, the droplet evaporation rate, dm/dt, has to be calculated. For droplet evaporation, the following equation, which is valid for both micro and nano-sized droplets, is used (Eslamian et al. 2006):

$$\frac{dm}{dt} = 4\pi R\rho_g D_g \ln(1+B)\left[1 + Kn\frac{\sqrt{\pi}Kn\xi + \zeta}{1 + Kn\xi}\right]^{-1} \tag{3}$$

where

$$Kn = \frac{\lambda}{R} \tag{4}$$

In above equations, Kn is the Knudsen number, $\xi = 1.3330$ and $\zeta = 1.0161$ are matching parameters, ρ_g is the gas density, D_g is the diffusion coefficient of solvent vapor in the surrounding gas and B is the Heat Transfer number. In equation (3), the term in the last bracket accounts for the non-continuum effects. Note that in derivation of equation (3), it has been assumed that the evaporating species (solvent) on the droplet surface is in thermodynamic equilibrium with its vapor in the gas phase. In the absence of combustion, B is calculated from following equation:

$$B = \frac{W_{g\infty} - W_{gR}}{W_{gR} - 1} \tag{5}$$

where $W_{g\infty}$ and W_{gR} are the mass fraction of the solvent vapor far from the droplet and on the droplet surface, respectively. With evaporation of the solvent, vapor concentration of the carrier gas increases in the reactor. This increase is governed by *(i)* the solvent vapor introduced into the gas by droplet evaporation and *(ii)* vapor loss by diffusion from the gas to the reactor wall. The change in oxygen mass fraction used in the calculation of Transfer number, B, can be related to the droplet evaporation rate via the following equation:

$$\frac{dW_{g\infty}}{dt} = -\frac{N_0}{\rho_g}\frac{dm}{dt} - \frac{2k_m}{R_r}(W_{gR} - W_{g\infty}) \tag{6}$$

where N_0 is the droplet number density, R_r is the reactor diameter, and k_m is the mass transfer coefficient. The mass transfer coefficient k_m is calculated using available correlations (e. g. Incropera and DeWitt 1985).

The heat generated from the hot walls is partially consumed for droplet evaporation and the rest causes a continuous temperature rise in the carrier gas. Applying the energy balance on a circular element of the reactor perpendicular to the reactor longitudinal axis with thickness dx will lead to the following equation for the time rate of change of the carrier gas bulk temperature:

$$\frac{dT_b}{dx} = \frac{1}{\rho_g Q c_p}\left[2\pi R_r h_w (T_w - T_b) - 4\pi^2 R^2 R_r^2 h_s (T_b - T_l) N_0\right] \tag{7}$$

where h_w is the heat transfer coefficient from the tube walls to the carrier gas inside the pipe and can be obtained from correlations available in the literature for different flow and heat transfer conditions. Analytical correlations for constant wall temperature and constant heat flux cases ($Nu_T = 3.391$ for constant wall temperature and $Nu_H = 4.123$ for constant wall heat flux) are used to obtain the flow Nusselt Number ($= h_w 2R_r / k$) and consequently h_w for laminar flows inside tubes. For turbulent flows empirical correlations such as:

$$Nu_{Tu} = 0.023\,\mathrm{Re}^{0.8}\,\mathrm{Pr}^{0.4} \tag{8}$$

may be used, where Re is based on the reactor diameter (Bird et al. 2002). h_s in equation (7) is the heat transfer coefficient from the surrounding air to the droplet and is obtained assuming the droplet evaporation is symmetric and therefore, $h_s D / k = 2$. In equation (7), x is the distance along the reactor, Q is the carrier gas volume flow rate, and T_w is the reactor wall temperature.

The carrier gas bulk temperature variation along the reactor can be obtained using the chain law through the following equation:

$$\frac{dT_b}{dx} = \frac{dT_b}{dt}\frac{dt}{dx} \tag{9}$$

Assuming that the carrier gas is ideal, the droplet residence time (the time t required for droplet to travel distance x) along the reactor can be found as follows (Jayanthi et al 1993):

$$\frac{dt}{dx} = \frac{\pi R_r^2}{Q}\left(\frac{T_0}{T_b}\right)\left(\frac{1 - Y_{g\infty}}{1 - Y_{g\infty}^0}\right) \tag{10}$$

where t here is the residence time, T_b and T_0 are the temperatures of the carrier gas at distance x from the reactor inlet, and at the reactor inlet, $x = 0$, respectively, and $Y_{g\infty} = W_{g\infty} / M_s$ and $Y_{g\infty}^0 = W_{g\infty}^0 / M_s$ are the solvent mole fraction in the carrier gas at x and $x = 0$, respectively. M is the solvent vapor molecular weight. The correction factors in brackets in equation (10) account for the changes in the vapor content of the carrier gas as a result of droplet evaporation. Note that in equation (10) the gravitational settling component to the droplet residence time is neglected, which is fairly justified for small droplets.

Transition from Shrinkage to Constant-Diameter (Induction Period)

It has been assumed that once the solute concentration on the droplet surface reaches the CSS, the diameter of the drying droplet at this time can be considered as the final particle size (Jayanthi et al. 1993). However, in reality, there is a time delay between the onset of solute precipitation on the droplet surface and the time at which sufficient amount of solute required for the formation of a thin rigid shell is precipitated. This time period is called the induction time or induction period. The following empirical equation shows the dependence of the induction time on temperature (Lancia et al. 1999):

$$t_{ind} = \tau \exp(E_{act} / \mathrm{R}T) \tag{11}$$

where τ is a constant, E_{act} is the activation energy for the process, and R is the universal gas constant. The values of τ and E_{act} must be obtained individually for each material synthesis process. It was found that the induction time for spray pyrolysis of zirconia from zirconium hydroxychloride was short compared to the shrinkage period (8%). It was also found that the induction time is a weak function of temperature (Eslamian et al. 2006). Therefore, for the synthesis of a powder, its process induction time should be obtained by comparing the numerical and the experimental data.

From the phase diagram for a binary system of solute-solvent, as the droplet evaporates, its composition changes from one liquid phase (solute + solvent) to liquid (solute + solvent) and solid phases. Within the temperature range of each particular material synthesis the anhydrous or hydrated compounds may form. These hydrated compounds have different latent heat of crystallization, which influences the particle morphology in the drying process. In addition, hydrates may cause the formation of a high porosity shell. Material that releases more energy during the crystallization process may lead to earlier fragmentation when its shell structure is inelastic and impermeable to liquid.

To account for the effect of the induction time on the final droplet/particle size, once the solute concentration on the droplet surface reaches CSS, computations must be continued for extra t_{ind} time. It is assumed that except for the heat balance equation on the droplet surface i.e. equation (1), other equations governing the shrinkage period are still valid for the induction period. Heat balance on the droplet surface should be modified to account for the heat of precipitation or crystallization (ΔH_{cr}) in induction and constant-diameter periods:

$$mc_p \frac{dT_l}{dt} - 4\pi Rk(T_b - T_l) - L\frac{dm}{dt} - \Delta H_{cr}\eta(\frac{4}{3}\pi R_0^3 C_0^{'})\frac{1}{t_{ind}} = 0 \tag{12}$$

where $C_0^{'}$ is the solute initial concentration in mole/m^3. The last term in equation (13) is the average rate of heat of crystallization released during the induction period; η is the mass fraction of the solute which precipitates in the induction period (t_{ind}), a parameter which is not known, a priori. The amount of the solute precipitated during the induction period can be calculated using the solute concentration distribution profile at the time the outer particle diameter becomes constant, i.e. the end of the induction period (Eslamian et al. 2009):

$$\eta = \frac{\int_{r_{ES}}^{r_{CSS'}} 4\pi r^2 C dr}{\frac{4}{3}\pi R_0^3 C_0} = \frac{3}{R_0^3 C_0}\int_{r_{ES}}^{r_{CSS'}} C r^2 dr \tag{13}$$

where *CSS'* is the concentration on the droplet surface at the end of the induction period. Calculations for the induction period should be repeated several times using the new values for η until no more substantial changes in η is observed.

Constant-Diameter Period

In the constant-diameter period, droplet/particle outer diameter remains unchanged, while evaporation of the trapped liquid causes continuous increase in the particle wall thickness and inner diameter. Since the total volume of the droplet during this stage is constant, the continuous depletion of the liquid due to evaporation must create a continuously expanding, vapor-saturated space within the droplet core interior to the shell. Although the precise configuration of this vapor space is not clear, it is reasonable to postulate that the solvent will continuously wet the inner surfaces of the shell such that the vapor core will not be in direct contact with the shell. Evaporation rate in this period is different from that in the shrinkage period and is obtained from the following equation (Lee and Law 1991):

$$\frac{dm}{dt} = 2\pi d_{s,c} \frac{k}{c_p} \ln(1 + B) \tag{14}$$

where $d_{s,c}$ is the final particle/droplet outer diameter obtained at the end of the induction period. Assuming a constant droplet evaporation rate in this stage (B is nearly constant), from equation (14), with the initial mass of the solvent given, the duration of the constant period (t_C) can be calculated (the final mass of the solvent is zero).

As the evaporation of the trapped liquid in the inner core continues, more solute precipitates in the inner wall of the particle. The particle outer and inner diameters, $d_{s,c}$ and d_f and the total void volume within the particle (V_p), are interrelated through the following equation.

$$\rho \frac{\pi}{6}(d_{s,c}^3 - d_f^3) - \rho V_P = m_s \tag{15}$$

where the solute mass within the droplet, m_s, is conserved during the process. The shell thickness, δ, for the case of hollow particles assuming zero porosity ($V_p = 0$) is obtained from the following equation:

$$2\delta = d_{s,c} - d_f \tag{16}$$

Equations (1)-(16) can be solved using numerical methods or a commercial software. Eslamian et al. (2009) have outlined the steps one should follow in order to solve these equations.

3.3. Effect of Operating Conditions

Limited research is available on the effect of operating conditions on the morphology of powders produced by SP and SD. Jayanthi *et al.* (1993) modeled the effect of reactor temperature and initial concentration on solute distribution within the droplet. Lenggoro *et al.* (2000) performed a theoretical and experimental investigation on the morphology of powders produced by SP. They considered the effects of temperature, reactor flow rate, droplet number density, and initial droplet size on powder morphology. They found that at low temperatures (~ 200°C) the prepared zirconia particles are non-disrupted and the material is relatively amorphous. On the other hand, they found that at higher temperatures (~ 500°C), the particles are disrupted and the material is crystalline. Seydel et al. (2004) studied the solute precipitation and crust formation in aqueous solution droplets of sodium chloride, ammonium sulphate, and suspensions of silicon dioxide in water. In their study, partly dried droplets were caught in an oil bath and were characterized by an optical microscope. They observed several single cubic crystals of sodium chloride on the edge of the caught droplet. On the other hand, an amorphous crust formed on the ammonium sulphate droplet and the droplet containing silica sols lead to the formation of a similar crust. They did not provide any information on the morphology of the final dried particles. Zhou et al. (2001) investigated the effect of the solvent on the morphology of spray dried polymethyl methacrylate (PMMA) particles. They stated that the product PMMA particles derived from the PMMA-acetone dilute solution have a smaller particle size than those from PMMA-tetrahydrouran dilute solution. They attributed this to the better solubility of PMMA in acetone than in tetrahydrouran. They observed that, when the solvent is changed, several particle morphologies, including solid, porous, and honeycomb spheres can be obtained. Lin and Gentry (2003) investigated the effect of several parameters, including the drying temperature, initial concentration and solubility of the solute, and the heat of solute crystallization on the morphology and shape of the particles produced from suspended solution droplets. They employed several solutes including NaCl and NH_4Cl. They observed that the crust formation first occurred at the two poles of the droplet. In the case of NaCl solution droplet, the final particle shape was irregular cube-like, whereas, in the case of NH_4Cl, the final particle was an inflated irregular sphere.

In the following section, the effects of the most important processing parameters on the morphology and other characteristics of powders produced by SD and SP are reviewed.

3.3.1. Effect of Atomization Technique

Effects of atomization methods and solute concentration on the morphology of spray dried $MgSO_4$ powders were systematically investigated by Eslamian and Ashgriz (2007). They employed four different spray generators: (a) a vibrating mesh nebulizer, which generates low velocity droplets of about 1.5 m/s with a size range between 1 to 13 μm; (b) a splash plate nozzle, which generates droplets with an average velocity of about 17 m/s and a size range between 5 to 110 μm; (c) an air mist atomizer, which generates droplets with an average velocity of about 42 m/s and a size range between 5 to 40 μm; (d) a pressure atomizer, which generates droplets with an average velocity of about 17 m/s and a size range between 5 to 110 μm. These spray generators were used to produce spray dried $MgSO_4$ powder.

They identified several particle morphologies including smooth-surface uniformly disrupted particle, smooth-surface disrupted particle from a single opening, non-smooth-surface contracted particle, smooth-surface non-disrupted particle and non-uniformly disrupted particles. The particles prepared by the vibrating mesh nebulizer were mostly non-disrupted thick shells. They observed that increasing the initial solution concentration resulted in thicker-walled particles. The particles produced by the splash plate nozzle and the air mist atomizer had a variety of morphologies, however, the majority were contracted shell-like particles. They also observed that increasing the initial solute concentration resulted in a reduction in the number of disrupted particles, which was attributed to an increase in the particle wall thickness and strength due to increase of solute concentration. They also observed that the wall thicknesses of $MgSO_4$ particles were nearly spherically symmetric for most cases. Non-spherically-symmetric evaporation was expected for high velocity droplets generated by air mist and splash plate nozzles. However, no obvious non-symmetrical crust formation was observed.

They found that the spray characteristics, which have considerable effect on the particle drying rate and morphology, include droplet number density, droplet velocity, atomizing air, and droplet size. Low droplet number density, low droplet velocity and size, and accompanying atomizing air favors rapid drying of the droplets.

Figure 3 summarizes different possible final morphologies for hollow particles formed in spray drying or pyrolysis, most of which were observed in the study by Eslamian and Ashgriz (2007). Depending on the relative velocity of the droplet with respect to the surrounding air, the evaporation process could be either spherically symmetric or asymmetrical. If a relative velocity between the air and the droplet exists, a boundary layer will form around the droplet (see Figure 3), which results in a non-uniform solvent evaporation and, therefore, a non-uniform crust formation. Depending on the nature of the solute and the process conditions, the crust could be either permeable or impermeable. If the crust is impermeable and the surrounding air is still hot, the trapped solvent inside the shell continues to evaporate, and the pressure increases inside the shell-like particle. This pressure buildup may cause several final particle morphologies. If the crust is uniform and contains no defects, the uniform stress applied on the internal wall of the shell causes a uniform disruption of the particle and the particle will be cut into several orderly pieces (type *a* particle in Figure 3). If the particle wall thickness is not uniform, it may break from the weakest part, and a particle with a small hole will form (type *b* particle in Figure 3). As another scenario, if the crust is strong enough to resist the pressure, the vapor trapped in the particle may eventually condense, and the shell may contract by the depression (type *c* particle in Figure 3). However, if the shell is fully dried, it is possible that the condensation can not cause any permanent deformation on the particle, and a hollow smooth-surface particle forms (type *d* particle in Figure 3). In addition, if the crust is thin and weak, or the pressure buildup is substantial, or the particles collide in the reactor, it is possible that the drying particle bursts to form a bunch of particle flakes (type *e* particle in Figure 3). On the other hand, for the case of a permeable crust, the evaporated solvent leaves the particle, without increasing the internal pressure, and a non-disrupted porous particle forms (type *f* particle in Figure 3).

3.3.2. Effect of Temperature

The most important parameter affecting the morphology of powders produced by SD and SP is the process temperature. This is because the temperature has a great influence on the solvent evaporation rate. As discussed by Jayanthi et al. (1993), for a particular precursor, at given operating conditions, the morphology of particles and whether they are solid and fully-filled, or hollow and disrupted, depends on the concentration distribution within the droplet. The concentration distribution is a strong function of the process temperature. Employing the model of Eslamian et al. (2009) the effect of the process temperature on the final particle morphology is discussed below.

Figure 4 shows the effect of reactor temperature on the solute concentration profile at the onset of precipitation within a droplet with 5 μm initial diameter for a given initial droplet number density, N_0, carrier gas flow rate, Q, and initial relative humidity, RH_0, and initial solution concentration, C_0. The reactor inside diameter is 10 mm. The concentration profile inside the droplet depends on the operating conditions and the reactor geometry. For the reactor conditions of their study as specified above, and for the wall temperatures up to about 1000 °C, the concentration profile entirely lies above the Equilibrium Saturation (ES) line, which is favorable for the production of fully-filled particles (Jayanthi et al. 1993).

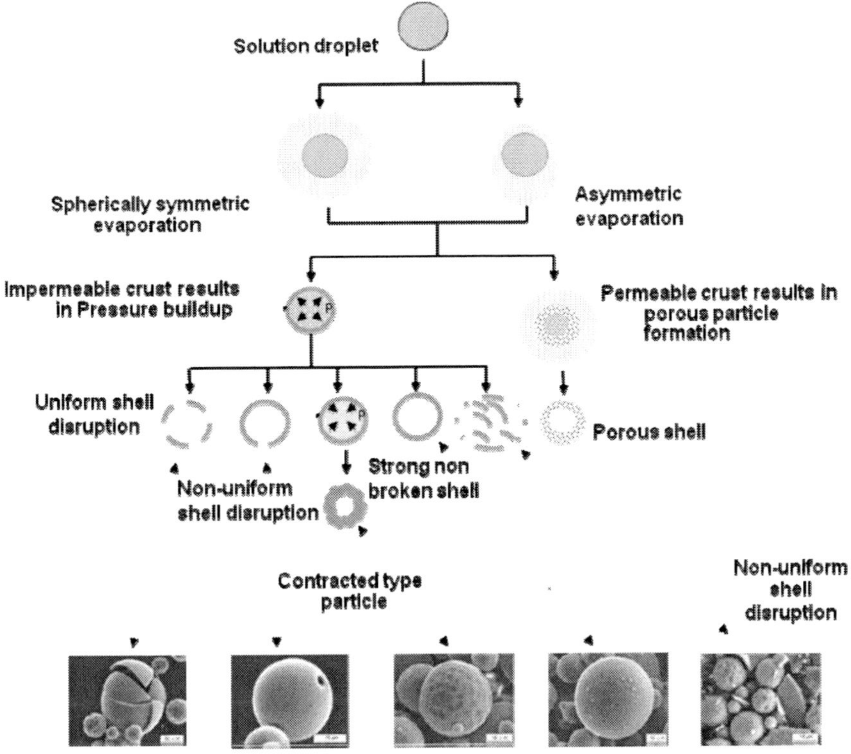

Figure 3. Different possibilities for hollow particle formation during spray drying/pyrolysis. Type (a) uniform shell disruption, type (b) non-uniform shell disruption, type (c) contracted shell particle, type (d) non-disrupted smooth surface particle, type (e) non-uniform shell disruption, and type (f) porous particle (Eslamian and Ashgriz 2007).

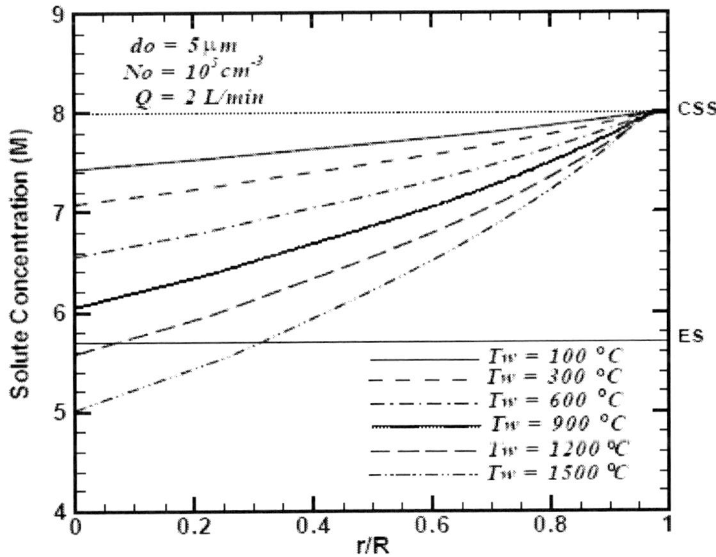

Figure 4. Solute concentration profile within the droplet for various wall temperatures for given initial droplet size d_0,droplet number density N_0, carrier gas flow rate Q, initial relative humidity, RH_0= 10%, and initial solute concentration C_0= 2 M (Eslamian et al. 2009).

The gradient of the concentration within the droplet depends on the characteristic time of the droplet evaporation and solute diffusion from the droplet surface to the droplet center. As the reactor wall temperature and therefore the reactor bulk temperature increase, the solvent evaporation rate from the droplet surface increases and the characteristic time of evaporation compared to the characteristic time of solute diffusion decreases. As a result, at high temperatures, at the onset of precipitation on the droplet surface, the concentration in the vicinity of the droplet center falls below the ES line. The occurrence of this type of concentration profile is an indication of having hollow particles as postulated by Jayanthi et al (1993). Note that, because of the symmetry of evaporation, the slope of the concentration profiles at the droplet center is theoretically zero; however, due to the finite number of the numerical nodes within the droplet, the slopes shown on Figure 4 do not seem zero. Nevertheless, this does not affect the accuracy of the analysis.

3.3.3. Effect of Droplet Number Density, Carrier Gas Flow Rate and Droplet Initial Size

Eslamian et al. (2009) also investigated the effect of other parameters on particle characteristics. Figure 5 shows the solute concentration profiles within the droplet for various droplet number densities and initial droplet diameters respectively, while other parameters are kept constant. From Figure 5 it is deduced that a high droplet number density favors the formation of a less steep concentration profile and as a result the synthesis of fully-filled particles. This is because an increase in droplet number density leads to a decrease in evaporation rate. Using droplet number densities below what is shown in Figure 5 did not show any significant change in the concentration profile. This is because, at low droplet number densities, where the distance between the droplets is large and the carrier gas effectively removes the moisture, droplets essentially evaporate without influencing each other's evaporation. It should be noted that the concentration profile is not a function of the carrier gas flow rate. Also note that smaller initial droplet sizes favor the formation of solid, fully-filled particles. As the initial droplet size increases the likelihood of the formation of hollow particles increases.

3.3.4. Effect of Pressure

Eslamian and Ashgriz (2006*b* and 2006*c*) systematically investigated the effect of pressure on powder morphology and other powder characteristics. Particle shape and morphology depends on the precursor properties and precipitation mechanism, as well as on the droplet evaporation rate. Droplet evaporation rate is a function of the reactor pressure and temperature. Evaporation rate controls the solute distribution profile within the droplet, and determines whether the particles are solid or hollow. Eslamian and Ashgriz (2006*b*) showed that, when the pressure is reduced to 60 Torr, the decrease of the evaporation rate due to the non-continuum effects is about 60% of that of the continuum-based evaporation rate.

In SP, a low evaporation rate is favorable to the production of less hollow and more fully-filled particles. On the other hand, depending on the nature of the precursor, a relatively high reactor temperature, which causes a high evaporation rate, is essential for a chemical decomposition to occur within the precursor. Hence to increase the likelihood of forming solid particles, it is advantageous to conduct the SP process at reduced pressures.

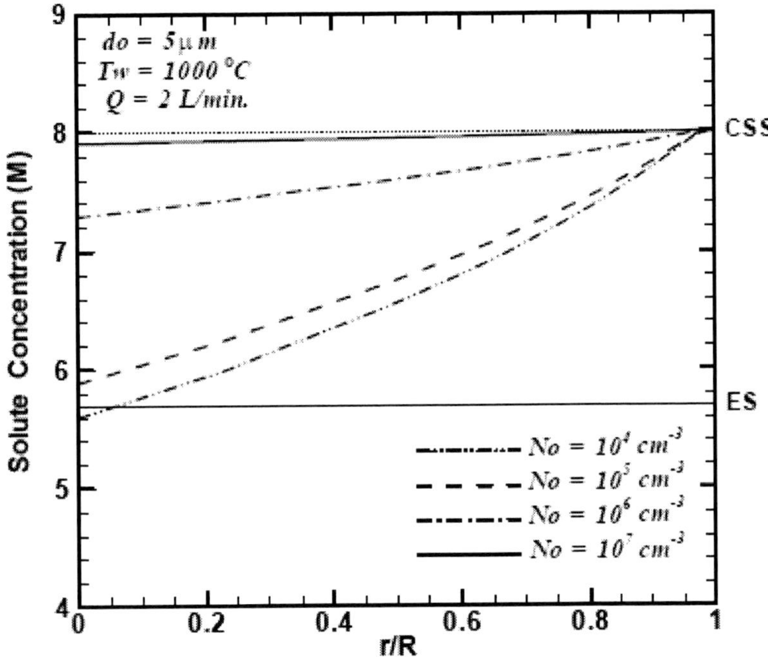

Figure 5. Solute concentration profile within the droplet for various droplet number densities for given d_0, T_w, Q, $RH0$= 10%, and C_0= 2 M.

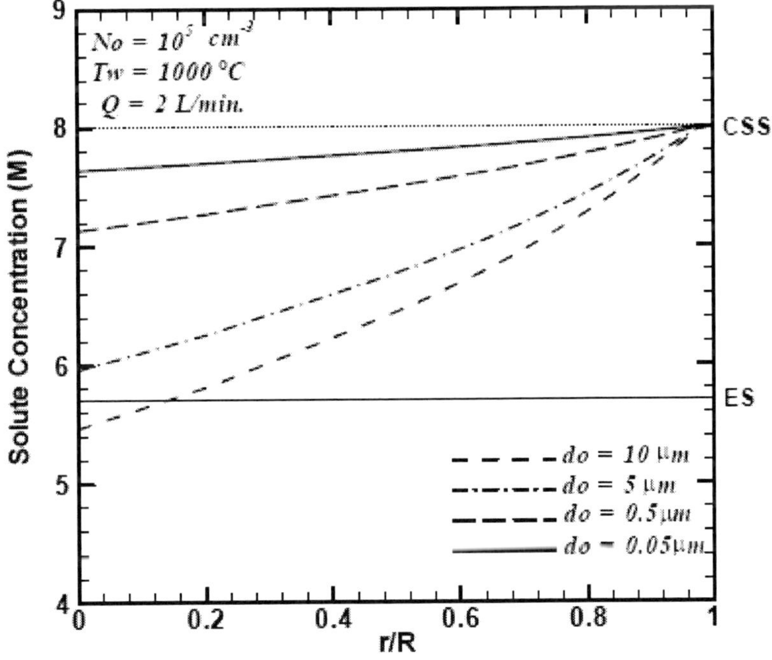

Figure 6. Solute concentration profile within the droplet for various initial droplet sizes for given N_0, T_w, Q, RH_0= 10%, and C_0=2M.

In addition to the effect of pressure on evaporation rate and particle morphology, the X-ray Diffraction (XRD) tests on NaCl, $MgSO_4$, and ZrO_2 powders, have shown that reducing the operating pressure leads to the formation of powders with enhanced crystallinity (Eslamian and Ashgriz 2006*c*).

3.4. Recent Advances and Future Trends

Some of the major challenges in powder production by SP technique include the production of nano-sized particles, improving the throughput of the process and making it economical.

Most of the powders prepared by SP even though large in size (micron-sized), but structurally are nanocrystalline. However, sometimes, the overall size of the particles is crucial for some applications. Regarding the production of nano-sized particles, electrospraying has attracted attention as a possible way to produce tiny solution droplets which are sufficiently small to allow production of sub-micrometer and nanoparticles without resorting to extremely dilute solutions or impractically small nozzle orifices. Also, electrospraying is a versatile tool for liquid atomization that has the advantage of uniform droplet generation. Jaworek and Sobczyk (2008) have reviewed and discussed the contribution of electrospraying in the production of nanoparticles, nano-thin films and also nano-encapsulation. In micro- and nano-particle production, electrospray allows the production of particles with uniform size over a wide size range. It seems that there are still a number of challenges to be faced before commercialization will be possible; nonetheless advances in electrospray applications in nanotechnology and biotechnology will certainly continue in near future, and new achievements in this field can be expected.

In most research studies on SP and SD in the lab scale, ultrasonic atomization has been used to generate droplets/sprays. To increase the powder production rate, other atomization methods have to be examined without affecting the particle size, size distribution and quality. Some researchers have already used high throughput nozzles such as pressure nozzles. For instance, Nimmo et al. (2003) used a twin-fluid atomization technique to produce lead zirconate titanate (PZT) powder using a starting solution composed of lead acetate, zirconium acetate, and titanium propoxide (stabilized by acetylacetone) dissolved in water. The effects on phase composition on the final product were examined as a function of temperature profile along the tube reactor, and on the amount of the excess Pb in the starting solutions. Commercialization of SP technique is closely interrelated to its throughput and strong evidence that SP is the best method for the production of some particular advanced powders.

It has been shown that SP is capable of producing composite powders with applications in emerging technologies. In the future, SP will be used to produce a variety of other new composite materials. Fukui et al. (2002) synthesized composite powders such as NiOSDC and La(Sr)CoO_3 by spray pyrolysis used as anode and cathode of a solid oxide fuel cells (SOFC), respectively. As another example of the application of SP in power production, Bakenov et al. (2008) reported the production of nano-structured lithium manganese oxide with spherical particles via ultrasonic spray pyrolysis technique. Rechargeable lithium-ion batteries (LIBs) have become the key components for a wide range of portable electronic devices and most promising energy supplier for electricity-powered transport. They observed that the material produced by SP showed a pronounced stability upon prolonged cycling at room temperature at high charge–discharge rates up to 10°C. They stated that the electrochemical performance of the nanostructured $LiMn_2O_4$ prepared was superior to the material prepared by conventional methods.

Another challenge is to address and elucidate the many complex physical and chemical phenomena involved in SP and SD. In SP and SD similar to many other industrial processes, it is important to be able to predict the characteristics of the final product and also to design the components of the equipment and trouble-shoot the processes. Although several attempts have been made to model some aspects of the physical and chemical phenomena involved in these processes, more work is needed in this area. In addition, these models such as that described earlier in this chapter, have not been incorporated into the commercial packages. Currently the commercial packages are adequate in modeling the flow patterns, but they are weak when it comes to the intra-particle phenomena. Intra-particle phenomena are more important when the final morphology of particles plays an important role in particle characteristics. Fletcher et al. (2006) have addressed the important issues regarding available CFD codes for the industrial spray dryers. They reviewed the fundamental flow behavior in dryers and their modeling using a commercial CFD code. They argued that the key point to emerge is the need to perform three-dimensional, transient calculations and to include hindered drying and wall interaction models. They also noted that coalescence and agglomeration models need to be validated and included in the simulations. In addition, new turbulence modeling approaches, such as Detached Eddy Simulation (DES), are under investigation to determine their capabilities in this area. This progress relies on both continued model development and experimental investigations to validate these models.

4. Flame Spray Pyrolysis

4.1. Introduction

The conventional way of particle production by flame/combustion technology is the vapor phase flame synthesis method. Flame synthesis of particles, in general, is a scalable technology that was developed mainly by research for manufacturing of some commodities such as carbon blacks, fumed silica and pigments. However, with the major advances in the scientific understanding of combustion and aerosol formation and growth, now optimal reactor design and flame production of sophisticated inorganic nanoparticles with controlled composition, size and morphology have become possible. This leads to a series of new products such as highly selective (TiO_2-spot-coated SiO_2) epoxidation catalysts or V_2O_5-coated TiO_2 for selective catalytic reduction of NO_x with NH_3 at lower temperatures than conventional catalysts that can lead to better fuel utilization and/or effective pollutant (e.g. Hg) removal during incineration (Pratsinis 2006).

Recently, the understanding and knowledge gained from the flame synthesis method has lead to the creation of new processes like Liquid-Fed Flame Reactors or Flame Spray Pyrolysis (FSP), making possible the manufacturing of a much broader spectrum of products.

In FSP, called also Liquid Flame Spray (LFS), the heat provided by the combustion of a gaseous or liquid fuel and the precursor results in evaporation of the entire solvent and the solid content of the precursor (Kammler et al. 2001). FSP has high potential for synthesis of composite nanoparticles that can be used for fabrication of catalysts, sensors and electroceramics. The FSP reactors are quite attractive as they can utilize a broader spectrum of liquid precursors than conventional flame reactors. Typical products synthesized by FSP are titania and alumina (Bickmore et al. 1998), $MgAl_2O_4$ (Bickmore et al. 1996), Fe_2O_3 (Grimm et al. 1997), and so on. Multicomponent oxide powders, such as beta-$SrMnO_3$ and $NiMn_2O_4$ (Kriegel et al. 1994), superconductors (Zachariah and Huzarewicz 1991), alumina (Tikkanen 1997) and $BaTiO_3$ (Brewster and Kodas 2004) have been made by pyrolysis of solutions containing inorganic precursors in oxy-hydrogen flames. Using organic solvents as liquid fuel, leads to a self-sustaining spray flame. This process has the potential to produce composite mixed-metal oxide powders in the size range of 1 to 200 nm from low cost precursors with production rates up to 250 g/h (Narayanan and Laine 1997, Laine et al. 2000). These processes clearly differ from the conventional spray pyrolysis (SP) as the precursor is released from the droplet environment undergoing gas phase reaction and subsequent particle growth by coagulation, surface growth and sintering.

A typical set-up for nanoparticle synthesis by flame spray pyrolysis consists of a unit for the generation and dispersion of droplets such a spray nozzle, a heat source for initial droplet evaporation and ignition (pilot flame or the spray flame itself for combustible liquids), and an oxidant (oxygen/air) to facilitate combustion (see Figure 7). The nozzle feed rate and spray characteristics, flow rate and type of the oxidant, and precursor combustion enthalpy are the parameters that control the temperature profile and gas phase reactions of the precursor followed by subsequent particle growth and sintering, taking place within and after the spray flame.

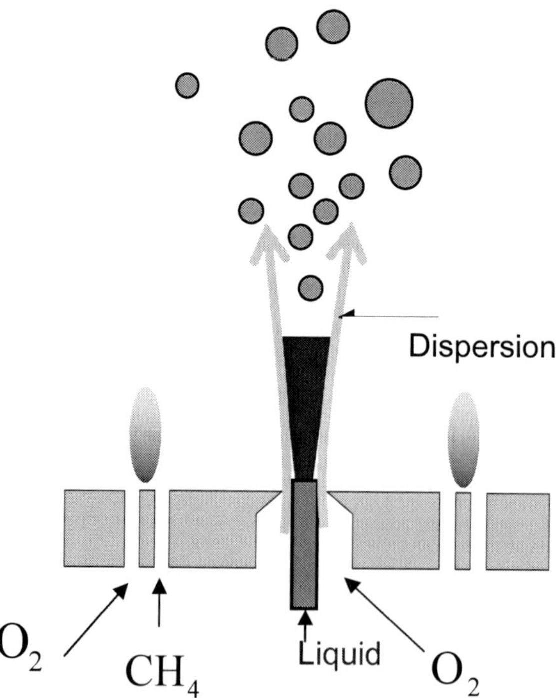

Figure 7. Schematic of a typical coaxial nozzle used for synthesis of nanoparticles by Flame Spray Pyrolysis.

The product powder is usually collected by filtration. Micron-sized liquid droplets can be generated, using various atomizers; however ultrasonic and gas-assist pressurized atomizers are most common in spray combustion. Ultrasonic atomizers show good performance with regard to droplet size and homogeneity and the droplets can be embedded in any gaseous flow for transportation into the reaction zone, where the droplet velocity can be controlled independently. Gas-assisted pressurized atomizers, however, are easy to incorporate into the spray flame apparatus, their operation is simple and reliable, and have a high rate of spray generation. High shear forces are needed to breakup the liquid jet, thus, high gas velocities are inevitable leading to the high quenching ability of such nozzles resulting in small (sub-micron) particles.

4.2. Mechanism of Particle Formation

Controlled synthesis of particles is of great importance in FSP, because small variation of the processing conditions alters the quality and properties of the particles. Parameters such as the spray and flame characteristics, gas phase reactions and processes such as particle nucleation, coagulation, agglomeration and sintering determine the quality of the powder (Eslamian and Heine 2008). While the majority of ceramic powders synthesized by FSP were made of dense solid nanoparticles e.g. (Tani et al. 2002, Mueller et al. 2004), in some cases it was found that the ceramic powders were an inhomogeneous mixture of large (~ 1 μm) hollow and solid dense nanoparticles (~ 10 nm) e.g. (Mädler et al. 2002*a*, Jossen et al. 2005*a*, Mädler et al. 2002*b*).

Jossen et al. (2005*b*) studied the effect of precursor and solvent characteristics on the morphology of flame-made ceramic powders. They developed a morphology map for FSP-made oxides. According to that map, given the density of the precursor combustion enthalpy, which is the total combustion of the solvent and precursor per unit mass of the atomizing gas, and also the ratio of the solvent boiling point to the precursor decomposition point, one can determine whether the particles will be either hollow and shell-like or dense and solid. Their results showed that for the data available in the literature, inhomogeneous ceramic particles are formed at low densities of combustion enthalpy (< 4.7 kJ/g_{gas}), and when the solvent boiling point is smaller than the melting or decomposition point of the metal precursor ($T_{bp}/T_{d/mp} < 1.05$).

Later Jossen et al. (2005*a*) produced yttra-stablized-zirconia (YSZ) synthesized by FSP of a solution of yttrium nitrate hydrate ($Y(NO_3)_3xH_2O$ = YN*x*), as the yttrium source, and zirconium n-propoxide ($Zr(OC_3H_7)_4$ = ZP), as the zirconium source. They observed that at constant density of combustion enthalpy, the powder produced from YN6 was inhomogeneous mixture of hollow and solid particles, whereas the powder produced from YN0.5 was homogeneous and composed of solid, dense particles. This observation indicates that the enthalpy of combustion could not adequately determine the final morphology of the flame-made-particles. The presence of more water molecules in YN6 compared to YN0.5 is believed to be the main reason for this discrepancy. The presence of water can lead to a decrease in the flame temperature, and as a result, the droplet vaporization rate changes. This indicates that the spray and flame characteristics also should have a role on the morphology of FSP-made-particles.

In a spray which contains high velocity droplets, or in a cloud of stationary droplets, the droplet vaporization and burning rate and the flame location is influenced by droplet characteristics, such as droplet size, velocity, spacing, etc. This is known as the group combustion theory, studied primarily by Chiu and his colaborators (Chiu and Liu 1977 and Chiu et al. 1982). In a spray flame, group combustion number or Chiu number G, interpreted as the ratio of the total heat transfer rate between droplets and the atomizing gas to the rate of energy transported by convection, is an indicator of the group combustion mode. Chiu number G is a function of spray characteristics, such as the total number of the droplets in the spray (*n*), droplet size (*d*), the average center-to-center spacing of the droplets (*L*), Reynolds number (Re) based on the droplet diameter, the relative velocity between the droplets and the atomizing or carrier gas, the liquid properties, gas Lewis number (*Le*), and Schmidt number (*Sc*) (Chiu and Liu 1977):

$$G = 1.5Le(1+0.276Sc^{1/3}\,\mathrm{Re}^{1/2})n^{2/3}\left(\frac{d}{L}\right) \tag{17}$$

Four principal combustion modes were identified, which are characterized as follows:

a) Isolated-droplet combustion mode, where the droplets vaporize and burn individually without interfering with the adjacent droplets ($G < 10^{-2}$).
b) Internal group combustion mode, where vaporization is occurring within the spray or the cloud core totally surrounded by a flame. Within the flame, each droplet is enveloped by individual flames ($10^{-2} < G < 1$).
c) External group combustion mode, where the droplets lying on the spray or cloud boundaries burn in the form of a flame and those within the spray core vaporize slowly without burning ($1 < G < 100$).
d) External sheath group combustion, in which only those droplets lying in a thin layer of the spray peripheries burn, while those in the spray core do not evaporate ($G > 100$).

Eslamian and Heine (2008) studied the influence of spray flame characteristics and group combustion mode on powders produced by FSP and proposed a mechanism for the formation of ceramic powders. They also modified the Chiu number G to provide a criterion to predict the morphology of flame-made ceramic particles.

They observed that Chiu number G increases with increase of the liquid feed rate. Higher liquid feed rates, while the spray geometry and volume is almost the same, results in the formation of a spray with higher liquid mass per unit volume of the spray. Their other observation was that Chiu number G decreases with increase of the distance from the nozzle exit. This is because close to the nozzle exit, the droplet number density and flux is high, and only a small fraction of the droplets evaporate. They concluded that close to the nozzle exit, the droplets in the spray core do not burn or burn partially and the majority of them merely evaporate. The fuel vapor is then transported to the radial and axial directions by means of diffusion and convection. Close to the nozzle exit, where the internal and external group combustion modes prevail, the droplets on the spray boundaries burn, while enveloped by a flame. Additional fuel vapor is supplied to the boundaries from the spray core.

The possible mechanism through which an inhomogeneous mixture of hollow microparticles and solid nanoparticles form is illustrated in Figure 8.

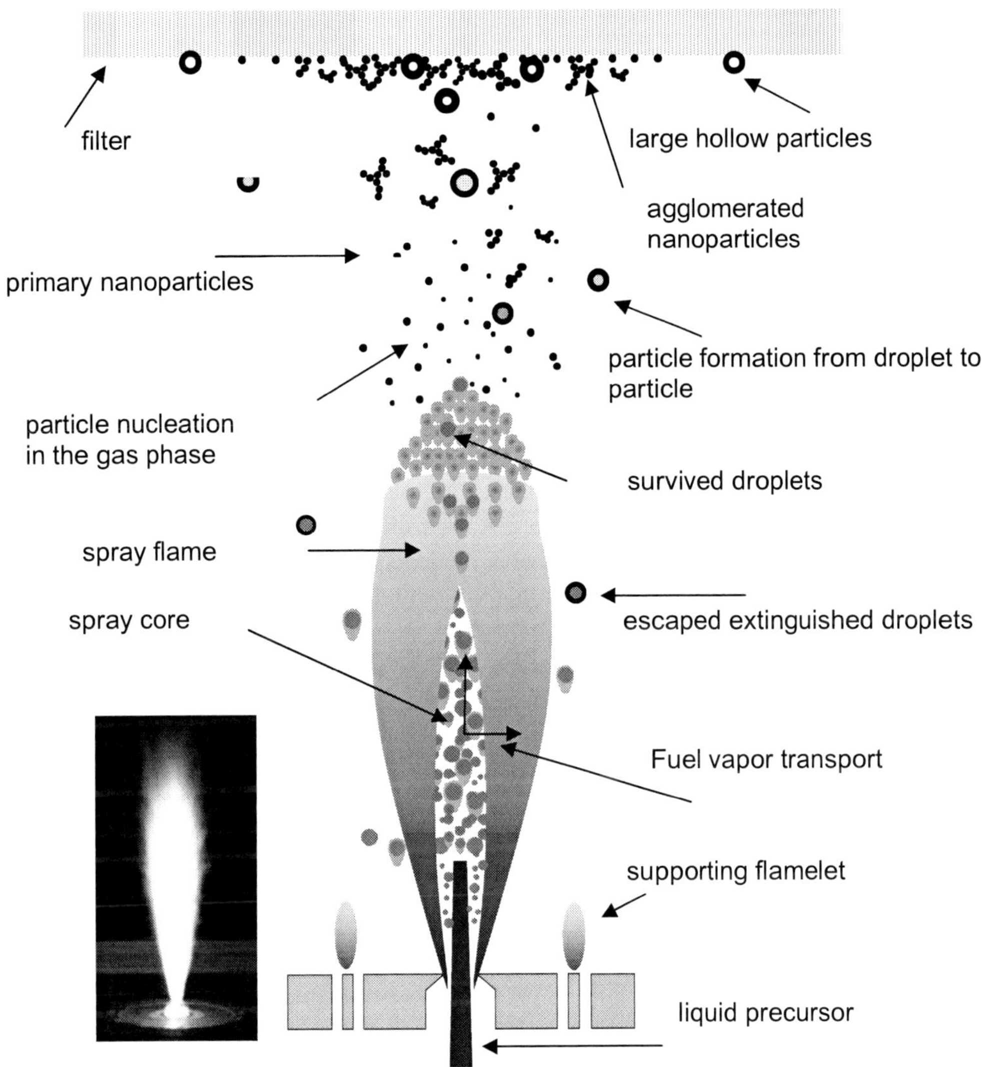

Figure 8. Mechanism of particle formation by flame spray pyrolysis (FSP), and the effect of group combustion mode on the morphology of flame-made particles (Eslamian and Heine 2008).

The precursor droplets that entirely evaporate in the hot flame are responsible for the formation of nanoparticles: The vapor species react, forming intermediate and product molecules and clusters that quickly grow to nano-sized particles. These primary spherical particles then grow by coagulation and sintering, finally forming agglomerates or even hard bonds before being deposited onto the particle collector. On the other hand, under certain conditions, a small number of the droplets particularly with large initial sizes may escape from the spray boundaries and become extinguished to simply produce large hollow particles. It is also possible that some of the large droplets present in the relatively cool spray core, do not completely evaporate. These survived droplets then simply form large particles which are usually hollow, due to the high temperature of the process, and may disrupt, owing to the continuous evaporation of the trapped liquid and, therefore, the pressure buildup inside the hollow particles (Eslamian and Ashgriz 2007).

If the Chiu number G, as defined in equation (17) is modified to take into account the effect of the flame temperature, which indeed has a profound influence on droplet vaporization and particle synthesis, a quantitative criterion for the formation of solid nanoparticles or inhomogeneous mixture of hollow, shell-like and solid nanoparticles will be obtained, which is more realistic than both Chiu number G and the enthalpy of combustion density. The modified Chiu number called FSP Chiu Number G_{FSP} is defined as follows:

$$G_{FSP} = 1.5Le(1+0.276Sc^{1/3}\,\mathrm{Re}^{1/2})n^{2/3}\left(\frac{d}{L}\right)\left(\frac{T_0}{T_{ad}}\right) \tag{18}$$

where T_{ad} is the adiabatic flame temperature, and T_0 is the precursor feed temperature employed to make G_{FSP} non-dimensional. Having a high adiabatic flame temperature, which is usually achieved when the flame enthalpy of combustion density is high, leads to a decrease in FSP Chiu number G_{FSP}. Therefore, according to this criterion, having a small FSP Chiu number G_{FSP}, favours the formation of homogeneous solid nanoparticles.

From practical or industrial point of view, it would be more convenient if just the flame temperature could have been used to predict the powder morphology. In general, a high temperature flame favours the formation of homogeneous nanoparticles. However, considering the flame temperature or adiabatic flame temperature as the only factor cannot adequately predict the morphology of the particles. This is because the flame temperature across the spray changes dramatically, both radially and axially. For instance, as discussed above, in external group combustion mode, the temperature on the spray boundaries could be very high, while the droplets within the spray core are not even evaporating.

4.3. Recent Advances and Future Trends

Specialty oxides like stable CeO_2 for catalytic or planarization applications as well as stable ZnO quantum dots that exhibit the blue shift of UV light can be readily made by FSP (Mädler et al. 2002*c*). Carbon-coated silica or titania nanoparticles have been made that can either better blend in a polymer without requiring functionalization with surfactants or facilitate the formation of electrodes for a Li-battery (Kammler and Pratsinis, 2003). Biomaterials have been made by these processes, creating some unprecedented opportunities for orthopedics. Non-agglomerated fumed silica particles (50-90 nm in diameter) that are blended with dimethylacrylate contributes to the development of nanocomposites for novel dental fillings that could replace current ones based on polymeric resins (Mueller et al. 2004). A new flame-nozzle process is developed that freezes particle growth and allows formation of non-agglomerated and even blue titania (Wegner and Pratsinis 2003). Scale-up relationships for flame aerosol reactors are developed and validated with a large body of data allowing a systematic design and operation of industrial units (Mueller 2003). With this technology inorganic nanoparticles with closely controlled morphology and composition are made exhibiting unique performance in this field that is dominated by wet chemistry.

To summarize, FSP offers a promising and rather inexpensive route for large-scale production of nanoparticles. The field is expanding from synthesis of simple oxides to more complex, functional nanoparticles. Materials produced by FSP emerge into other engineering areas, such as heterogeneous catalysis, biomaterials, dental materials, fuel cell membrane production and electroceramics fabrication. Early results in heterogeneous catalysis, in particular, indicate the potential of flame-made catalysts. Microelectronics and even medical applications will profit from these developments. New questions arise and underline the need for basic research in synthesis of mixed oxides with precisely controlled characteristics. Some of them include scale-up for synthesis of particles with controlled functionalities; mesoscopic chemistry relationships that can be verified and used, and nano-thin coated nanoparticles made in large quantities. However, many new discoveries are awaiting on this rather unexplored interface of material and engineering science (Pratsinis 2006).

Despite the great potential of nanoparticles, there is a significant concern for their negative adverse health effects, such as their penetration to body's organs. Also, environmental issues are becoming an increasingly important area of research. Until recently the potential negative impacts of nanomaterials on human health and the environment have been rather speculative and unsubstantiated. However, within the past number of years several studies have indicated that exposure to specific nanomaterials, e.g. nanoparticles, can lead to adverse effects in the lungs and the brain of test animals (Lam et al. 2004, Oberdorster 2004). However, the data acquired from the academic and industrial nanoparticle laboratories are scattered and inadequate to comment on possible adverse effects of nanoparticles on human bodies. Some organizations seem tend to treat this technology as harmful and even try to impose restrictions on nanoparticle research. There is, however, plenty of information on the effects of nanoparticles on human health that has been completely overlooked. Humans have been in contact with nanoparticles for centuries and even with their manufacturing for over a hundred years. Clearly there is a great demand to place the health effects of nanoparticles on a firm scientific basis to better protect the national investment in this field and, most importantly, guide researchers and even investors. Given the current advances of aerosol science in characterizing nanoparticles and the large body of anecdotal data regarding exposure to nanoparticle commodities (carbon black, fumed silica, pigmentary titania), there is also a great opportunity for systematic research on the health effects of nanoparticles at an international level (Pratsinis 2006).

5. Other Methods

5.1. Emulsion Combustion Method

Takatori (1997) developed a spray method, called emulsion combustion method (ECM), to produce micro and nano sized ceramic particles with desired properties. This method is basically a combination of the emulsion (Akinc and Richardson 1986) and flame spray pyrolysis (FSP).

In ECM method, an aqueous solution of metal salts is stirred with a fuel such as kerosene and a small amount of emulsifier to obtain water-in-oil (W/O) type emulsion. Using a spray nozzle, the solution is then atomized to produce a spray of droplets. The size of the emulsion droplets depends on the atomization conditions and nozzle type and is on the order of 10 μm for air-assist nozzles; the size of the dispersed micro-solution droplets depends on the string process and is about 1 μm in diameter (Takatori et al. 2004).

The ECM is classified into liquid phase syntheses, but it also has the features of gas phase syntheses, so it is a liquid/gas phase method. If the process temperature is high enough so that the droplets including their solid content entirely evaporate, then the vapor species nucleate and grow to produce nanoparticles in the gas phase. If this is the case, the ECM is similar to Flame Spray Pyrolysis (FSP). However, if during the ECM, the droplets partially evaporate, i.e. only the solvent evaporates, then this process would be similar to the conventional Spray Pyrolysis (SP). The most significant differences between ECM and SP are the size of the reaction fields and the reaction period. Although in laboratory SP experiments, one can spray a small amount of aqueous solution into very fine droplets, whose sizes are the same as those of the dispersed droplets in ECM, this cannot be done easily in a large-scale industrial process. An isolated small reaction field of about 1 μm is easily prepared using the emulsion process. The reaction period of the SP with average droplet sizes of about 5 μm is longer than that of the ECM.

ECM is relatively a new process and as such very few physical or mathematical models are available for this process. A mathematical model that explains the mechanisms of particle formation by ECM has been developed by Eslamian et al (2009). This has been done by combining a model for the burning of W/O emulsion droplets (Leite and Lage 2000) and a model for particle formation by SP (Eslamian et al 2009). The model has been also employed to investigate the morphology and size of the particles. Different scenarios of particle formation are explained in Figure 9. Depending on the process conditions, the relative size of the dispersed and emulsion droplets, and the properties of the fuel, the final particles may be either micro or nano-sized and either hollow or solid fully-filled.

In lab scale, several powders have been produced by ECM including silica and zinc oxide nanoparticles (Tani et al., 2003), and alumina particles (Tani et al. 1998). However, to our knowledge, this method has not yet been commercialized.

5.2. Spray Freeze Drying

Spray Freeze Drying (SFD) is a method of producing biopharmaceutical powders that are sensitive to high process temperatures usually experienced in spray drying (SD). The detail of the process is described in a book by Costantino and Pikal (2004). This method involves the atomization of a precursor solution such as protein plus a suitable substance that carries the droplets into the liquid nitrogen. The ice is removed from the frozen droplets by sublimation under vacuum. As this process involves no heat for drying, the denaturation associated with the spray drying process can be avoided. Still, aceptic powder handling is needed and the production yields are low.

A variation of this process is spray freezing into liquid, where the impingement of the feed solution onto the cryogenic liquid results in intense atomization into micro-droplets which freeze instantly. The microparticles can then be separated by sieving or evaporation of the cryogen and the sublimation of the solvent. Spray freezing into liquid allows particularly good size control and fast freeze, but it still involves aseptic powder handling

Recently SFD has attracted the attention of formulation scientists. Spray-freeze drying of aqueous solutions of pure proteins or protein/sugar combinations (Maa et al. 1999) produced larger, more porous particles than those prepared by SD. In addition, the SFD protein particles had superior aerosol performance than the SD particles, indicating their suitability for pulmonary delivery.

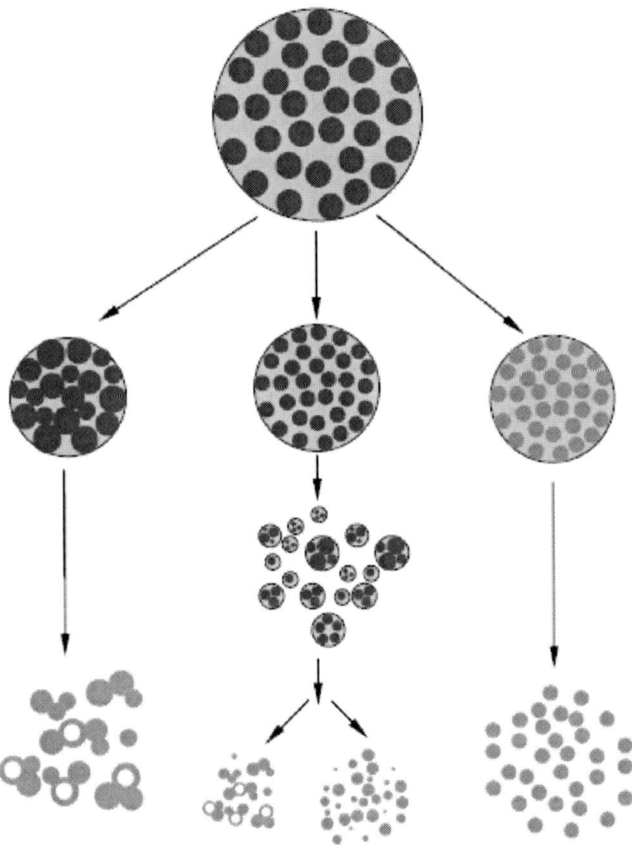

Figure 9. Particle formation mechanism in emulsion combustion method (ECM) .

Costantino et al. (2000) examined the friability of SFD protein particles and found to be dependent on their size. It was proposed that because small particles freeze in a shorter time than do large ones, a finer microstructure, and hence higher friability and also greater protein aggregation, is the result. In another study, it was found that the degree of aggregation of bovine serum albumin was directly proportional to the specific surface area of SFD trehalose or mannitol particles (Costantino et al. 2002).

When the modeling of SFD process is concerned, a mathematical model based on a steady-state heat-transfer condition has been derived by Maa and Prestrelski (2000) that can be used to estimate the freezing time in liquid nitrogen, for instance.

6. SUMMARY

In this chapter, the most popular methods of powder production that exploit spray technology, were reviewed. This included melt atomization for powder metallurgy, spray drying, spray pyrolysis, and flame spray pyrolysis. Spray freeze drying and emulsion combustion method were considered, briefly.

The chapter was started with a general introduction to powder technology and its importance. Then the principle of powder metallurgy by melt atomization and the effects of process conditions on powder morphology were considered. Particle formation by melt atomization consists of atomization of a melt followed by solidification of spray droplets. Therefore, powder morphology is dependent on the atomization process, and then the thermal history and solidification of droplets. Also, the geometry and dimensions of the nozzle are important in maximizing the delivery of the gas energy to the liquid metal stream. Droplet dynamic and thermal behavior is strongly dependent upon droplet diameter. Smaller droplets solidify faster. Another controlling parameter is the initial gas velocity. Increasing the initial axial gas velocity has three effects on individual droplet dynamic and thermal behavior. There is a reduction in flight time, an increase in mean heat transfer coefficients, and an increase in mean cooling rates. Increasing the melt superheat has no effect on individual droplet dynamic behavior. Increasing the melt mass flow rate does not have any effect on individual droplet dynamic and thermal behavior either, but affects the spray cooling and solidification. The main challenges in melt atomization process are mostly related to the atomization process; therefore, development of nozzles with stable operation, simple design, and with high wear and corrosion resistance against the hot melt stream is desirable.

Then, fundamentals of spray drying and spray pyrolysis processes were elucidated. A rather comprehensive model, explaining the intra-particle reactions and their effects on particle morphology and overall reactor design was introduced. Using the experimental evidence and modeling results, the effects of operating conditions on characteristics of powders produced by these methods were reviewed. One of the most important characteristics of powders is their morphology and, whether under certain process conditions, they are hollow or solid fully-field. This depends on the condition of the concentration profile developed within the solution droplets, at the onset of solute precipitation on droplet surface. Parameters such as, low process temperature, large droplet number density, small initial droplet size, and high initial solution concentration favor the development of a rather uniform concentration distribution within the solution droplet, and the formation of solid fully-filled particles. It was pointed out that some of the major challenges in powder production by spray pyrolysis technique include the production of nano-sized particles, improving the throughput of the process, and also process commercialization. Spray drying, however, is commercialized with industrial units already installed in food and biotechnology factories. However, reliable models that can be used for effective design of spray dryers have to be developed. For both spray drying and pyrolysis more research is required to understand the physics and chemistry of particle formation.

In contrast to spray pyrolysis, that produces micron-sized ceramic powders, powders produced by flame spray pyrolysis are nano-sized. This is because, in spray pyrolysis one solution droplet directly is converted to one particle, whereas in flame spray pyrolysis the entire droplet evaporates completely, and then the gaseous species nucleate and grow to form nanoparticles. Both conventional and flame spray pyrolysis can be used to produce functional powders; however, since flame spray pyrolysis produces nanoparticles, it is attracting more attention. Commercialization of the process, reducing the cost of powders by using cheap precursors, and the need for fundamental research to understand the complex phenomena involved in this process are the major challenges in flame spray pyrolysis. Additional research on the adverse effects of nanoparticles on health is required as well.

References

Akinc, M. and Richardson, K. 1986. Preparation of ceramic powders from emulsions, Mat. Res. Soc. Symp. Proc., 73.

Anderson, I E. and Terpstra, R. L. 2002. Progress toward gas atomization processing with increased uniformity and control, *Materials Science and Engineering A*, 326(1), 101-109.

Antipas, G.S.E. 2006. Modelling of the break up mechanism in gas atomization of liquid metals. Part I: The surface wave formation model, *Computational Materials Science,* 35(4), 416-422.

Bakenov, Z., Wakihara, M., Taniguchi, I. 2008. Battery performance of nanostructured lithium manganese oxide synthesized by ultrasonic spray pyrolysis at elevated temperature, *J. Solid State Electrochem.* 12, 57–62.

Bickmore, C. R., Waldner, K. F., Treadwell, D. R. and Laine, R. M. 1996. Ultrafine spinel powders by flame spray pyrolysis of a magnesium aluminum double alkoxide. *Journal of American Ceramic Society*, 79(5), 1419–1423.

Bickmore, C. R., Waldner, K. F., Baranwal, R., Hinklin, T., Treadwell, D. R. and Laine, R. M. 1998. Ultrafine titania by flame spray pyrolysis of a titanatrane complex. *Journal of the European Ceramic Society*, 18(4), 287–297.

Bird, R. B., Stewart, W. E., and Lightfoot, E. N. 2002. Transport Phenomena, 2nd edition, John Wiley, New York.

Brewster, J. H., Kodas, T. T. 2004. Generation of unagglomerated, dense, $BaTiO_3$ particles by flame-spray pyrolysis, *AIChE Journal*, 42(S11), 2665 – 2669.

Chau, A., Eslamian, M., Ashgriz, N. 2008. On production of non-disrupted particles by spray pyrolysis, *Particle and Particle Systems Characterization*, 25, 183-191.

Chiu, H. H. and Liu, H. H. 1977. Group combustion of liquid droplets, *Combustion Science and Technology*, 17, 127-142.

Chiu, H. H., Kim, H. Y. and Croke, E. J. 1982. Internal group combustion of liquid droplets, Proceedings of Nineteenth Symposium (International) on Combustion/ The Combustion Institute, 971-980.

Costantino, H., Firouzabadian, L., Hogeland, K., Wu, C., Beganski, C., Carrasquilla, K., Cardova, M., Griebenow, K., Zale, S., Tracy, M.. 2000. Protein spray freeze drying. Effect of atomisation conditions on particle size and stability. *Pharmaceutical Research*, 17, 1374-1383.

Costantino H, Firouzabadian L, Wu C, Carrasquillo K, Giebenow K, Zale S, Tracy M. 2002. Protein spray freeze drying. 2. Effects of formulation variables on particle size and stability. *Journal Pharmaceutical Science*, 91, 388-395.

Costantino, H. R. and Pikal, M .J. 2004. Lyophilization of Biopharmaceuticals, Chapter 13: Spray Freeze Drying of Biopharmaceuticals: Applications and Stability Considerations. American Association of Pharmaceuticals Scientists; Springer.

Eslamian, M., Ahmed, M., and Ashgriz, N. 2006. Modeling of nanoparticle formation during spray pyrolysis, *Nanotechnology*, 17, 1674-1685.

Eslamian, M. and Ashgriz, N. 2006*a*. Modeling of particle formation during spray pyrolysis using droplet internal circulation, *International Communications in Heat and Mass Transfer*, 33, 863-871.

Eslamian, M. and Ashgriz, N. 2006*b*. The effect of pressure on the morphology of spray-dried magnesium sulfate powders. *Canadian Journal of Chemical Engineering*, 84(5), 581-589.

Eslamian, M. and Ashgriz, N. 2006*c*. The effect of pressure on the crystallinity and morphology of powders prepared by spray pyrolysis, *Powder Technology*, 167, 149–159.

Eslamian, M. and Ashgriz, N. 2007. Effect of atomization method on the morphology of spray generated particles. *ASME Journal of Engineering Materials and Technology*, 129(1), 130-142.

Eslamian, M., Rak, J. and Ashgriz, N. 2008. Preparation of aluminum/silicon carbide metal matrix composites using centrifugal atomization, *Powder Technology*, 184-11-20.

Eslamian, M. and Heine, H. C. 2008. Characteristics of spray flames and the effect of group combustion on the morphology of flame-made nanoparticles, *Nanotechnology* 19, 045712.

Eslamian, M., Ahmed, M. and Ashgriz, N. 2009. Modeling of Particle Formation via Droplet-to-Particle Spray Methods, *Drying Technology*, 27, 1–11.

Eslamian, M. Ahmed, M. and Ashgriz, N. 2009 Modeling of micro- and nano-particle formation produced during emulsion combustion method. To be published.

Farid, M. 2003. A new approach to modeling of single droplet drying, *Chemical Engineering Science*, 58, 2985-2993.

Fletcher, D. F., Guo, B., Harvie, D.J.E., Langrish, T.A.G., Nijdam, J. J. and Williams, J. 2006. What is important in the simulation of spray dryer performance and how do current CFD models perform? *Applied Mathematical Modelling,* 30(11), 1281-1292.

Fukui, T., Ohara, S., Murata, K., Yoshida, H., Miura, K., Inagaki, T. Performance of intermediate temperature solid oxide fuel cells with La(Sr)Ga(Mg)O-3 electrolyte film, *Journal of Power Sources,* 106(1-2), 171-176.

Grant, P. S. Cantor, B. and Katgerman, L. 1993. Modelling of droplet dynamic and thermal histories during spray forming—II. Effect of process parameters, *Acta metall, mater.* 41(11), 3109-3118.

Grimm, G., Schultz, M., Barth, S. and Muller, R. 1997. Flame pyrolysis—a preparation route for ultrafine pure gamma Fe_2O_3 powders and the control of their particle size and properties. *Journal of Matererails Science*, 32(4), 1083–1092.

Heine, M. C. and Pratsinis S. E. 2005. Droplet and particle dynamics during flame spray synthesis of nanoparticles, *Industrial Engineering Chemical Research*, 44, 6222-6232.

Huttunen-Saarivirta, E. 2004. Microstructure, fabrication and properties of quasicrystalline Al–Cu–Fe alloys: a review, *Journal of Alloys and Compounds*, 363(1-2), 154-178.

Huang, Z., Czisch, C., Schreckenberg, P., Büllesfeld, F., Fritsching, U. 2005. Powder production by gas atomization of bioglass melt, *Particle and Particle Systems Characterization.*, 22, 345–351.

Incropera, F. P., DeWitt, D. P. 1985. Fundamentals of Heat and Mass Transfer, Wiley Incorporated, New York.

Jaworek, A. and Sobczyk, A. T. 2008. Electrospraying route to nanotechnology: An overview, *Journal of Electrostatics*, 66(3-4), 197-219.

Jayanthi, G. V., Zhang, S. C., and Messing, G. L. 1993. Modeling of solid particle formation during solution aerosol thermolysis, *Aerosol Science and Technology*, 19, 478-490.

Jossen, R., Mueller, R. Pratsinis, S. E., Watson, M. and Akhtar, M. K. 2005a. Morphology and composition of spray-flame-made yttria-stabilized zirconia nanoparticles, *Nanotechnology*, 16(6), 609-617.

Jossen, R., Pratsinis, S. E., Stark, W. J. and Mädler, L. 2005b. Criteria for flame-spray synthesis of hollow, shell-Like, or inhomogeneous oxides, *Journal of American Ceramic Society*, 88(6), 1388-1393.

Kammler, H. K. Mädler, L., Pratsinis, S. E. 2001. Flame Synthesis of Nanoparticles, *Chemical Engineering and Technology*, 24(6), 583 – 596.

Kammler, H. K. and Pratsinis, S. E. 2005. Carbon-coated titania nanostructured particles: Continuous, one-step flame-synthesis, *Journal of Materials Research*, 18(11), 2670-2676.

Kearns, M. 2004. Development and applications of ultrafine aluminium powders, *Materials Science and Engineering* A, 375-377, 120-126.

Kriegel, R., Töpfer, J., Preuß, N., Grimm, S., Böer, J. 1994. Flame pyrolysis - a preparation route for ultrafine powders: metastable b-$SrMnO_3$ and $NiMn_2O_4$. *Journal of Materials Science Letters*, 13, 1111 – 1113.

Kuo, K. K. 2005. Principles of Combustion, 2nd Edition, John Wiley, New York.

Laine, R. M., Hinklin, T., Williams, G., and Rand, S. C. 2000. Low-cost nanopowders for phosphor and laser applications by flame spray pyrolysis. In Eckert, J., Schlörb, H., and Schultz, L. (Eds.), Metastable, mechanically alloyed and nanocrystalline materials, Parts 1 and 2 (pp. 500–510).

Lam, C. W., James, J. T., McCluskey, R. Hunter, R. L. 2004. Pulmonary toxicity of single-wall carbon nanotubes in mice 7 and 90 days after intratracheal instillation. *Toxicological Sciences*, 77 (1), 126-134.

Lancia, A., Musmara, D., and Prisciandaro, M. 1999. Measuring induction period for calcium sulfate dehydrate precipitation, *AIChE Journal*, 45(2), 390-397.

Lee, A., and Law, C. K. 1991. Gasification and shell characteristics in slurry droplet burning, *Combustion and Flame*, 85, 77-93.

Leite, L. F. T., Lage, P. L. C. 2000. Modeling of emulsion droplet vaporization and combustion including microexplosion analysis, *Combustion Science and Technology*, 157, 213-242.

Lenggoro, I. W., Hata, T. T., Iskandar, F., Lunden, M. M. and Okuyama, K. 2000. An experimental and modeling investigation of particle production by spray pyrolysis using a laminar flow aerosol reactor, *Journal of Material Research*, 15(3), 733-743.

Lin, J.-C. and Gentry, J. W. 2003. Spray drying morphology: experimental study, *Aerosol Science and Technology*, 37, 15–32.

Liu, Y. B., Lim, S. C., Lu, L. and Lai, M. O. 1994. Recent development in the fabrication of metal matrix-particulate composites using powder metallurgy techniques, *Journal of Materials Science*, 29(8), 1999-2007.

Maa, Y-F., Nguyen, P., Sweeney, T., Shire S., Hsu, C. 1999. Protein inhalation powders: Spray drying vs. spray freeze drying. *Pharmaceutical Research*, 16, 249-255.

Maa, Y-F., Prestrelski, S. 2000. Biopharmaceutical powders: Particle formation and formulation considerations. *Current Pharmaceutical Biotechnoly*, 1, 283-302.

Mädler, L., Pratsinis, S. E. 2002a. Bismuth oxide nanoparticles by flame spray pyrolysis, *Journal of American Ceramic Society*, 85, 1713-1718.

Mädler, L., Stark, W. J. and Pratsinis, S. E. 2002b. Flame-made ceria nanoparticles, *Journal of Materials Research*. 17(6), 1356-1362.

Mädler, L. Stark, W.J. and Pratsinis, S. E. 2002c. Rapid synthesis of stable ZnO quantum dots, *Journal of Applied Phyisics*, 92, 6537-6540.

Mueller, R., Mädler L. and Pratsinis, S. E. 2003. Nanoparticle synthesis at high production rates by flame spray pyrolysis, *Chemical Engineering Science,* 58(10), 1969-1976.

Mueller, R., Kammler, H. K. Pratsinis, S. E., Vital, A., Beaucage, G. and Burtscher, P. 2004. Non-agglomerated dry silica nanoparticles, *Powder Technology*, 140(1-2), 40-48.

Mueller, R., Jossen, R., Pratsinis, S. E., Watson, M. and Akhtar, M. K. 2004. Zirconia nanoparticles made in spray flames at high production rates, *Journal of American Ceramic Society*, 87(2), 197-202.

Narayanan R. and Laine, R. M. 1997. Synthesis and character-. ization of precursors for group II metal aluminates, *Appllied Organometallic Chemistry* 11(10–11), 919–927.

Nimmo, W., Ali, N. J., Brydson, R. M., Calvert, C., Hampartsoumian, E., Hind, D. and Milne, S. J. 2003. Formation of lead zirconate titanate powders by spray pyrolysis, *Journal of American Ceramic Society*, 86(9), 1474–80.

Oberdorster, E. 2004. Manufactured nanomaterials (Fullerenes, C-60) induce oxidative stress in the brain of juvenile largemouth bass. *Environmental Health Perspectives*, 112(10), 1058-1062.

Orban, R. L. 2004. New research directions in powder metallurgy, *Romanian Reports in Physics*, 56(3), 505-516.

Ortiz, M. B. 1996. Centrifugal atomization for production of aluminum particles, Department of Mechanical and Aerospace Engineering, State University of New York at Buffalo, Master's Thesis.

Pratsinis, S. E. 2006. Overview - Nanoparticulate Dry (Flame) Synthesis and Applications, Technical Proceedings of the 2006 NSTI Nanotechnology Conference and Trade Show, Volume 1, Chapter 4: Nanoparticle Processes and Applications, 301 – 307.

Rohatgi, P. and Asthana, R. 1991. The solidification of metal matrix particulate composites, *JOM*, 43, 35-41.

Schaefer, R. J. and Kushner, B. G. 1990. Intelligent Processing of Materials (Eddited Wadley, H. N. G. and Eckhart, W. E. Jr). Minerals, Metals and Materials Society, Warrendale, PA.

Stefanescu, D. M., Moitra, A., Kacar, A. S. and Dhindaw, B. K. 1990. The influence of buoyant forces and volume fraction of particles on the particle pushing/entrapment transition during directional solidification of Al/SiC and Al/Graphite composites, *Metallurgical Transactions A* 21A, 231-239.

Sydel, P., Sengespeick, A., Blomer, J. and Bertling, J. 2004. Experimental and mathematical of solid formation at spray drying, *Chemical Engineering Technology,* 27(5), 505-510.

Takatori, K. 1997. R and D Review of Toyota CRDL (in Japanese), 32, 1–12.

Takatori, K., Tani, T., Watanabe, N. and Kamiya, N. 1999. Preparation and characterization of nano-structured ceramic powders synthesized by emulsion combustion method, *Journal of Nanoparticle Research*, 1(2), 197-204.

Tani, T. Mädler, L. and Pratsinis, S. E. 2002. Homogeneous ZnO Nanoparticles, *Journal of Nanoparticle Research*, 4, 337-343.

Tani, T. Watanabe, N. and Takatori, K. 2003. Emulsion combustion and flame spray synthesis of zinc oxide/silica particles, *Journal of Nanoparticle Research*, 5(1-2), 39-46.

Tani, T. , Takatori, K., Watanabe, N., Kamiya, N. 1998. Metal oxide powder synthesis by emulsion combustion method, *Journal of Materials Research*, 13(5), 1099-1102.

Tikkanen, J., Gross, K. A., Berndt, C. C., Pitkänen, V., Keskinen, J., Raghu S., Rajala, M., and Karthikeyan, J. 1997. Characteristics of the liquid flame spray process, *Surface and Coating Technology*, 90, 210-216.

Uslan, I., Saritas, S. and Davies, T.J. 1999. Effects of variables on size and characteristics of gas atomized aluminum powders, *Powder Metallurgy*, 42(2), 157-163.

Wegner, K. and Pratsinis, S. E. 2004. Nozzle-quenching process for controlled flame synthesis of titania nanoparticles, *AIChE Journal*, 49(7), 1667-1675.

Wu Y and Lavernia, E. J. 1991. Spray-atomized and co-deposited 6061 Al/SiC composites *JOM*, 43(8), 16-23.

Wu, Y. and Lavernia, E. J. 1992. Interaction mechanisms between ceramic particles and atomized metallic droplets, *Metallurgical and Materials Transactions A* 23A(10), 2923-2937.

Xiong, Y. and Kodas, T. T. 1993. Droplet evaporation and solute precipitation during spray pyrolysis, *Journal of Aerosol Science*, 24(7), 893-908.

Yingxue, Y., Shengdong G. and Chengsong, C. 2004. Rapid prototyping based on uniform droplet spraying, *Journal of Materials Processing Technology*, 146(3), 389-395.

Yule, A. J. and Dunkley, J. J. 1994. Atomization of Melts: For Powder Production and Spray Deposition, Oxford University Press.

Zachariah, M. R. and Huzarewicz, S. 1991. Aerosol processing of YBaCuO superconductors in a flame reactor, *Journal of Materials Research*, 6(2), 264-269.

Zhou, X. D., Zhang, S. C., Huebner, W. and Ownby, P. D. 2001. Effect of the solvent on the particle morphology of spray dried PMMA, *Journal of Materials Science*, 36, 3759–3768.

In: Powder Metallurgy Research Trends
Editors: L. J. Smit and J. H. Van Dijk ISBN: 978-1-60456-852-3

Chapter 3

GOVERNING FACTORS OF PHYSICAL AND CHEMICAL BEHAVIOR OF REACTIVE POWDER MATERIALS

Vladimir N. Leitsin, Maria A. Dmitrieva and Tatiana V. Kolmakova

ABSTRACT

Factors determining kinetics of physical and chemical conversion, structure of reaction product, and parameters of dynamically loaded reacting powder body surface radiation are investigated by the computer simulation method. The model takes into account structural parameters of the powder material capable of exothermal chemical reactions, an opportunity for phase transfers, formation of a product of chemical reaction, change of an aggregative state of the reacting medium, mechanical activation of the reacting components, the forced filtration of the molten component of a mixture, and change of parameters of a condition on each step of physical and chemical conversions. The algorithm of the numerical solutions of imitating simulation problems of physical and chemical behavior of reactive powder materials has been developed on the basis of the created model, providing the solution of the related problems of a macrokinetics of chemical transformations, mechanical modification of powder materials during dynamic loading, thermal balance and a filtration of a liquid phase. All parameters of the model are repeatedly specified during each step of time. The method for interpretation of experimental results of investigation of physical and chemical processes in the multicomponent reacting systems received by optical pyrometry methods for an estimation of a degree of realization of various stages of physical and chemical conversions, structure modification and other state parameters of reacting powder body, registration of changes of chemical reactions stages, both on the surfaces and in the bulk of the model compact has been developed. The governing factors of researched processes have been investigated during the computer simulation results analysis.

INTRODUCTION

The research focused on processes of chemical compounds synthesis by powder metallurgy methods of the powder metallurgy resulting in formation of new materials during reaction, have been intensively developed after A.G.Merzhanov, V.M.Shkiro and I.P.Borovinskaya patented the method of synthesis of refractory inorganic compounds by combustion in a gasless oxygenless system [1]. Further research of materials synthesis by methods of technological combustion of powder mixtures in the USSR and Russia are made by scientists A.G.Merzhanov, I.P.Borovinskaya, A.P.Aldushin, B.I.Khaikin, K.G.Shkadinskii, V.I.Itin, Y.S.Naiborodenko, Y.M.Maksimov, N.Z.Lyakhov, V.V.Aleksandrov, M.A.Korchagin, G.A. Nersisyan, S.L.Kharatyan, etc. [2 - 11]. The reviews of research efforts made in this field are reported in [12 - 14]. Experimental research has shown that a characteristic feature of reactive powder mixture behavior is a many-stage and multiphase character and variety of physical-chemical processes. The essential part of practically significant reactions between powder components can be included into a gasless class. Many researchers have found the ability of melting for the fusible component of a reactive powder mixture [9, 15, 16]. Solid-phase combustion when the temperature obtained from synthesis is lower than temperature of melting all the components of mixture and provides conservation of material structure to be set at the development stage of initial powder compact. However, this combustion can be obtained experimentally only after intensive mechanical activation [6, 17].

Intensive mechanical compression of reacting components of mixtures can result in an increase of reactivity - decrease of reaction initiation threshold and reduction of components interaction duration. This effect is called mechanical activation and applied for a certain synthesis regime and the opportunity to obtain new materials.

The effect of mechanical activation on regimes and conditions of interaction in various powder systems is investigated by N.Z.Lyakhov, M.A.Korchagin, V.V.Boldyrev, E.G.Avvakumov, N.S.Enikolopyan, Y.A.Gordopolov, V.S.Trofimov, A.S.Steinberg, S.S.Batsanov, M.A.Meyers, V.F. Nesterenko and N.N.Tadani [6, 18-26], etc. Besides formation of dense composites [6], general factors of mechanical activation in mechanical loading of reacting powder compacts can be plastic deformation of material crystal structure and removal of oxide and adsorbed layers from the surface of powder mixture particles [24, 25, 27]. Compaction of powder body at intensive mechanical loading allows increasing the components reactivity in a wide range and provides the conditions of chemical conversions realization, i.e. provides obvious technological advantages.

It has been found experimentally that during preparation of reacting powder mixture the components which differ by specific density, plasticity, etc., mix poorly, hence, it is practically impossible to get homogeneous distribution of particles of one component in another one; and the mixture always shows the formation of agglomerates of particles of one sort [4, 6]. The subsequent pressing of powder mixture results in the formation in volume of a heterogeneous powder material of porosity structure – heterogeneity of specific pore volume in local volumes of powder compact [28]. Thus, reacting powder mediums are structurally inhomogeneous materials characterized by presence of macroscopic structure of concentration heterogeneity.

Estimation of properties of structurally inhomogeneous powder materials is possible in the view of composite material micromechanics [29-32]. The element of periodicity for periodic structure material can represent the behavior of all the material.

The development of modern mechanics of deformable reacting powder mediums occurs on the frontier of mathematical modelling, powder materials mechanics, mechanics of reacting mediums, micromechanics of composite materials, heat and mass transfer theory, chemical kinetics [33].

1. Model of Reacting Powder Layer

The powder mediums able to undergo gasless exothermal chemical conversions are considered. Conditions of chemical transformation processes in conditions of self-propagating high-temperature synthesis (SHS) (in combustion wave) and mechanochemical synthesis at dynamic (shock) loading of initial powder compacts (green compacts) of reacting components mixture and inert filling agent are studied.

The dynamic loading impulse can be generated at detonation of condensed explosive being in contact with initial compact or at collision of high-speed plunger with it. Notice that pressure jump occurs on compact's boundary and propagates into the compact at shock wave velocity [34]. The matter compression is accompanied by collapse of pores and intensive mixing of the components, friction on the surface of particles and their deformation. In aggregate, these processes determine mechanical modification of powder materials. Combination of defective conditions arising in powders at dynamic compression of pores, plastic deformation, current and mixing is possible only at shock compression [35].

The powder mixture of reacting components and inert filling agent is a preformed granular structure [36]. Preforming is formation of powder compacts by application of pressure to a powder in a closed form or shell. At pressing in closed volume the coalescence of particles occurs, and required forms and sizes are preformed. Change of powder body volume occurs as a result of packing the voids between powder particles and binding (mechanical adhesion) of particles due to displacement and deformation of separate particles [34, 37]. In case of plastic components, the compaction occurs due to deformation of particles. At first deformation is localized by bonding areas and then propagates deeply into particles. At brittle materials pressing, deformation is shown in destruction of ledges on the particles surface. Rear surface structure of the sample is characterized by surplus of a plastic component of powder composition and small porosity. And powder mixture structure in near-wall regions is characterized by heavy porosity.

In preparation of reacting compact a mixture of initial components undergoes premixing, activating, and pressing. Reacting powder mediums are characterized by presence of macroscopic structure of concentration heterogeneity formed in a process of powder compact preparation due to synergetic processes of discrete systems self-organization followed by formation of powder body's internal structure [38] under intensive mechanical loading conditions. Formed at the stage of mixture preparation the structure of the mixture concentration heterogeneity can be modified in combustion wave and in a process of shock compression.

Real powder body is a modelling heterogeneous composition of reacting components A and B with inert filling agent having determined structure parameters, physical and chemical characteristics. Powder mixture is modelled by set of component particles. Material of particles of one sort is considered to be homogeneous and isotropic with set physical properties. The structure of initial mixture is characterized by form and sizes of the particles and their units, their location, concentration of components and porosity. It is supposed that before shock compression the powders mixture of reacting components and inert filling agent has been preliminary pressed to state of close packing with defined values of medium porosity, characteristic dimensions of particles agglomerate and component concentration dispersion. At loadings a powder compact behaves as a porous body [33].

Powder compact is considered to be a reacting layer if at each moment of time it is possible to assign arbitrary set of reacting powder sections which are perpendicular to some direction and on which microscopic characteristics of mechanochemical processes, effective medium state parameters, concentration and phase characteristics of structure can be considered as constants. The direction perpendicular to these sections is a direction of mechanochemical processes development. The layer of powder medium can be presented in a form of packing for elements of macroscopical structure of concentration heterogeneity. An example of geometrical model of such packing of structure elements in the form of rectangular prisms with sizes $a{\times}a{\times}b$ is shown in figure 1.

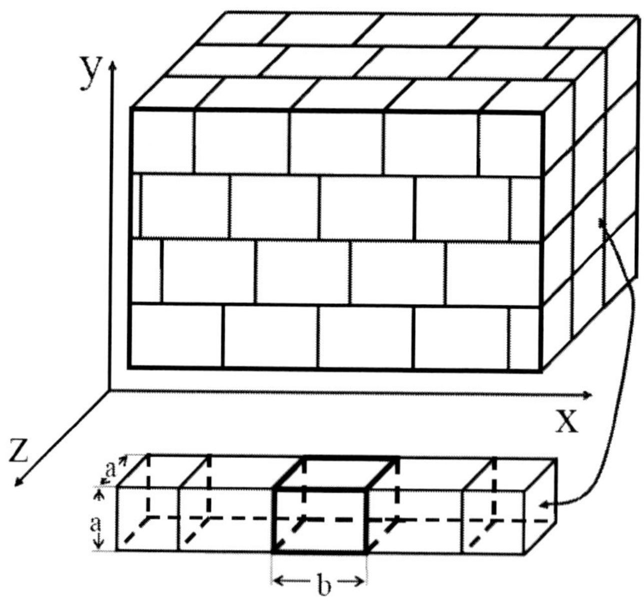

Figure 1. Geometrical model of powder body structure.

The considered direction of mechanochemical processes development is a mainstream in evolution of all characteristics and parameters of a reacting powder compact. Such approach is helpful in setting a problem of physical and chemical processes modelling in one principal direction, and therefore, considering the processes of mechanochemical conversions distribution to be quasi-one-dimensional.

The processes of SHS at initiating combustion wave on compact's out-side in principal direction of mechanochemical synthesis at loading powder composition layer by macroscopically plane shock pulse which spreads with specified amplitude and velocity satisfy the conditions to be mentioned above.

The layer model structure is considered to be a solid skeleton from powder particles with through porosity which forms regular structure of cells of concentration heterogeneity in space. Macroscopic concentration heterogeneity of mixture is supposed to have regular spatial structure. Mixture with specified at an average concentration of components is inhomogeneous on volume of periodicity cell with size $a \times a \times b$ – reaction cell. Concentration heterogeneity of the element of macroscopic structure of mixture (cell) is set by changing the concentration of components in direction b in the assumption that required share of a low-melting component δ is concentrated at left side of the cell $a \times a$ in its part of size *do* to be defined by characteristic size of particles agglomerates (figure 2). For specified character of functions of distribution of mixture's volumetric concentration components (step-function or parabolic, for example) the parameter b/a can be used as the characteristic of modelled macroscopic structure of mixture concentration heterogeneity. Side b of periodicity cell is chosen perpendicularly to the surface of mechanochemical conversions initialization. This direction is a princepal direction of changing the reacting mixture structure in a process of mechanochemical conversions [39].

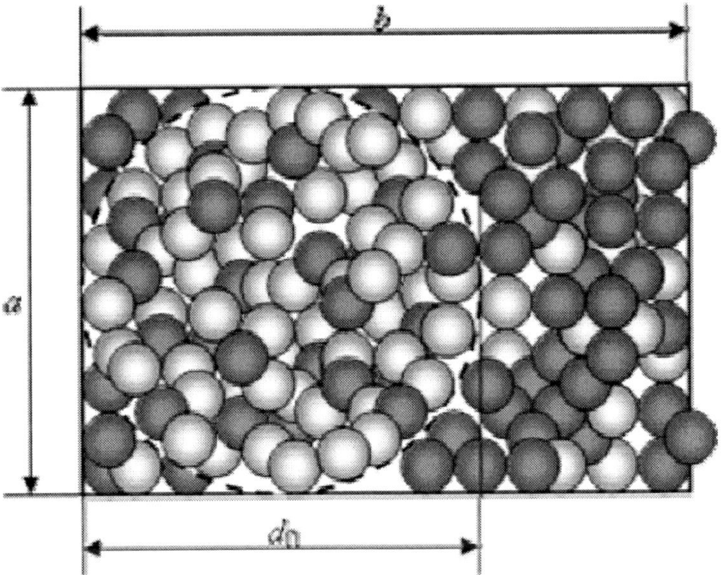

Figure 2. Concentration heterogeneity structure.

An element of macroscopic structure of concentration heterogeneity, reacting cell, is assumed to be representative volume of reacting powder medium considering structurally inhomogeneous powder material as a micro-inhomogeneous composite [31].

Reacting layer is modelled by sequence n of reaction cells and each of them represents an element of concentration heterogeneity macroscopic structure of the reacting powder mixture. The choice of reaction cells number (n) which is necessary for adequate representation of mechanochemical processes in a reacting powder material is determined by conditions of modelled experiment:

1 To model distributing the fronts of chemical conversions initiation and combustion of powder mixture layer (half-space) which is thick enough, the set of reactionary cells should provide stationary conditions of state parameters in depth of a powder material;
2 Modelling of mechanochemical processes in a layer of reacting powder mixture of finite thickness requires to take into account the ratio of cell dimensions and layer's thickness.

In each section of a reactionary cell which is perpendicular to principal direction the statement about thermal homogeneity is accepted. The statement having proved in [6] allows formulating fundamental conservation laws for reacting cells' microvolumes relating components' local features and effective parameters of powder material micro-layers in reacting layer sections.

2. Thermal Processes in Reacting Powder Body

For heat conduction processes the energy conservation law has form of the first law of thermodynamics which is written for unit of moving medium volume as follows [40]:

$$Q_v dt + L_v dt = \rho\left(du + d\left(w^2/2\right)\right), \qquad (1)$$

where Q_v is quantity of heat in unit of medium volume for time unit, L_v is work done by external forces on medium volume unit for time unit, u is internal energy of one kg of medium, and w is medium motion (flow) velocity.

Internal energy is connected with its enthalpy di:

$$du = di - d(p/\rho) \qquad (2)$$

Heat balance equation per time for volume V limited by surface F in a considered body is as follows:

$$\int_V Q_v dV + \int_F \mathbf{q} dF = \int_V q_v dV. \qquad (3)$$

The first member of equation is change of enthalpy for considered volume, the second member is quantity of heat losses through surface F due to heat conductivity, the third one – the quantity of heat evolution by internal sources.

Here $q_v = q_{v+} + q_{v-}$

where q_{v-} — heat sinks determining heat loss on phase transitions; q_{v+} – internal sources of heat resulted from volume chemical reactions and viscous-plastic deformation.

Taking Biot-Fourier law $q = -\lambda\ grad\ T$ and Gauss-Ostrogradski formula we can deduce from equation (3) the following relation:

$$Q_v = div\ (\lambda\ grad\ T).$$

Taking into account all this transformations, the differential heat-balance equation is as follows:

$$div(\lambda\ grad\ T) + q_v + L_v + \frac{dp}{dt} - \frac{p}{\rho}\frac{d\rho}{dt} = \left[\frac{di}{dt} + \frac{d(w^2/2)}{dt}\right]\rho. \quad (4)$$

Energy conservation law (4) for thermal processes in a reacting powder layer has the form of one-dimensional heat transfer equation with variable coefficients:

$$\frac{\partial(\rho_s c_s T)}{\partial t} = \frac{\partial}{\partial x}\left(\lambda_s \frac{\partial T}{\partial x}\right) + q_{v+} + q_{v-} \quad (5)$$

The possibility of heat loss on phase transitions and change of parameters with change of temperature are also considered. All coefficients of equation (5) are local effective characteristics of heterogeneous medium.

Initial conditions:

$$t=0,\ T = T_0(x), \quad (6)$$

where $T_0(x)$ – initial distribution of temperature on layer thickness.

The number of reaction cells in a model of powder mixture reacting layer is determined by conditions of stationarity achievement for state parameters in powder depth or by ration between thickness of a modelled powder layer and size b of a reactionary cell. Vanishing temperature gradient can be condition of state parameters stationarity on the boundary of modelling sequence of reactionary cells lying away from modelling reacting layer face. In the processes modelling in a thin layer of the reacting powder components mixture which has heat insulation on the boundary being opposed to face the condition of equal-zero temperature gradient determines zero flow of heat on heat-insulated boundary. In modelling mechanochemical conversions in a reacting powder layer we can consider the layer's heat loss on its far boundary to be negligible. Thus, for all specified cases of the reacting powder layer behaviour the research of which is practically essential the modelling of thermal processes can be done under condition of setting zero temperature gradient on back boundary.

Boundary conditions are:

$$x=0,\ T=\widetilde{T}, \qquad x=n\cdot b,\ \frac{\partial T}{\partial x}=0, \tag{7}$$

where n – number of reactionary cells on reacting layer thickness,

$\widetilde{T}$ - value of temperature on face boundary of a powder layer, the surface of application of mechanical load pulse.

Intense heat can cause melting of the low-melting component of a mixture since some point of time t^*.

In this case the reacting layer is considered to be the liquid-filled skeleton formed by refractory components and reaction products, in which movement of liquid-phase component is possible.

We offer to use two-temperature equations of thermal balance [41] instead of (5) to describe thermal processes in reacting powder medium saturated by liquid phase of a low-melting component.

Temperature regime is determined by temperature profiles T_s - of solid-phase skeleton (consisting of solid reagents and products of synthesis) and T_l - for liquid phase. The system of heat balance equations in this case is represented in as follows:

$$(1-\Pi_l)\frac{\partial(\rho_s c_s T_s)}{\partial t}+\alpha_v f_T(T_s-T_c)=\frac{\partial}{\partial x}\left[\lambda_s(1-\Pi_l)\frac{\partial T_s}{\partial x}\right]+q_{v+}+q_{v-}$$

$$\Pi_l\frac{\partial(\rho_l c_l T_l)}{\partial t}-\alpha_v f_T(T_s-T_l)=\frac{\partial}{\partial x}\left[\lambda_l\Pi_l\frac{\partial T_l}{\partial x}\right]-c_l G_l\frac{\partial T_l}{\partial x}, \tag{8}$$

where ρ_s – effective density of skeleton material, ρ_l – density of a liquid phase of the powder mixture fusible component, c_S – effective thermal capacity of the solid-phase powder skeleton material in the absence of deformation, c_l – a thermal capacity of the low-melting component melt in the absence of deformation, Π_l – effective porosity of skeleton, α_v – coefficient of volume internal heat exchange between liquid phase and skeleton , λs, – effective heat conductivity coefficient of the skeleton material, λ_l, – heat conductivity coefficient of liquid phase, $G_l=\rho_l v$ – liquid phase consumption, f_T – relative surface of heat exchange in powder layer section, v – velocity of liquid phase motion.

Initial conditions:

$$t=t^*,\ T_l=T_S(x), \tag{9}$$

where $T_S(x)$ – distribution of temperature on reacting layer thickness at a point in time t^*.

Boundary conditions:

$$x=0,\ T_S=T_l=\widetilde{T},\ x=nb,\ \frac{\partial T_S}{\partial x}=0,\ \frac{\partial T_l}{\partial x}=0. \tag{10}$$

Parameter $0<f_T<1$ included in equation (8) represents specific surface of heat transfer in powder layer sections and corrects the value of volume internal heat exchange coefficient deduced from solution of model problem on thermal interaction of unit particle and fluid flow. Its value changes during mechanical compaction of powder layer.

It is considered that action of energy sinks q_{v-} is initiated at achievement of temperature of phase transfer threshold values that compensates the sources of mechanical and chemical character q_{v+} till completion of phase transfers work.

Basic Fourier heat conduction law $q = -\lambda \ grad \ T$ is often applied for granular layer and considered as quasi-homogeneous medium [36]. The problem is in finding the value λ, which consists of contribution of various components which operation is nonadditive.

Reacting powder body is complex system with various mechanisms of thermal interaction. The defining mechanisms of thermal action are:

1 Heat conductivity and heat stresses in a solid-phase skeleton.
2 Convective heat exchange between a skeleton and flow of low-melting component liquid phase.
3 Heat conductivity of stationary liquid phase in a layer.

These boundary problems of modelling processes of transfer in saturated granular layer are considered in [35] for a separate particle or for particle ensembles interacting with flow or stationary liquid phase. The decision of a problem on thermal conductivity and stresses inside spherical particle with internal thermal emission provides the estimations of possible temperature drop on a particle which is defined by particle durability in relation to thermal stresses. Postulation of nondestruction of solid-phase skeleton of reacting powder compact at interaction with liquid-phase of low-melting component provides dependence of acceptable temperature drop on a particle from tensile strength of particles material, which can be used to estimate coefficient of particle's volume heat exchange with liquid phase flow. Limiting value of volume heat exchange coefficient can be deduced from heat balance equation for a particle which interacts with flow:

$$\alpha_v = 0.25\psi\rho_l c_l v \Pr^{-2/3}, \ \psi = 1.09\left(1-(1-\Pi)^{2/3}\right). \quad (11)$$

To define effective heat conductivity of powder medium λ_{S2} under condition of liquid phase absence the dependences of heat conductivity coefficient in a form of binomial function can be applied [42]:

$$\lambda_{S2} = \begin{cases} \lambda_k\left(1-\dfrac{3}{2}\Pi\right), & \Pi \le 0.1 \\ \lambda_k(1-\Pi)^{3/2}, & \Pi > 0.1 \end{cases}, \quad (12)$$

where λ_k – effective coefficient of skeleton heat conductivity.

We will estimate heat conductivity coefficient in a granular body saturated with liquid phase taking into account the solution of the problem on definition of effective heat conductivity of dispersion medium with identical spherical particles excluding frame conductivity, convection and radiation [43]. :

$$\lambda_{S1} == \lambda_g \frac{\chi(12-11\Pi)+11\Pi-6}{7\chi\Pi+24-7\Pi}+$$
$$+\frac{\sqrt{\left[\chi(12-11\Pi)+11\Pi-6\right]^2+(7\chi\Pi+24-7\Pi)(\chi+17-10\Pi)}}{7\chi\Pi+24-7\Pi}, \quad (13)$$

where $\chi = \lambda_g / \lambda_l$, λ_g - effective coefficient of grain heat conductivity, λ_l - coefficient of heat conductivity of liquid component.

3. The Processes of Modification of Powder Body at Compression

With the help of the shock waves the phase transfers, the decomposition of different components and chemical reactions in powder mixtures with the formation of numerous ceramic and intermetallic compounds were initiated [37]. Chemical reactions in solids at the shock compression of previously pressed powders are characterized by the fast rates and low temperatures of reaction initiation (in comparison with chemical conversions without dynamic loading) [20]. The possibility of the reaction induced by a wave in different powder mixtures is shown in [23, 44 – 48]. The existence of the threshold value of the pressure of the shock initiation of chemical reaction is revealed. The presence of such threshold shows that such mechanisms as plastic deformation, mass transfer and the behavior of particles like liquid, not melting energy and components expansion influence on the initiation of the reaction. These effects are not connected with the thermal activation of the synthesis process due to the heating of the system in a whole or in local points; it means that they are caused by the processes of mechanical activation. Qualitative characteristic of mechanical activation can be the temperature of the reaction start.

The analysis of the experimental data [16, 49] showed that the grain refining of the shock compressed powder was mainly connected with the activity of unloading waves if the pressure previous to unloading was lower than some threshold value.

As it follows from [17], the removal of oxide and adsorbed surface layers of the particles of the reactive components provides the rise of the level of mechanical activation. The processes of the removal of the surface layers of powder particles into the pores is observed in the shock loaded porous powder compacts at the some threshold value of the shock pulse amplitude [27].

So, at the simulation of the mechanochemical processes in the reactive powder materials at the shock loading it is important to take into account such reasons for mechanical activation as plastic deformation, fracture of oxide and adsorbed layers on the surface of particles during the shock pulse [50]. At the simulation of the mechanochemical processes in the adjacent layers of the rear surface (relating to loading) it is necessary to take into account the interaction of initial shock wave with the reflected ones. The fracture of particles will cause extra mechanoactivation of powder mixture on this surface.

At the loading of powder layer with macroscopically plane shock pulse, macroscopic stress state in powder layer is close to the condition of uniform compression. This circumstance allows making a conclusion about the possibility of presenting the heterogeneous powder medium as the mixture of homogeneous mediums with the interpenetrative continia each of which follows the separate components of the powder mixture [51].

Averaging the volume, using the method based on the local conservation equations for each system with the followed averaging by the control volume and the compression it to zero, one can move from the porous medium to the solid one with some average density [52]. The law of conservation of mass, wave and energy after averaging are:

$$\frac{\partial \rho}{\partial t} + \nabla_i \rho v_i = 0, \quad \rho = \rho_s (1 - \Pi),$$
$$\rho \frac{\partial v_i}{\partial t} = \nabla_j \sigma_{ij}, \quad \frac{d}{dt} = \frac{\partial}{\partial t} + v_i \nabla_i, \quad \nabla_i = \frac{\partial}{\partial x_i}, \tag{14}$$
$$\rho \frac{dW}{dt} = \sigma_{ij} \dot{\varepsilon}_{ij}, \quad \dot{\varepsilon}_{ij} = \frac{1}{2}\left(\nabla_i v_j + \nabla_j v_i\right), \quad \sigma_{ij} = -P\delta_{ij} + S_{ij},$$
$$\Pi = \frac{4}{3}\pi a_n^2 \zeta, \quad i, j = 1,2,3,$$

where a_n, ζ – radius and concentration of pores; ρ – average density; σ_{ij}, $\dot{\varepsilon}$ – averagad stress tensors and rate of deformation; v_i – i-rate component; W -internal energy density; P – pressure; S_{ij} – deviator of stress tensor; Π – relative volume of pores (porosity).

For the description of micromechanics of the plastic deformation process of porous medium the laws of conservation of mass, impulse and energy are to be add in the equations of the state of the porous solid. We use the single-pore model of V.F. Nesterenko [53] for deriving the equation of the state, determining the behavior of the material point of the effective medium.

The porous medium (figure3, a)) is modelled by the cell which is α-in a radius pore surrounded by the b-in the radius sphere with the central spherical c-radius (figure3, b)) [53]. The material in a hollow sphere is considered to be viscoplastic.

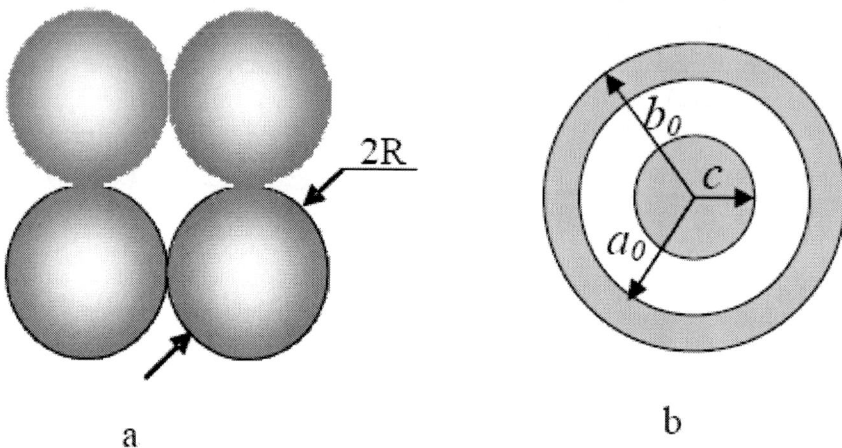

Figure 3. «Real» material (a), Nesterenko model (b).

Peripheral part of the model cell is considered to undergo the viscoplastic deformation and its mechanical characteristics are the temperature functions.

Nesterenko model satisfies the following requirements:

1 the same chracteristic sizes of model and powder specified by the equality of cell mass and the real particle;
2 equality of the specific volumes of the powder and the model;
3 at the porosity $\Pi_0 > \Pi^*$ the coincidence of the volumes of a plastically deformed material and hollows in the powder at the porosity Π^* (Π^* corresponds to the minimal porosity of the initial powder achieved due to the replacing of particles); at $\Pi_0 < \Pi^*$ the equality of the material volume under plastic flow and hollows in the initial state;
4 dependence of yield stress and viscosity on temperature;
5 self-consistency of the model following the conditions of the equality of pressure P (t) growth time t_g on surface of a model cell and time of collapsing of a pore t_α, $t_\alpha = t_g$.

Geometrical parameters of the modified model are:

$$b_0 = \frac{d}{2}\sqrt[3]{\frac{1}{1-\Pi_0}}; \quad a_0 = \frac{d}{2}; \quad c = \frac{d}{2}\sqrt[3]{2 - \frac{1}{1-\Pi_0}}; \quad \frac{1}{1-\Pi} = \left(\frac{a_n}{a_0}\right)^3 + \left(\frac{1}{1-\Pi_0} - 1\right),$$

where d – diameter of particle, Π – current value of porosity, a_n – current value of internal radius (figure 3).

The following equation was used for the description of the dependence of yield stress on the temperature in a modified model:

$$\sigma_T = \begin{cases} \sigma_{T1}\left(1 - T/T_m\right) & T \le T_m \\ 0 & T > T_m \end{cases},$$

where σ_{T1} –value of yield stress at low temperatures, T_m – melting temperature.

The dependence of viscosity on temperature was accepted as:

$$\eta = \eta_m \exp B\left(\frac{1}{T} - \frac{1}{T_m}\right),$$

where η_m – viscosity of melt.

Nesterenko model is represented as a boundary problem:

Equation of motion

$$\rho_s \ddot{r} = \frac{\partial \sigma_{rr}}{\partial r} + \frac{2}{r}(\sigma_{rr} - \sigma_{\Theta\Theta}), \quad (15)$$

Viscoplastic flow equation

$$\sigma_{rr} - \sigma_{\Theta\Theta} = \sigma_T + 2\eta\left(\frac{\partial v}{\partial r} + \frac{v}{r}\right) = \sigma, \quad (16)$$

Incompressibility equation

$$r^3 - r_0^3 = a_n^3 - a_0^3, \quad (17)$$

Boundary conditions

$$\sigma_{rr}(r = a_n) = 0;\ \sigma_{rr}(r = b_n) = -P(t), \quad (18)$$

Initial conditions

$$a(0)=a_0;\ \dot{a}_n(0) = 0;\quad T(r_0,0) = T_0. \quad (19)$$

Using the dependency of σ_T and η on the temperature, the solutions of the boundary problem were received [54]:

$$P(t) = 2\int_{a_n}^{b_n}\left\{\sigma_{T1}\left(1 - \frac{T}{T_m}\right) - 6\eta_m \exp\left[B\left(\frac{1}{T} - \frac{1}{T_m}\right)\right]\left(\frac{\dot{r}}{r}\right)\right\}\frac{\partial r}{r} -$$

$$-\rho_s\left[(a_n \cdot \ddot{a}_n + 2\dot{a}_n^2)\left(1 - {a_n}/{b_n}\right) - \frac{1}{2}\dot{a}_n^2\left(1 - \frac{a_n^4}{b_n^4}\right)\right];$$

$$\rho_s c_s \dot{T}_N = -2\sigma_{T1}\left(1 - \frac{T}{T_m}\right)\frac{\dot{r}}{r} + 12\eta_m \, exp\left[B\left(\frac{1}{T} - \frac{1}{T_m}\right)\right]\cdot\left(\frac{\dot{r}}{r}\right)^2; \tag{20}$$

$$b_n = \left[a_n^3 + a_0^3 \cdot \left(\frac{1}{1-\Pi_0} - 1\right)\right]^{1/3}; \quad r = \left(r_0^3 - a_0^3 + a_n^3\right)^{1/3}$$

$$a_n \geq c = \frac{d}{2}\sqrt[3]{2 - \frac{1}{1-\Pi_0}}; \quad \Pi_0 \leq \Pi^*.$$

The solution (20) can be used as the determining equation for viscoplastic compression of the porous powder medium. The second equation represent the formula for the heating rate of particle with the radius of the model sphere. It is important to point out that in the Nesterenko model the dimension of particles is introduced as a characteristic dimension. It allows simulating adequately the real powder mixture in which the dimension of pores can be different at similar particles dimensions and porosity. The model takes into account the dependence of thermal characteristics on the temperature and allows taking into account the inhomogeneity of the plastic deformation of particles and, particularly, the possibility of the localization of intensive plastic deformations at the boundaries of particles.

For heterogeneous powder medium it is offered to use the Nesterenko model separately for each component. Such method allows to investigate the local behavior of the powder compact during the shock compression, i.e. at the microscopic level [55]. In the powder body microvolumes it is offered to define the decrease of porosity, plastic deformation and fracture of the surface layers of the powder medium particles in each moment of the mechanical loading according to the possibility of the dissipation of shock pulse energy quota on plastic deformation and fracture. Even so, the possibility of the plastic flow of the components of the powder mixture, the fracture of the surface layers of particles are determined by exceeding of the average shock pressure with the threshold flow conditions and the realization of the jet stream of the particles material [53].

The single-pore model of the porous reactive powder medium can be simulated on the base of Nesterenko model if the spherically symmetric characteristic of the pore flowing is considered with increments. Let's consider that the shock pulse influences on the powder medium with the definite value of porosity, temperature, phase concentration, and other state parameters at each moment. The realization of different nonlinear and asymmetrical mechanisms of plastic deformation is possible when the threshold values of the plastic deformations of the components of the powder mixture and shock pulse pressures are achieved.

So, the modification of Nesterenko model for the microvolumes of the reactive powder medium lies in the following [33]:

1 component-wise use of the model for the components of heterogeneous mixture;
2 the use of the current values of the state parameters of the powder medium as the initial parameters for the use of the model in the following loading time step.

For each powder component, the shock pulse energy is considered to be wasted for the plastic flowing of pores until the size of the deformability of powder particles material achieves the boundary value or the material is pressed to the limiting state.

In the macrolevel it is offered to use the average state parameters of the porous medium which were received with the help of the Tuvinin model without using the formal average density value behind the shock pulse front. The particles are believed to be dynamically loaded by the shock pulse and then to be unloaded into the surrounding pores. Such conception allowed calculating the parameters of the final state of the shock compact powder medium behind the shock pulse front [56]:

$$\begin{aligned}
&U_f = v_{imp} + [a_f + (1-\Pi_0)D_p]/2b_f + \\
&\{[a_f + (1-\Pi_0)D_p]^2/4b_f^{\ 2} + (1-\Pi_0)v_{imp}D_p/b_f\}^{1/2}, \\
&P_f = (1-\Pi_0)\rho_0 D_p U_f, \\
&\rho_f = (1-\Pi_0)\rho_0\, D_p/(D_p - U_f), \\
&W_f - W_0 = U_f^2/2,
\end{aligned} \tag{21}$$

where D_p - velocity of the shock pulse in the porous medium, U_f- mass velocity, U_f- pressure behinde the shock pulse front, W_0 and W_f - specific internal energies of the mediums before and after the shock loading, ρ_0 and ρ_f – material density before the shock compression and after, a_f and b_f – parameters of the shock adiabats, v_{imp} – the rate of the «equivalent» plunger.

The parameters of the shock adiabats can be determined analytically for the mixture of heterogeneous materials. Postulating that the proportion $U_{S1,2}=a_f+b_fU_{P1,2}$, is valid for the mixture components, constants a_f and b_f can be determined according to the additivity concept [57]:

$$\begin{aligned}
a_f &= a_1a_2[\rho_1(1-x)+\rho_2x]/[a_1\rho_1(1-x)+a_2\rho_2x]; \\
b_f &= b_1b_2[\rho_1(1-x)+\rho_2x]/[b_1\rho_1(1-x)+b_2\rho_2x].
\end{aligned}$$

Here x – mass concentration of the first component, ρ - density, a_i, b_i – parameters of the shock adiabats of components, indexes i=1 and 2 mean the mixture components.

There is a conversion zone between the shock pulse front and the area of the final states. Its width is determined by the decay time of the compression and unloading waves circling in particles, and by the time of the thermal relaxation of particles. The dissipation of kinetic energy of the material particles vibration leads to the activation of the mixture components and to thermal component appearance in the energy balance equation.

The problem of mechanical modifications of reactive powder materials under dynamic loading is solved by an energy-balance method [54]. It is considered that in each microvolume the kinetic energy of the shock compression can dissipate with the constant velocity during shock loading .

Thermodynamics of the shock compression of the powder body is represented by the proportion for the average values of the dissipated energy at the plastic and viscous flowing of components, the energy of motion at the pores collapse, fracture of the surface layers of powder particles.

Using the Nesterenko model, the equation for the deformation specific energy dissipated into the heat around a pore on the stage of its collapsing can be presented as [53]:

$$W_d = \frac{3c_v}{R^3}\int_{a_n}^{b_n}(T - T_0)r^2 dr,$$

where R – powder particles radius.

For the case when the processes of the viscous dissipation is neglible the average yield stress value σ_{Tm} is introduced and the equation for W_d is represented as:

$$W_d = \frac{2\sigma_{Tm}}{3\rho_s}\left|F(x_0) - F(x_1)\right|, \quad F(x) = \ln x - 1,$$
$$x_0 = \frac{1}{\Pi_0}; \quad x_1 = \frac{1-\Pi_0}{(1-\Pi_1)\Pi_0}; \tag{22}$$

If only the processes of viscous dissipation mainly contribute to W_d, so, introducing the average viscosity η, this eqaution is represented as [53]:

$$W_d = \frac{4\eta'}{3\rho_s}\int_{a_n}^{b_n}\frac{(\partial(1/(1-\Pi))/\partial t)d\alpha}{(1/(1-\Pi))[1/(1-\Pi) - 1/(1-\Pi_o) + 1]} \approx \frac{\eta'}{\rho_s t} \approx \frac{2\eta'}{\rho_s d}\sqrt{\frac{P}{\rho_s}} \tag{23}$$

The difference between the specific energy of the compression W and the specific energy W_d dissipated about the spherical pore on the stage of its collapsing represent the microkinetic energy: $W_k=W-W_d$. It is spent for fracture of the surface contact layers of particles and, probably, the compression of powder material due to the liquid way behavior of the material in the surface layers of particles [53]. With the increasing of the loading pulse the part of the shock pulse energy dissipated to the plastic deformation and viscous pores flowing decreases, it means that the part of energy, which can be spent for the fracture of particles surface layers, increases.

The work required for the fracture of the surface layers of particles can be calculated by the Irvin formula [58]:

$$A_* = \pi(1-\mu^2)E^{-1}K_{IC}^2, \tag{24}$$

where A_* – minimal required work for increasing the area of crack per unit, K_{IC} – specific value of stress intensity, E - Young modulus, μ - Poisson's ratio.

Let the area of the formed surface increase from 0 to the area of the surface of a particle S_p, so the energy required for fracture of the whole surface of one particle will be equal to:

$$W_{pi} = \frac{2\pi\left(1-\mu^2\right)}{E} K_I^2 S_p.$$

In this case the total energy required for the fracture of the surface layers from all the particles of microvolume is represented as:

$$W_p = \sum_{i=1}^{M} \frac{2\pi\left(1-\mu_i^2\right)}{E_i} K_{Ii}^2 S_{pi}, \tag{25}$$

where M – number of particles.

The work done at the fracture of the surface layers of particles provides the activation of the reactive components of the powder mixture and in the addition to the work of the plastic deformation determines the sources of the heat in the heat balance equations.

It is offered to determine the sequence of the mechanisms of the dissipation of the mechanical loading energy for each component of the powder mixture: at first, the plastic flowing is realized until it achieves the threshold value of deformation, then the transfer to the dynamic regime of deformation leading to the formation of jet streams of surface layers of powder particles and to the fracture of oxide and adsorbed layers is possible [50, 55].

Fast stress relaxation causes the recovery of equilibrium interatomic distances in a matter with the rate close to the speed of sound in this material. Undissipated mechanical energy of the compression transfers into the kinetic energy of some elements of the material substructure (microvolumes). If this energy is higher than some threshold value, the unloading leads to the internal bonds breakage in the most defected areas and to the dispersion of particles. The unspent energy of the shock pulse required for the dispersion of particles can be calculated according to the balance of solid body fracture energy

$$W = K\left(\sigma_T \varepsilon + \frac{12\gamma}{d} + \frac{\rho V^2}{2}\right), \tag{26}$$

where σ_T – yield stress, ε - ultimate strain, γ - specific fracture work of the surface, d – average size of particles, ρ - density, V – average rate of particles after fracture, K - empirically determined coefficient.

In the work [27] it is offered to use a pressure scale for characterizing the powder solid body relative to the shock pulse, i.e. to calculate the pressures of the shock pulse corresponding to the transfer to qualitatively different variants, macroflow, regimes of deformation and heat of particles:

1. The pressure under which the existence of the shock transfer is not possible:

$$P_{s1} = P_{\min} = \frac{2}{3}\sigma_T \ln\left(\frac{\rho_s}{\rho_s - \rho_0}\right), \tag{27}$$

Here ρ_0 – initial density of powder, ρ_s – density of solid material.

2. The lower limit of «jet» regime of shock compression (localization of deformation at the surface layers of particles):

$$P_{s2} = 0{,}36\frac{\rho_0}{\rho_s}\frac{Hv}{(1-\rho_0/\rho)}, \tag{28}$$

where Hv=2,58$\sigma_в$ – hardness of particles material, $\sigma_в$ – ultimate strength [27].

Pressure P_{s2} and density ρ after the shock pulse front are connected by shock adiabat:

$$P_{s2} = \frac{2}{3}\sigma_T \, ln\left(\frac{\rho_s}{\rho_s - \rho}\right), \tag{29}$$

and it is necessary to solve these two equations together.

Taking into account the possible presence of chemical conversions in the investigated powder medium, the scale of the critical pressure levels determines the conditions of the reactive abilities increase of the reactive components.

3. The result [49] allows to supplement the pressure scale [27] with the critical pressure value P_{s3} providing the fracture of particles during the shock pulse unloading. The pressure of uniform compression P required for fracture of solid body can be calculated according to the energy-balance method:

$$P_{s3} = \left[\frac{2Ek}{3(1-2\mu)}\left(\sigma_T\varepsilon + \frac{12\gamma}{d} + \frac{\rho V^2}{2}\right)\right]^{1/2}, \tag{30}$$

The start of plastic deformation of the powder mixture components is determined by achieving pressure after the shock pulse front of threshold values (27). In this case the equation (22) determining the work of plastic pores flow in the *i*-level of loading is represented as:

$$\Delta W^{(j)}{}_{iT} = \frac{2\sigma^{(j)}{}_{Tm}}{3\rho^{(j)}{}_{s}} \left| F(x_0) - F(x_1) \right|; \quad F(x) = \ln x - 1;$$

$$x_{0i} = \frac{1}{\Pi_{(i-1)}}; \quad x_1 = \frac{1 - \Pi_{(i-1)}}{\left(1 - \Pi_i^j\right)\Pi_{(i-1)}}; \tag{31}$$

where $\Pi_{(i-1)}$ and $\Pi^j{}_i$ – values of relative volume of pores in the last two levels of plastic deformation (by time and component-wise), $\sigma^{(j)}{}_{Tm}$ и $\rho^{(j)}{}_s$ – average yield stress and density of j plastic mixture component.

The equations (31) are solved relative to unknown value of porosity $\Pi^j{}_i$ .This value is taken as initial value of porosity $\Pi_{(i-1)}$ in each following step of using these equations (31).

If the condition of «jet» regime of deformation of powder medium components (28) is achieved, the scale of the fracture of the surface layers of powder particles is calculated after the plastic deformation complete. For j-component of the powder mixture the degree of the surface layers fracture is determined by the part of surface $\Delta s_p^{(j)} / s_p^{(j)}$, cleaned due to the jet streams of material, and can be obtained from the equation (25) for the work of surface layers fracture:

$$\Delta W^{(j)}{}_p = \sum_{i=1}^{M} \frac{2\pi\left(1 - \mu^{(j)2}_i\right)}{E^{(j)}{}_i} K^{(j)2}_{Ii} \Delta S^{(j)}{}_{pi}, \tag{32}$$

where $\Delta W^{(j)}{}_p$ – residual shock loading energy, unspent for plastic deformation of particles on the loading step, or fracture of the surface layers of particles of other components. After fracture of the part $\Delta S^{(j)}{}_{pi}$ of the surface layers this part of the shock pulse energy is spent by powder medium for heat. In the equation of the heat balance this part of energy will determine the increase of the mechanical heat source.

In the process of the powder solid body compression the relative surface of the heat exchange in the section of powder layer determining the relative surface of the volume heat exchange between the solid-phase skeleton and liquid phase (equations (8)) is changed. Taking into account the results of the work [59], the value of relative surface can by calculated:

$$f_t = 3.5\left(\Pi/\Pi^*\right)^{4/3} - 2.5,$$

with the assumption that the powder particles of high-melting components of mixture form tetrahedral periodic packing.

4. Filtrational Processes in the Saturated Porous Medium

Mechanical and reaction heating of heterogeneous powder medium can result into melting of low-melting component. In this case, powder medium is considered to be solid-phase skeleton with through porosity, which is saturated with low-melting liquid-phase component. Relative pores volumes filling by melt define the degree of skeleton saturation with liquid phase.

The forced filtration of melt with velocity v in principle direction of mechanochemical processes in the reacting layer is considered. This motion is possible if there are open porosity and porous pressure gradient before heat zone.

The filtration equation for slow movements of viscous liquid through a powder body micro-layer described by Darcy's law [58]:

$$F(v) = \left(\frac{M(T)}{K}\right)v, \tag{33}$$

where $M(T)$ – kinematic viscosity, K – permeability of the porous medium

The law of filtration can be used to estimate the velocity of liquid phase filtration in each section of a reacting layer.

In porous medium with quiescent solid phase the average tensor of viscous stresses in liquid phase is neglected (i.e. viscosity is considered only in force of interfacial interaction). Thus, as it is shown in [60], inertial forces, pulsating momentum transfer and kinetic energy of pulsating motion is negligible too.

Having considered all these assumptions from momentum conservation equation and continuity equation we have the following system:

$$\nabla P_{ж} = -\frac{M(T)}{K}v, \quad \nabla v = 0. \tag{34}$$

For porous model body consisting of homogeneous solid spherical particles with diameter d, the value of permeability is

$$K = \Pi_l^2 d^2 / 180(1 - \Pi_l)^2.$$

We offer to consider forced filtration of liquid phase under thermocapillary forces P_{tk} [61] and porous stresses during mechanical compression taking into account bearing capacity of solid-phase skeleton P_f:

$$P_l = P_{tk} + P_f,$$

$$P_{tk} = \left[C_b\alpha_b - (C_a\alpha_a + \Pi)\right]\frac{(T - T_m)}{1 - 2\nu}E, \tag{35}$$

where C_a and C_b – volume concentrations of solid-phase and liquid components, α_a and α_b - coefficients of thermal expansion of skeleton and liquid component, T_m – melting temperature for a low-melting component of powder system.

To consider bearing ability of solid-phase skeleton which depends on temperature we offer to use stiff-plastic model of continuous medium. In this case bearing ability of solid-phase skeleton will be defined by dependence of effective yield stress on temperature. Taking into consideration the equation for average pressure of shock pulse (21), porous stresses during action of mechanical loading momentum are set by equation:

$$P_f = (1 - \Pi_0)\rho_0 D_p U_f - C_a \sigma_T(T). \tag{36}$$

5. Macrokinetics of Chemical Conversions

The aim of the macroscopic chemical kinetics is to study the chemical reactions in the real conditions, taking into account physical processes imposed on the main chemical process.

The rate of the chemical reaction implies the amount of the matter reacting in the unit of volume per time unit [62]. This value is the function from the temperature and the concentration of reactive matter. For the description of the dependence of the reaction rate on the concentration of reactive matter, the power law usually is used:

$$w = k_r C_A^{nA} C_B^{nB} \ldots,$$

where w – reaction rate, C_A, C_B – concentration of matter A, B, …, taking part in the reaction, k_r –reaction rate constant depending on the temperature, n is a power-low exponent (by the matter A, B, …). The dependence of the constant rate on the temperature is given by Arrhenius equation:

$$k = k_0 \cdot exp(-E_a / RT),$$

where T – temperature, R – universal gas constant, E_a – activation energy, k_0 – pre-exponential factor.

The reaction rate can be determined by the change of the conversion of one of the components. As the low-melt component in the liquid phase can move in the porous skeleton, it is very difficult to determine adequately the conversion, that is why we will determine the reaction rate by the conversion z of high-melting component: $w = \partial z / \partial t$

Macrokinetic equation for the conversions z in differentiated form is represented as [2, 3, 10]:

$$\partial z / \partial t = k_0\, exp(-E_a / RT)\varphi(z) = \Phi(T, z), \tag{37}$$

where φ – inhibiting function. which is determined by the reaction type

In the case when the reaction product and at least one of reagents at combustion do not melt, the type of the kinetic function is determined by the mechanism of reagents transfer and by the structure of the growing layer of products [63]:

$$\frac{\partial z}{\partial t} = k_0 e^{-E_a/RT}\left(e^{-mz} z^{-n}\right) \tag{38}$$

The common laws are:

m=0, n=0 - linear
m=0, n=1 - parabolic
m=0, n=2 - cubic
m>0, n=0 – exponential

It is known that the reaction rates with the liquid phase essentially exceed the known values for solid-phase conversions [10]. Deriving the macrokinetic equation which determined the interaction in the powder system by the solid-liquid type in [10], the following assumptions were made: the formation of the front of the conversions on the particles of high-melting component happens faster than the complete interaction; all solid particles have the same size.

In this case macrokinetic inhibiting function φ (z) presents is represented as [10]:

$$\phi(z) = (1 - z)^{m - 1/n}, \tag{39}$$

where n=1, 2, 3 depending on one - , two-, or three- dimensional front of the reaction transfers into the high-melting skeleton, m=1,2. The case m=1 gives the equation of the diminishing surface and is realized with the excess of low-melting component; m=2 is for stoichiometric mixtures taking into account the changes of the concentration of low-melting component in the reaction.

Postulating that the power law of the reactive diffusion in the solid phase [64] works for the powder reactive layer in the macrolevel, the type of the inhibiting function and the dependence of the pre-exponential factor on the characteristic size b of the element of the structure can be determined [65].

The kinetics of the reactive diffusion at the interaction of the diffusion pair in the solid phase is described with the help of the law [64]:

$$y^{n+1} = K_r t, \tag{40}$$

where y - reaction product layer thickness, K_r - constant of growth rate of this compound, t – time. For the model cell of concentration inhomogeneity the rate of the conversion of the high-melting component can be determined as:

$$z = \frac{V_A C_A}{V C_A} = \frac{V_A}{V},$$

where C_A - volume concentration of high-melting component V_A – the reacted material volume, V – total material volume in the cell.

Supposing that the conversions was in the layer with the size $a \times a \times y$, we get:

$$z = \frac{a \cdot a \cdot y}{a \cdot a \cdot b} = \frac{y}{b}.$$

Taking into account (2.47) we get in [33] for the reactive layer $a \times a \times b$:

$$\frac{\partial z}{\partial t} = \frac{K_r}{(n+1) z^n b^{n+1}} \tag{41}$$

With $K_r \approx k_0$ the ratio (41) determines the power law of macrokinetics (38) of chemical conversions (m=0; n), and n can be fractional [64]. Particularly, for the reactions of powder materials of Ni-Al, Ti-Al (n=1) type by solid phase type the macrokinetic inhibiting function is represented as $\varphi(z)=0{,}5z^{-1}$, corresponding to the parabolic law.

Pre-exponential factor k_0 in the equation of macrokinetics is determined by the experimental values of the conversions rate of the investigated reactive mixture of «standard» structure. It is natural to examine the structure corresponding to the homogeneous distribution of the mixture components concentration and the specific volume of pores as «standard» for the laboratory sample. Given the fact that the thickness of the reaction-product layer in the reactive mixture under study follows the reactive-diffusion power law, the pre-exponential factor is assumed to be depended on the size b of the discrete structure element of the powder-concentration inhomogeneity and we have [66]:

$$k_0 = k/b^{n+1}, \tag{42}$$

where $k = k_0$ for unit- thickness reactive cell.

The calculation of the barrier reduction of activation processes by the decrease of energy activation parameter is given in [67] for the creep rate and the long-term strength of metals on the temperature and stresses, and in [68] for solid phase conversions.

The change in the reactivity of the powder mixture under mechanical loading can be calculated as [66]:

$$E_a = E_0 \quad H\left(P - P_i^*\right)\alpha_i A_i, \tag{43}$$

where E_0 is the activation energy for conversions of the reactive mixture with no mechanical loading; H is Heaviside function, P_i^* is the critical shock pressure $P=P(t)$ responsible for involvement of a particular activation mechanism; α_i are the parameters defining the contribution of the mechanical loading A_i, that changes the reactivity of the powder mixture by the i-th-mechanism.

It is considered that the contribution of different mechanisms of activation in E_a is additive. The coefficients α_i in the ratio (43) are the matching parameters of the reactive cell model and can be determined at the simulation of laboratory experiments. The value of the coefficient k in the ratio (42) also refers to the matching parameters.

The change of the threshold criterion for the reaction initiation which is reached by the introduction of the local condition of reactive equivalence of mixture is taken into account. It is offered to consider the condition of the reactive equivalence of the mixture states, which are different by the values of macrokinetic parameter of activation energy, as [65]:

$$\beta = \frac{E_a}{RT} = \frac{E_0}{RT_0} = \mathrm{B}, \tag{44}$$

where T_0 is an experimentally determined temperature of the conversions initiation corresponding to the activation energy E_0.

The condition (44) is used to determine the values of energy activation reached during the mechanical activation: $E_a = \frac{T}{T_0} E_0$. The numerical evaluation of A_i, spent on plastic deformation and the fracture of the surface layers of the reactive components particles, allowed to determine the matching coefficients α_i in (43).

Under the assumptions done for the reactive powder layer the problem of macrokinetics of mechanochemical synthesis can be solved consistently for different macrolayers of elements of macroscopic structure of concentration heterogeneity.

It is considered that in each section (microlayer) of the reactive layer the condition of the thermal, concentration, and phase homogeneity is given. After the initiation of chemical conversions the conversions rate $\dot{z}$ and the power of heat sources in the heat balance equation $q_{v+} = Q\frac{\partial z}{\partial t} + \varphi$ is determined by the ratio (38) where the pre-exponential factor, the inhibiting function and the activation energy are determined by the powder medium state. These macrokinetic parameters are determined with taking into account the modifications of powder material of the microlayer and the whole reactive cell [55, 65].

The rate of the conversions of high-melting component in each microlayer is determined by the numerical integration of the conversions rate $\dot{z}$.

So, in the examined model of the reactive cell the pre-exponential factor in the macrokinetic equation for the conversions rate is considered to be structurally dependent and determined for the whole representative volume taking into account its deformation, but the parameter of energy activation is considered to be local and determined in the microlevel by the processes of the activation of the initial components in the microvolumes of the mixture.

In the microlevel the type of macrokinetic inhibiting function is determined by the type of reaction in each moment of time depending on phase and concentration state of microvolumes of the mixture. The possibility of solid-phase chemical conversions and the reactions with the liquid phase presence is taken into account.

6. The Scheme of Computer Simulation of Mechanochemical Processes

The scheme of computer simulation is a general set of nonstationary problems of heat balance (I – look 2), of the shock modification of powder material (II – look 3), forced filtration (III – look 4), macrokinetics of chemical conversions (IV – look 5), modelling of the modification conditions of the rear surface of powder compact at the exit of the shock pulse (V) and modelling of the characteristics of heat and luminescent light of the side and rear surfaces of reacting powder compacts (VI).

In the model of the reactive material this set is to provide the performing of a computing experiment helping to reveal the governing factors of absolutely different conditions of the mechanical chemical gasless conversions in combustion wave without loading as well as in shock-compressed reactive powder systems. To such conditions one should refer:

1) Self-propagating high-temperature synthesis in the combustion wave. This regime is characterized by the heat initiation of chemical conversions without mechanical modification of green compact by dynamic loading.

2) Shock-assisted conversions. This condition is characterized by conversions, initiated directly after shock pulse in the local volumes of the reactive layer

In such regime the heating of reactive material till the melting temperature of a low-melting component is possible. In the last case in the scheme of modelling the considering of convective heat- and mass- transfer processes is nessesary.

3) The regime of thermo-activated conversions in mechanically activated reactive powder layer. This regime is characterized by the constitutive delay of the start of chemical conversion after the shock-wave pass. The initiation of the chemical conversions is possible after the heating of the reactive powder material from the external source of heat. In the extreme case such regime of conversions is realized at only heat initiation of chemical conversions without use of applied shock.

4) The regime of ultrafast solid-phase conversions. Chemical low-temperature conversions ending completing during the shock compression may be initiated in some powder mixtures at high activation of reactive components in the process of shock-wave modification in the local microlayers of the reactive layer.

The alternation of the regimes 2-3-4 of shock synthesis by the thickness of the reaction-product layer, local volumes of reactive layer with different concentration, thermodynamic and other state parameters as well as by the time of mechanochemical reactions is possible.

The aim of the work is to scheme the computer simulation providing, on the one hand, the division of the processes by different physical mechanisms, on the other hand, calling for the development of algorithm of calculating experiments adequately reflecting mechanochemical processes in absolutely different regimes of mechanochemical gasless reactions.

The first regime of thermo-activated reactions in the green compact is realized at the initiation of the combustion wave in the outer surface of the reactive layer. Such regime is well studied experimentally. The simulation of this regime in the process of calculating experiments is carried out for each reactive powder mixture. The value of the matching parameter k in macrokinetic equation (42) for chemical reactions rate is being chosen after matching the initial parameters of the model and known conditions of laboratory experiments for homogeneous stoichiometric powder mixture of reactive components and inert filling agent. The criterion for choosing k may be the coincidence of experimental and calculating values of the combustion wave rate. Validity of the sampling k is defined in the process of further computer simulation of laboratory experiments for examined powder mixtures with other structural characteristics (particle size, porosity) and other concentration of inert filling agent. The k value is used as a matching parameter of macrokinetics of chemical reactions at the simulation of mechanochemical processes.

Computer model for physicochemical processes of synthesis in the combustion wave requires co-solution of tasks I, III, IV.

The second investigated regime of mechanochemical reactions is characterized by asynchronous stages: 1) shock-wave modification and activation of powder medium, 2) initiation and realization of mechanochemical processes.

These stages are characterized by different rates of processes that allows to optimize the calculating experiments due to the variation of time and space steps. Computer simulation of mechanochemical processes on the first stage – mechanical modification of a powder solid body – calls for the solution of the problems I, II, and probably IV, and on the second stage of mechanochemical reactions – the solution of problems I, III and also probably IV.

The necessity of solving the problem IV on the first stage appears at the simulation of probable filtration of the low-melting liquid-phase component in the subsurface layer of porous compact which was heated due to dynamic loading.

On the second stage of simulation there appears the necessity of solving the tasks of filtration after the melting of the liquid-phase component of mixture if the open porosity of the solid-phase skeleton and the pore-pressure gradient are presence.

Following these criteria conditions results in using the two-temperature equation of the heat balance (8) with convection term in the task I.

There is principle difference between the task solution of the heat balance I at the simulation of fast and slow mechanochemical processes. Following the conditions of heat balance in fast processes, the change of enthalpy of local microvolumes of reactive powder system due to the heat conductivity may be little in comparison with the local changes of medium enthalpy due to the exothermal processes of plastic deformation and chemical reactions, which, in their turn, compete with endothermic processes of the phase transfers. Considering the changes of the heat conductivity of powder solid body due to the components temperature rise, the changes of concentration and phase state is actual for studying the conditions of initiation and kinetics of reactions in powder medium, formation of the structure of reactive products in slow processes realized after the shock pulse front.

The second regime of mechanochemical reactions is realized in insufficiently activated powder materials. The presence of delay of reactions after the completion of mechanical loading in local microvolumes of reactive powder medium is a characteristic of this regime.

So, in such regime of reactions there is a stage of heat activation determined by the processes of the heat conductivity of powder medium and the phases transfers of components between the stages of mechanical and mechanochemical modification of powder medium described in the previous case.

The second regime of mechanochemical reactions is characterized by three stages: the first is mechanical modification of powder medium (tasks I, II, (IV)); the second is heat activation (tasks I, (IV)); the third is mechanochemical reactions (tasks I, III, IV)).

The third regime of ultra-fast mechanochemical reactions is characterized by the connected processes of mechanical modification of powder solid body, initiation of chemical reactions and conversions during the dynamic loading. Such processes are initiated, as a rule, in a solid-phase regime. The character of ultra-fast mechanochemical reactions doesn't allow to simulate the whole process of the shock-wave synthesis in the sequence of stages different in time. In the sequence of microlayers the connected processes of mechanical modification, the heat balance and macrokinetics of chemical reactions of reactive powder solid are modelled (tasks I, II, III).

For all investigated regimes of mechanochemical processes in reactive powder mediums the solution of the task II – the task of the simulation of the processes of dynamic modification of powder medium, - is divided into the solution of the following problems:

a) defining the macroscopic parameters of dynamic loading and
b) micromechanics of powder microlayer.

The problems a) is suggested to be considered separately from the solution of the problem of powder medium modification during mechanochemical reactions [33]. Let's consider that mechanical loading pulse is spread into the porous quasi homogeneous medium with effective determined parameters of unloaded porous solid body. Solutions (21) allow to estimate the average state parameters during the macroscopically plane loading pulse in microvolumes of powder layer: the velocity of the shock pulse in porous medium, pressure after the front of shock pulse, internal specific energy increase after shock loading. These solutions allow to estimate effective parameters of mechanical loading by the cells of microscopic structure of concentration heterogeneity on average as well as by the sections of powder layer on average.

It is offered to consider the step-by-step loading of microlayers of powder solid body, supposing that on each step of loading the particles of powder medium get the equal quota from internal energy increase due to the shock compression [33]. The problem b) of shock modification of medium is offered to be solved by energy method, following the law of energy conservation at each moment of the loadings in all microlayers of powder medium in the form

$$\Delta W_f = \mathrm{H}(\Pi - \Pi_{\min}) \begin{bmatrix} \mathrm{H}(P_f - P_{s1}) \sum_i \Delta W_{iT} + \\ + \mathrm{H}\left(\Delta W_f - \sum_i \Delta W_{iT} \right) \mathrm{H}(P_f - P_{s2}) \sum_i \Delta W_{pi} \end{bmatrix}, \quad (45)$$

where H is the Heaviside's asymmetric single function, $\Pi_{\min}$ is the minimum allowed value of specific microlayer pore volume Π, P_f – pressure after the front of shock pulse (21), P_{s1} and

P_{s2} - critical levels of pressures (28) and (29), ΔW_f - internal specific energy increase at the shock loading step (21), ΔW_{iT} – work (31) of spherical symmetric plastic flowing of pores of *i*th component of powder mixture at the shock loading step, ΔW_{pi} – work (32) of the jet stream leading to the fracture of the surface layers of particles of *i*th component during the shock loading.

Any of the mechanisms of mechanical modification of powder mixture can be realized for each component when the necessary condition achieves the corresponding value of threshold pressures (27 – 30). For each step of dynamic mechanical loading the possible changes of porosity and the parts of the fracture surface layers of particles are estimated.

The simulation of the conditions of the rear surface modification at the exit of the shock pulse (task V) consist in the estimation of the consequences of the fracture of powder mixture particles at subsurface layer because of interaction between the initial shock wave and the reflected one with the use of the energy-balance method. Energy W_p which was not dissipated during the shock pulse due to the plastic deformation of the component and the fracture of oxide and adsorbed layers from the particles surface is considered to be possible to be spent for the fracture of particles after the shock pulse front [69].

The specific work necessary for plastic deformation of particles at stretching can be defined as

$$A_{pk}=C_k\,\sigma_{Tk}(T)\,\varepsilon_k,$$

where A_{pk} is the work for plastic deformation of *k*th component, C_k is the volumetric concentration of *k*th component, ε_k is the ultimate strain of *k*th component.

The work necessary for fracture of all particles on the microlayer is represented by the proportion [70, 71]:

$$A_{bk} = M_k \gamma_k i_k\, S_{ok},$$

where M_k is the number of particles of *k*th component in the microlayer, i_k is the number of parts formed as a result of the fracture of one particle, S_{ok} is the area of the surface formed at the particle fracture, γ_k is the specific energy of the of new surface of *k*th component formation.

At the end of the shock pulse in each microlayer the volume of the reactive cell, its size and concentration of powder components and specific volume of pores are calculated. According to the results of the estimation of plastic deformation work and surface layer fracture, the heat mechanical sources in the equation of the heat balance and the degree of mechanical activation of reactive components of powder mixture are determined.

Carrying capacity of powder skeleton is determined by the effective yield stress depending on the temperature and concentrations of the solid-phase components. The pore pressure (36) of liquid-phase component is determined by the difference between the pressure of dynamic mechanical loading (21) and effective yield stress of the skeleton.

So, the connection of the problem of mechanical activation II with the problems of the heat balance I, chemical macrokinetics III and filtration IV are provided. The problem of the estimation of the rear surface modification V is solved for the simulation of the characterisitcs of the radiation of the rear surface of reactive powder compacts.

Different analogues of nonstationary equation of the heat balance (5) or (8) are used for all microlayers of powder solid body. All factors at the equation depend on thermodynamic and concentration-phase state of microlayers. In each time step these factors are specified iteratively. All parameters of the state of the reactive powder layer are also specified [33].

In each microlayer of the powder solid body these equations simulate changes of the enthalpy at the interaction of the sources and sinks of heat, processes of volume heat exchange between solid-phase skeleton and melting component, heat transfer due to the heat conductivity and convections of liquid phase. Using the equations (8) with the corresponding initial and boundary conditions only, one may take into account the changes of the enthalpy of the microlayers of powder solid body in different steps of mechanochemical reactions.

Temperature rise of the microlayers of powder solid body after each time step of mechanochemical reactions results in changes of physical characteristics of the material of powder particles and the product of reaction as well as the heat activation of reacting components. The connection of the problem of the heat balance I with the problems of mechanical modification of powder medium II, filtration IV and chemical macrokinetics III is determined by this.

For simulation of mechanochemical processes in the reactive powder layer, registration of changes of the structural parameters and convective heat transfer it is enough to model the filtration through the sequence of microlayers of reactive cells in the principal direction [33].

It is offered to investigate the necessary conditions of the possible filtration of liquid-phase component, calling convective mass transfer, for each microlayer of powder solid body. The rate of convections of the liquid phase is determined according to the filtration law (33). The requirement to obey the necessary conditions of sufficient smallness of time step allows to take into account the convective mass transfer by recalculation of the concentration of powder components and porosity only in two adjacent microlayers. The calculation of the velocity of the filtration of melt and structural parameters of powder solid changed due to the mass transfer provides iterative correction of the state parameters of powder medium in each time point as well as connection of the problem of filtration IV with the problems of shock modification II and the heat balance I.

It is considered that for each microlayer of powder solid body the problem of macrokinetics of chemical transformations consist in the check of the conditions of the possibility of initiation of chemical reactions (44), choice of inhibiting function φ (38), (39) and its parameters depending on concentration and phase state, calculation of local value of the energy activation parameter E_a (43), measuring the rate of reactions $\dot{z}$ (38), calculation of the conversion degree of high-melting component of the mixture z in the investigated time point. Macroscopical dependence (42) of a pre-exponential factor on the size of reactive cell, which is recalculated after the shock wave pass through each microlayer, is used. Chemical reactions change the concentrations of components and it means that they change all effective physical characteristics of powder layer. The rate of chemical reactions determines the power of the heat sources in the heat balance equations. All these determine the connection of the macrokinetics problem III with the problems of mechanical modification II and the heat balance I.

In the scheme of the computer simulation of mechanochemical processes the visco-plastic behaviour of the components of powder solid body heat exothermic and endothermic processes in multiphase powder medium, filtration processes and kinetics of chemical reactions are taken into account. The effectiveness and the possibility of computer simulation of connected physical and chemical processes are determined by the use of the wide range of boundary problems.

In the considered scheme of the simulation the solutions of the following model boundary problems are used:

1. The solution of the model problem (21) of the plane shock pulse pass through the porous medium with the effective parameters of shock adiabat. These solutions allow to estimate the effective parameters of the shock pulse, which are used for the identification of the position of the pulse in each time moment, and to obtain the average value of impulse pressure and internal specific energy increase after the pulse front.
2. Solutions (20), (22) of the model boundary problem of the visco-plastic behaviour of the shock loaded powder medium for spherical - symmetric behaviour of the single cell of porous solid body. For effective parameters of the dynamic loading in each time step these solutions are used for the estimation of the degree of the porosity change, energy distribution of the shock pulse for work on plastic deformation and dynamic regime of pore collapse, determination of the power of heat sources due to the plastic deformation.
3. Solutions (24), (25) of the model problem of the surface layers fracture are used for the estimation of the required energy of mechanical loading for fracture of the surface layers of powder particles. In the model of calculation, this estimation is used for the determination of the possibility of the surface layers fracture, a part of clean surface of the particles of reactive components and the activation of the reactive mixture due to these processes.
4. The solution of the model problem of the calculation of volume heat exchange factor (11) in the saturated porous mediums. The factor value obtained for the separate particle, is corrected by specifying structural parameter $f_\alpha<1$, determining the part of free boundaries of particles in compacted powder solid body.
5. The solution of the model problem for the single cell of porous visco-plastic powder medium, representing the solution (20) modified for the registration of exothermic effects of chemical reactions in the surface layers of the particles of reactive components of the medium.

The scheme of computer simulation of mechanochemical processes in the reactive powder layer, allows to make the computational algorithm realizing the solution of the connected problems of the mechanical modification of the powder medium, the heat balance, the filtration of liquid–phase component and the macrokinetics of chemical reactions on the base of the numerical model of nonstationary physically and geometrically nonlinear two-temperature boundary problem of the heat balance. The simulation of the characteristics of the heat and luminescent radiation (task VI) is realized from the position of the approach taking into account the peculiarities of the realization of the model of the reactive layer: homogeneity postulate of all local parameters of the model in each time point (temperature, components concentration, porosity, degree and the rate of chemical reactions, the parameters of the activation energy and so on) by microlayer of reactive cell $a\times a\times dx$ (look figure 4) [71]. The sections of the microlayer of the reactive cell which are the constituents of the radiant surface researched by the optical pyrometry method have an area: for the rear surface $a\times a$, for the lateral face $a\times dx$.

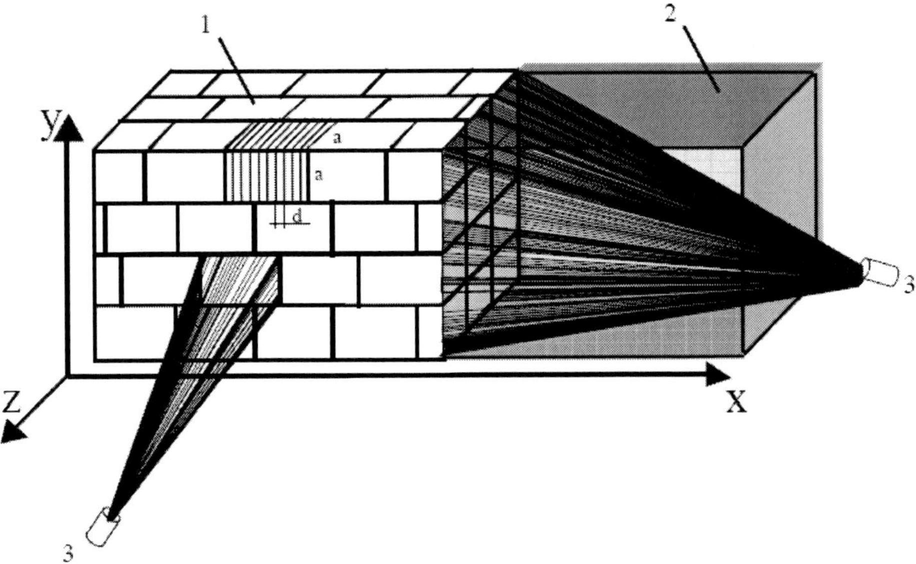

Figure 4. The measurement scheme for the radiation parameters research: 1- powder compact, 2- "window material", 3- radiation detector.

The radiation temperature for a lateral face of the reactive powder mixture is represented by a thermal component and a fluorescent one: $T_b = T_t + L$, where T_l is the thermal radiation temperature on the surface of the sample and L is the luminescent component of the surface emission. The thermal radiation temperature for the surface of the sample is estimated from calculated values of the thermodynamic temperature, structure characteristics of the emitting subsurface layer, and effective thermophysical parameters, using relations derived from the laws of thermal radiation from solids. The thermal radiation temperature on the surface of the sample is calculated from the relation derived from the Stefan-Boltzmann law for solid materials to give

$$T_t = \varepsilon_t^{1/4} T^t,$$

where ε_t is the total emissivity. The thermophysical characteristics and parameters of state of the emitting layer are averaged, using a mixture-ratio law. The thermodynamic surface temperature at each instant of time is determined by the temperature of the subregions making up the subsurface layer of the sample. The surface emissivity is defined by the emissivity of the subregions of the surface sublayer and is dependent on the powder-mixture ratio. The burning of the powder-mixture particles gives rise to chemiluminescent flashes. The flash frequency and intensity are estimated from the calculated rate of increase of powder component enthalpy [70] that determines the chemiluminescence regime under consideration:

$$I_{ch} = \rho c_p \Delta T C_i,$$

where ρ is the powder density, c_p is the heat capacity, ΔT is the temperature increment, and C_i is the concentration of one of the powder components.

So, the area of the rear surface is

$$S^t = \sum_{i=1}^{N^t} S_i^t = \sum_{i=1}^{N^t} a_i \cdot a_i ,$$

the area of lateral face is

$$S^b = \sum_{i=1}^{N^b} S_i^b = \sum_{i=1}^{N^b} a_i \cdot dx_i .$$

All heat physical parameters and state parameters of the reactive powder solid body determining the characteristics of the heat and luminescent radiation are considered as effective averaged surface values S^t and S^b. For example, temperatures of the rear and lateral face are:

$$T^t = \sum_{i=1}^{N^t} T_i \cdot S_i^t / S^t, \, T^b = \sum_{i=1}^{N^b} T_i \cdot S_i^b / S^b .$$

Results and Discussion

The powder systems of Zr-B, NiO-Al, and Ni-Al are chosen as subjects of investigation. The behavior of the powder mixtures compacts, in which the reactive components was preliminary mechanical activated and capable to self-propagating high temperature synthesis (SHS), is simulated. The physical and chemical mechanisms of conversion in the powder mixtures of reactive components and the inert filling agent with the macroscopic ratio of reactive components corresponding to stoichiometry of investigated reactions are researched; the powder particles of the corresponding products of chemical conversions are considered as an inert filling material.

SHS in the powder system Ni-Al is characterized by the activation energy of chemical conversions E_0 = 75,4 kJ/mole and starts to interact with the temperature close to the temperature of aluminum melting [4] (925 K), but the adiabatic temperature of mixture burning can reach the temperature of melting of the formed nickel aluminide. The reaction of the synthesis is represented by the equation [4]:

Ni + Al = NiAl + Q,
Q = 1377,983 kJ / kg.

Without intensive mechanical loading the powder mixture NiO-Al is characterized by the energy of activation E_0 = 146,1 kJ/mole and initiate with the temperature higher than the temperature of aluminum melting [71, 72] (1200 K). The reaction of synthesis is represented as [71, 72]:

$$3NiO+2Al=3Ni+Al_2O_3+Q,$$
$$Q=3391{,}308e3 \text{ kJ / kg}.$$

The synthesis in the powder system Zr-B without the intensive loading is characterized by the energy of the activation of chemical conversions E_0 = 309,32 kJ/mole and initiated at the heating with the temperature 1270 K [73]. The reaction of the zirconium diboride synthesis is represented by the equation:

$$Zr+2B = ZrB_2 + Q,$$
$$Q = 3005 \text{ kJ / kg}.$$

All constants and functions corresponding to physical, thermophysical and mechanical features of the components of powder mixture are taken from the literature [74-78].

For determining the dependence of the activation energy of chemical conversions in the local microvolumes of reactive powder layer on the energy consumption for the plastic deformation of crystal structure and the fracture of the surface layers of particles of reactive components, let's determine the matching coefficients α_i in equation (43) by known experimental facts. Such experimentally proved peculiarities of the behavior of reactive powder systems are following:

1 Mechanical activation of reactive powder components allows to decrease the temperature of the initiation of chemical conversions by 300-800 K and the synthesis temperature – by 600-650 K [16, 79-82].
2 In the absence of oxide and adsorbed films the reaction of nickel and aluminum with the formation of intermetallic compounds is initiated with T_{k2} = 493 K [17].

Using the equation (43) and the condition of reactionary equivalence (44), we will derive the equation for determination of matching parameters α_i:

$$\alpha_i = \frac{E_0}{A_i}\left(1 - \frac{T_{ki}}{T_0}\right).$$

The following values of coefficients α_i satisfying the experimental results were obtained:

α_1 = 1400,0; α_2 = 240,0.

According to the model the SHS-processes can be investigated due to the computing experiment simulating the physicochemical processes in the reactive powder mixture in the absence of the mechanical loading.

The simulation of the mechanochemical processes in the diluted powder mixture with the initial parameters agreed with the parameters of a laboratory experiment allows to estimate the matching parameter *k* of the reactive cell model. Further use of computer simulation method for the model powder mixture with the initial parameters agreed with the parameters of other laboratory experiments different from the first one by initial values of the concentration of inert filling agent and temperature, can serve as a test for checking of the certainty of the model and allows to research the physical mechanisms of the processes.

There was carried out a computing experiment by simulation of mechanochemical processes of self-propagating high temperature synthesis of intermetallic compound NiAl for the values of the initial concentration of inert filling agent (NiAl) 30 mass %, 48 mass % and initial temperatures T_0 298 K, 740 K, respectively.

The combustion process of the powder system Ni – 31,5 mass % Al diluted by the reaction product was simulated. The reactive thickness layer of 0,3 mm consisting of the particles of 40 μm. The concentrations of the components by the whole volume of the powder mixture are homogeneous.

The results of the computing experiment are represented in the table 1 in the comparison with the experimental data.

Table 1.

Initial temperature, concentration of inert filling agent	Calculation		Experiment [4]	
	Burning rate, m/s	Peak temperature, K	Burning rate, m/s	Peak temperature, K
298 K, 30 mass %	0,005	1640	0,0055	1680
740 K, 48 mass %	0,0044	1700	0,005	1720

Figure 5 shows the results of the computing experiments of simulation of mechanochemical processes of self-propagating high temperature synthesis in the powder mixtures Zr-B and NiO-Al with different values of the initial concentration of inert filing agent (C_{ZrB2} and C_{Al2O3} respectively.

The combustion of the powder mixture Zr-B-ZrB_2 in the SHS regime is studied experimentally [73]. The combustion rate and the peak temperature depend on the initial temperature values and the concentration of inert filling agent (reaction product). In figure 5 a the results of the simulation of the processes of powder system burning Zr – 19 mass % B, diluted by the reaction product with porosity Π_0=0,4 are shown. The reactive layer with the thickness of 0,01 m consisting of the particles of 50 μm is investigated. The concentration of the components of the powder mixture is homogeneous (b/a=1,1). The results of the computing experiment are shown in the figure 5 a in the comparison with experimental data [73].

In figure 5 b the results of the SHS processes simulation for the powder mixture NiO-Al with the different value of the initial concentration of inert filling agent (Al_2O_3) are shown. The reactive layer with the thickness of 0,01 m, consisting of the particles of 5 μm is investigated. The porosity of the powder compact is Π_0=0,3. The concentration of the components is homogeneous (b/a=0,9). The results of the computing experiment are shown in the figure 5 b in the comparison with experimental data [71].

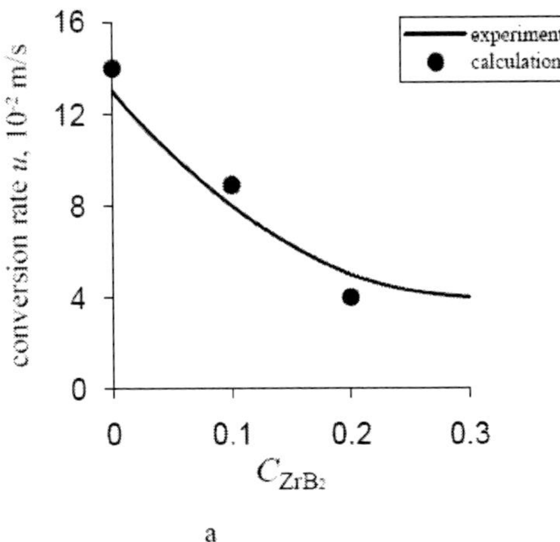

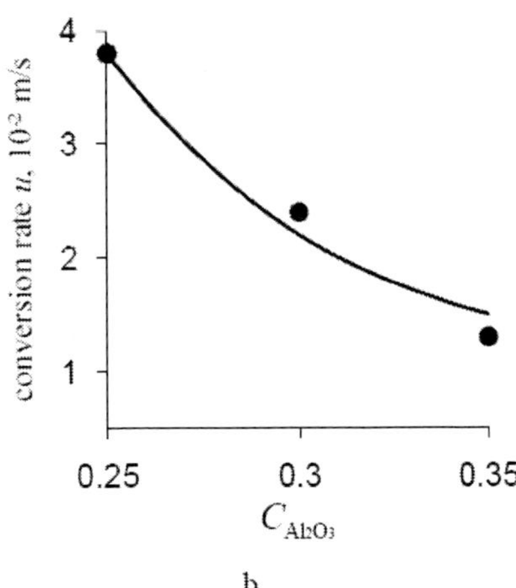

Figure 5. Combustion rate dependence on inert filing agent concentration.

In the table 2 the values of k coefficients (equaion (42)) obtained from the matching experimentally known definite values of the combustion rate with the results of computing experiments for investigated powder systems are given.

Table 2. Values of *k* coefficients for different powder systems

System	Zr-B	NiO-Al	Ni-Al
k	$9{,}0 \cdot 10^3$	$3{,}4 \cdot 10^4$	$1{,}125 \cdot 10^3$

The conditions of the shock initiation of the chemical conversions in the model sample of the powder mixture Zr-B, characterized by the presence of the macroscopic structure of the concentration heterogeneity of the initial components are investigated. The mixture of the reactive components with the particles size of 50 μm which was previously pressed till the average value of the relative volume of pores Π_0=0,4 with the initial temperature of a layer T_0=293 K is considered. The shock loading of the layer of the powder mixture is modeled by the macroscopically plane shock pulse with the amplitudes P_f=3GPa ÷7 GPa.

In the figure 6, the profiles of the initial concentrations of reagents (Zr and B) and the specific pores volume (Π) by the thickness of the reactive layer consisting of two reactive cells are shown.

The mechanical modification of the powder mixture due to the shock loading determines the compression of the material and the mechanical activation of the component. The distribution of the local values of the activation energy by two reactive cells are illustrated in the figure 7 by the curves 2-4 at the moment of the shock pulse complete. According to the condition of the reactive equivalence (44), the chemical conversions initiation in the powder mixture Zr-B with the initial temperature T_0=293 K (taking into account the temperature rise due to the visco-plastic compression) is possible when the local critical value E_{ac} of the activation energy pointed by the curve 1 in the figure 7 are reached. For the small amplitude of the shock pulse (curve 2) the mechanical activation of the mixture doesn't reach the level necessary for the chemical conversions initiation. The increase of the amplitude of the shock pulse till 5 GPa results in the formation of the areas in which the chemical conversions are initiated (curve 3). With the further loading increase this area of the shock-induced chemical conversions becomes wider (curve 4).

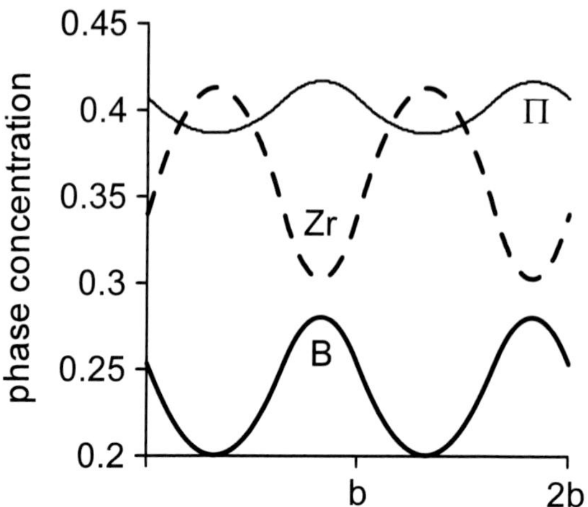

Figure 6. Initial bulk Zr and B concentration and porosity Π for Π_0=0.4.

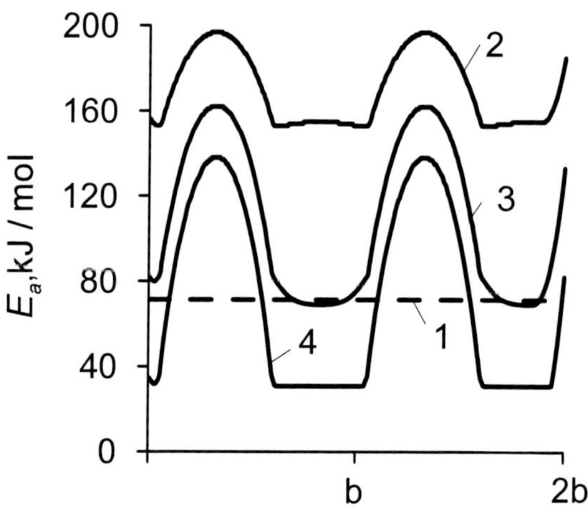

Figure 7. The local values of the activation energy: P_f= 3 GPa (2); P_f=5 GPa (3); P_f=7 GPa (4).

The computing experiments were carried out with the use of the methods of computer simulation described in [83, 84].

The combustion of the powder mixture $NiO+Al+25\%Al_2O_3$ with a particles size of 5 μm and the average porosity Π_0=0.3 was considered. The combustion wave is initiated by the ignition temperature at one of the sample side. The character of the heat and luminescent emission of the lateral face of the model sample is researched. The figure 8 shows the components concentrations and the specific volume of pores by the reactive cell. The combustion wave propagation in the sample is accompanied with the temperature gradient on its front that causes heat expansion and cracking of the powder mixture particles, which can be attended with the emission of light – mechanoluminescence in the subsurface layer of the lateral face of the sample.

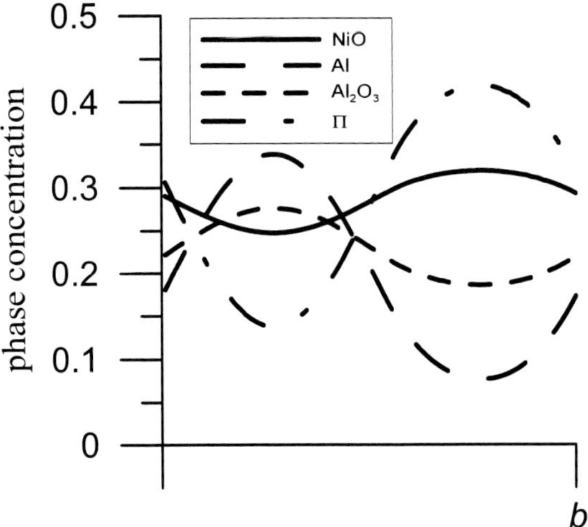

Figure 8. Initial bulk NiO, Al and Al_2O_3 concentration and porosity П for $П_0$=0.3.

The intensity I_m of this emission is proportional to the energy spent for the fracture of the particles as a result of unequal heat expansion. In the figure 9a the time distribution of the radiation temperature of the heat emission T_t with the intensity distribution of mechanoluminescence I_m of the lateral face of the model sample devided by its peak intensity I_{mm} is shown. The character of the changes of the radiation temperature reflects nonstationary regime of the physicochemical processes of synthesis and is determined by the structural phase transfers, the change of the regimes of the macrokinetics of chemical conversions and other factors. The figure 9 b) shows the intensity distribution of chemiluminescence I_{ch} of the lateral face of the model sample divided by its peak intensity I_{chm}. The behavior of chemiluminescence curve (figure 9 b) retraces the temperature curve (figure 9 a).

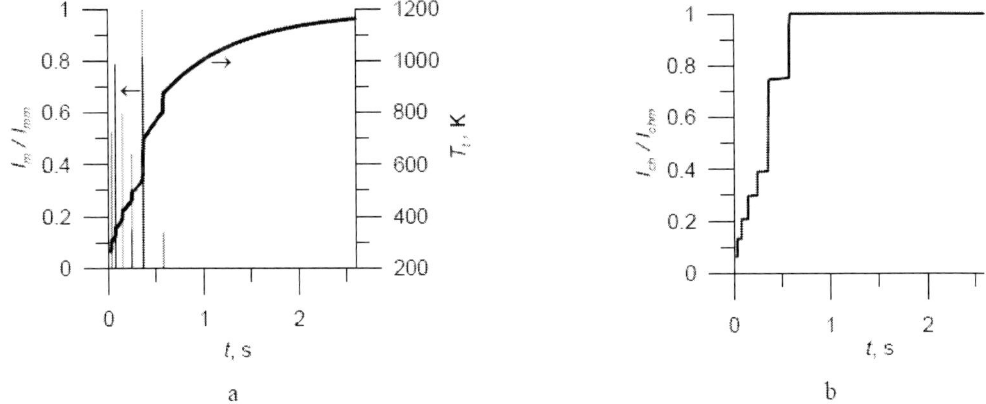

Figure 9. The distribution of the heat radiation the temperature and relative intensity of mechanoluminescence (a), and chemiluminescence (b) of the lateral face of the model sample NiO+Al+25 mass % Al_2O_3.

In the figure 10 the time distribution of the radiation temperature of the heat emission with the relative intensity distribution of mechano and chemiluminescence of the lateral face of the model sample Ni-Al with initial porosity $П_0$=0.25 and particle size of 40 μm are shown.

In contrast to NiO-Al mixture (figure 9), two main mechanoluminescence flashes are observed (figure 10 a). The chemiluminescence curve has an pulsing behavior (figure 10 b).

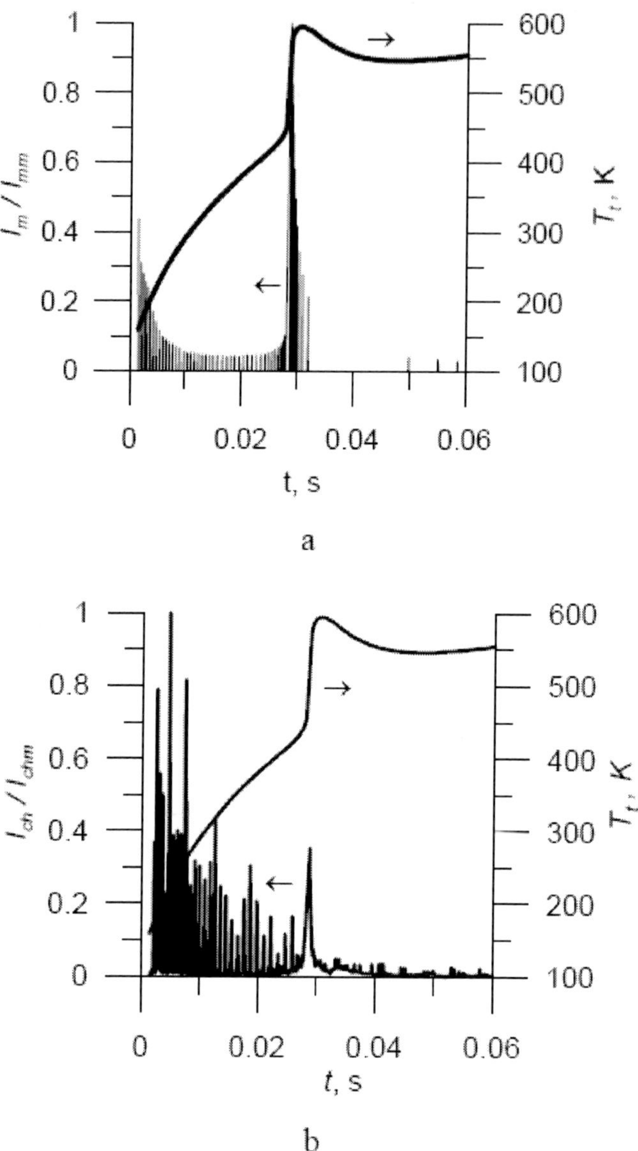

Figure 10. The distribution of the heat radiation the temperature and relative intensity of mechanoluminescence (a), and chemiluminescence (b) of the lateral face of the model sample Ni+Al.

The shock synthesis of the materials can be researched by the studying of the radiation of the rear surface of the sample of the reactive powder mixture loaded by macroscopically plane pulse.

The computing experiments on the research of the rear surface radiation parameters for the model sample NiO+Al+25%Al_2O_3 exposed to a shock-wave amplitudes P_f=2 GPa (figure 11a) and P_f= 8 GPa (figure 11b) was made. The character of chemiluminescence is similar for the investigated pressures of the shock wave. The maximum chemiluminescence flashes correspond to the temperature increase of the heat radiation accompanied with the step change of the intensity of chemical conversions. The first chemiluminescence flashes are observed at the first moments; it proves the initiation of the chemical conversions directly after the shock wave front in the most mechanically activated subareas of the rear surface. The flashes in the future periods are caused by the heat ignition of the low activated subareas. The time reduction between the bursts as well as the reduction of the total chemiluminescence for P_f=8 GPa in comparison with the results for P_f=4 GPa reflect the difference between the degrees of the mechanical activation of the reactive components.

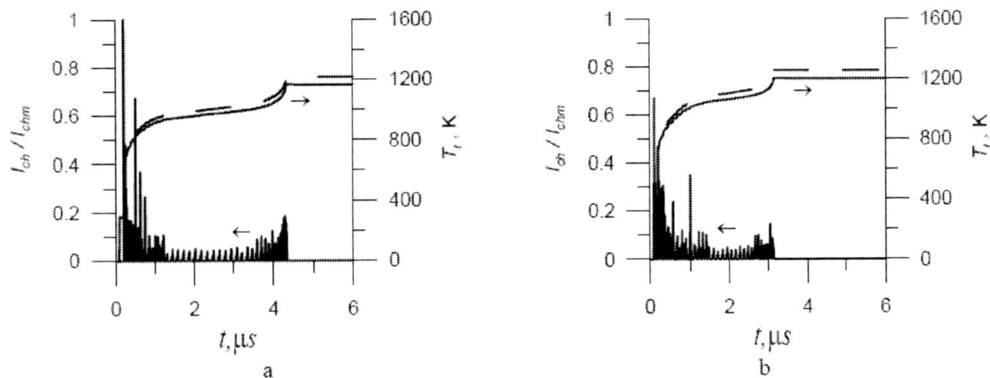

Figure 11. The distribution of the heat radiation temperature T_t on two wave-length ((- - -) – 420 nanometers, (——) – 720 nanometers) and relative intensity of chemiluminescence I_{ch}/ I_{chm} of the rear surface of the model sample.

The characteristics of thermal and luminescent emission from the surface of reactive powder compacts account for the parameters of state of reactive medium, attainable pressure levels, and kinetics of relevant chemical reactions. This means that the results obtained in the present work can serve as a basis for development of contactless methods for monitoring the physicochemical processes in reactive powder materials.

REFERENCES

[1] Refractory Inorganic Compounds Synthesis Method: Certificate of Authorship 255221 USSR, 1967 / Merzhanov, A.G.; Borovinskaya, I.P.; Shkiro V.M.; *Invention Bulletin,* 1971, 10.

[2] Merzhanov, A.G.; Borovinskaya, I.P. Doklady Akademii Nauk USSR 1972, 204, 366-369.

[3] Merzhanov, A.G. In: Fizicheskaya Khimia. Sovremennye Problemy (Physical Chemistry. Modern Problems); Ed. Kolotyrkin, Ya.M.; *Khimiya:* Moscow, 1983; pp. 5-45 [in Russian]

[4] Itin, V. I.; Naiborodenko, Y. S. *High-temperature synthesis of intermetallic compounds;* at. Tomsk University: Tomsk, RUSSIA, 1989; pp 1-214 [in Russian].

[5] Aldushin, A.P.; Khaikin, B.I. *Combustion, Explosions, and Shock Waves* 1974, 10, 273-280.

[6] Korchagin, M.A.; Grigorieva, T.F.; Barinova, A.P.; Lyakhov, N.Z. *Doklady Akademii Nauk* 2000, 372, 40-42.

[7] Aldushin, A.P.; Khaikin, B.I.; Shkadinskii, K.G. Combustion, *Explosions, and Shock Waves* 1976, 12, 725-731.

[8] Kirdyashkin, A.I.; Maksimov, Yu.M.; Merzhanov, A.G. *Combustion, Explosions, and Shock Waves* 1981, 17, 591-595.

[9] Korchagin, M.A.; Alexandrov, V.V.; Neronov, V.A. Izvestia Sibirskogo Otdeleniya Akademii Nauk SSSR. *Seria Khimicheskie Nauki* 1979, 6, 104-111.

[10] Alexandrov, V.V.; Korchagin, M.A.; Boldyrev V.V. *Doklady Akademii Nauk SSSR* 1987, 292, 879-881.

[11] Abovyan, L.S.; Nersisyan, G.A.; Kharatyan, S.L *Khimicheskaya Fizika.* 1994, 13, 127-133.

[12] Merzhanov, A.G. Combustion Processes and Materials Synthesis; Ed. Telena, V.T. and Chachoyan, A.V.; *Chernogolovka,* RUSSIA, 1998; 512. [in Russian]

[13] Carbide, Nitride and Boride Materials Synthesis and Processing; Ed. Weimer, A.W.; London – Weinheim – New-York – Tokyo – Melburne – Madras: Chapman and Hall, 1997; 671.

[14] SHS – Bibliography (1967 – 1995). *Int. Journal of SHS* 1996, 5, 513.

[15] Savitskii, A.P. Liquid-phase Sintering of Systems with interactive Components; Nauka. Sibirskoe Ontdelenie: Novosibirsk, RUSSIA, 1991; 184. [in Russian]

[16] Combustion Synthesis Chemistry; Ed. Koidzumi, M;. *Mir:* Moscow, RUSSIA, 1998; 247. [in Russian]

[17] Tarento, R. J.; Blaise, G. *Acta Metallurgia.* 1989, 37, 2305-2312.

[18] Boldyrev, V.V. *Izvestia Akademii Nauk SSSR.* Seria Khimicheskaya 1990, 10, 2228-2245.

[19] Avvakumov, E.G. Mechanical Methods of Chemical Processes Activation; *Nauka:* Novosibirsk, RUSSIA, 1986; 1-303. [in Russian]

[20] Enikolopyan, N.S. *Gurnal Fizicheskoi Khimii.* 1989, LXIII, 2289-2298.

[21] Gordopolov, Yu.A.; Trofimov, V.S.; Merzhanov, A.G. *Dokladi Akademii Nauk.* 1995, 341, 327-329.

[22] Shteinberg, A.S.; Knyazik, V.A.; Fortov, V.E. *Dokladi Akademii Nauk.* 1994, 336, 71-74.

[23] Batsanov, S.S. Combustion, Explosions, and Shock Waves. 1996, 32, 102-113.

[24] Meyers, M.A.; Batsanov, S.S.; Gavrilkin, S.M.; Chen ,H.C.; LaSalvia, J.C.; Marquis F.D.S. *Materials Science and Engineering.* 1995, A 201, 150-158.

[25] Nesterenko, V.F.; Meyers, M.A.; Chen, H.C.; LaSalvia, J.C. *Appl. Phys. Lett.* 1994, 65, 3069-3071.

[26] Thadhani, N. N. Shock-induced Chemical Reactions and Synthesis of materials; Progress in Materials Science; Ed. Christian, J.W.; Haasen, P. and Massalski, T. B.; Pergamon Press, Oxford, New York, Tokyo, 1993; vol. 37, No. 2; pp. 117-226.

[27] Shtertser, A.A. Combustion, Explosions, and Shock Waves. 1993, 29, 740-743.

[28] Balshin, M.Yu.; Kiparisov, S.S. *Fundamental Powder Metallurgy; Metallurgy:* Moscow, RUSSIA, 1978; 184. [in Russian]

[29] Vasiliev, V.V.; Protasov, V.D.; Bolotin, V.V. Composite materials: reference book; Ed. Vasiliev, V.V. and Tarnopolskiy, Yu.M.; *Mashinostroenie:* Moscow, RUSSIA, 1990; 512. [in Russian]

[30] Vanin, G.A. *Micromechanics of Composite Materials;* Naukova Dumka: Kiev, 1985; 304. [in Russian]

[31] Composite Materials. Ed. Broutman, L. J. and Krock, R. H.; Academic Press. New-York and London, 1974; A Subsidiary of Harcourt Brace Iovonovich, Publisher.

[32] Beran, M. *Statistical continuum theories;* N.Y.: Interci. Publ., 1968; 493.

[33] Leitsin, V.N.; Dmitrieva, M.A. *Simulation of Mechanochemical Processes in Reactive Powder;* NTL-Press: Tomsk, 2006; 188. [in Russian]

[34] Gordopolov, Yu.A. Proceedings of Russian Conference *"Combustion and Explosion Processes in Physics, Chemistry and Technology of Inorganic Materials"* 2002; Moscow; 74-81.

[35] Thadhani, N.N. *J. Appl. Phys.* 1994, 76, 2129-2138

[36] Goldshtik, M.A. *Transfer Processes in Granular Layer; Thermophysics Institut SB AS* USSR: Novosibirsk, 1984; 164. [in Russian]

[37] Batsanov, S.S. *Uspekhi Khimii.* 1986, LV, 579-607.

[38] Scherbakov, A.S. *Self-organization of Matter in Lifeless Nature: Synergy phylosophical Problems;* Moscow University Press: Moscow, 1990; 111. [in Russian]

[39] Shkadinskii, K.G.; Ozerkovskaya, N.I.; Chernetsova, V.V. *Khimicheskaya Fizika.* 1991, 2, 1437-1439.

[40] Kutateladze, S.S. *Heat Exchange Theory Base; Atomizdat:* Moscow, 1979; 416. [in Russian]

[41] Dmitrieva, M.A.; Leitsin, V.N. *Rus. Phys. J.* 1999, 42, 288-292.

[42] Skorokhod, V.V. *Poroshkovaya Metallurgiya; Naukova Dumka:* Kiev, 1977; 120-129. [in Russian]

[43] Buevich, Yu.A.; Korneev, Yu.A. *Inzhenerno-fizicheskii Zhurnal.* 1976, 31, 607-612.

[44] Tatsuhiko Aizawa; Yen, B. K.; Yasuhiko Syono Shock-indused reaction mechanism to synthesize refractory metal silisides; *Shock compression of condensed Matter,* 1997; pp. 651-654.

[45] Jiang, J.; Goroshin, S.; Lee,J. H. Shock wave induced chemical reaction in Mn+S mixture; *Shock compression of condensed Matter,* 1997; pp. 655-658.

[46] Yang, Y.; Gould, R. D.; Horie, Y.; Iyer, K. R. *Shock compression of condensed Matter,* 1997, 639-642.

[47] Meyers, M.A.; Batsanov, S.S.; Gavrilkin, S.M.; Chen, H.C.; LaSalvia, J.C.; Marquis, F.D.S. *Materials Science and Engineering.* 1995, A 201, P150-158.

[48] Fedorov, V.T.; Khokonov, Kh.B. *Tekhnicheskaya Fizika.* 1987, 1126-1128.

[49] Leitsin, V.N.; Dmitrieva, M.A. *Physical Mesomechanics.* 2002, 5, 55-65.

[50] Nigmatulin, R.N. *Multiphase Mediums Dynamics; Mir*: Moscow, 1987. [in Russian]

[51] Kiselev, S.P.; Ruev, G.A.; Trunev, A.P. Shock-wave Processes in Two-component and Two-phase mediums; *VO Nauka:* Novosibirsk, 1992; 261. [in Russian]

[52] Nesterenko, V.F. Impulse Loading of Heterogeneous Materials; *Nauka:* Novosibirsk, 1992; 200. [in Russian]

[53] Green, R. J. International Journal of Mechanical Science. 1972, 14, 215-224.

[54] Leitsin, V.N.; Dmitrieva, M.A.; Kobral, I.V. *Physical Mesomechanics.* 2001, 4, 43-49.

[55] Schetinin, V.G. *Shock waves in condensed matter;* Ed.. Birukov, A.L; Saint-Petersburg, 1998; Pp. 186-197.

[56] Batsanov, S.S. Effects of explosions on materials: modification and synthesis under high-pressure shock compression; N.Y.: Springer-Verlag, 1994; 194.

[57] Goldshtein, R.V.; Yentov, V.M. *Qualitative Methods in Continuum Mechanics; Nauka*: Moscow, 1989; 224. [in Russian]
[58] Skorokhod, V.V.; Solonin, S.M. *Poroshkovaya Metallurgiya.* 1972, 2, 43-49.
[59] Martynenko, O. G.; Pavlyukevich, N.V. *Inzhenerno-fizicheskiy Zhurnal.* 1998, 7, 5-18.
[60] Timokhin, A.M., Knyazeva, A.G. *Chem. Phys. Reports.* 1996, 15, 1497-1514.
[61] Frank-Kamenetskiy, D.A. Diffusion and Heat Transfer in Chemical Kinetics; *Nauka:* Moscow, 1987; 502[in Russian]
[62] Khauffe, K. *Reactions in Solids and on Their Surfaces;* Inostrannaya Literatura: Moscow, 1963; 275. [in Russian]
[63] Larikov, L.N. Metallofizika I Noveyshie Tekhnologii. 1994, 16, 3-27.
[64] Leitsin, V.N.; Dmitrieva M.A. *Chem. for Sustainable Development.* 2005, 13, 2691-274
[65] Leitsin, V.N.; Skripnyak, V.A.; Dmitrieva, M.A. Three-scale model for numerical simulation of mechano-chemical processes in shock-compressed powder bodies; Shock compression of condensed matter; ed. Furnish, M.D.; Thadhani, N.N. and Horie, Y; American Institute of Physics 0-7354-0068-7/02, 2002; Pp. 1093-1096.
[66] Pavlov, V.A. Basic Physics of Metal Plastic Deformation; Izdadelstvo Akademii Nauk SSSR: Moscow, 1962; 199. [in Russian]
[67] Benderskiy, V.A.; Filippov, P.G.; Ovchinnikov, M.A. Doklady Akademii Nauk SSSR. *Fizicheskaya Khimiya.* 1989, 308, 401-405.
[68] Leitsin, V.N.; Kolmakova, T.V.; Dmitrieva, M.A. *Physical Mesomechanics.* 2004, 7, 95-100.
[69] Leitsin, V.N.; Kolmakova, T.V.; Dmitrieva, M.A. *Physical Mesomechanics.* Special Issue 2004, 7, 78-81.
[70] Leitsin, V.N.; Dmitrieva, M.A.; Kolmakova, T.V.; Kobral, I.V. *Rus. Phys. J.* 2006, 49, 1198-1203.
[71] Filatov, V.M.; Naiborodenko, Yu.S. *Combustion, Explosions, and Shock Waves.* 1992, 28, 47-52.
[72] Shidlovskii, A.A.; Gorbunov, V.V. *Combustion, Explosions, and Shock Waves.* 1982, 18, 420-422.
[73] Borovinskaya, I.P.; Merzhanov, A.G.; Novikov, N.P.; Filonenko A.K. *Combustion, Explosions, and Shock Waves.* 1974, 10, 2-10.
[74] Chirkin, V.S. Thermophysical Properties of Materials for Nuclear Engineering; Atomizdat: Moscow, 1968; 484. [in Russian]
[75] Babichev, A.P., Babushkina, N.A., Bratkovskiy, A.M. in Physical Quantities: Guide; Ed. Grigoriev I.S. and Meynihov E.Z.; *Energoatomizdat:* Moscow, 1991; 1232. [in Russian]
[76] Gurevich, L.V.; Veits, I.V.; Medvedev, V.A. Thermodynamic Properties of Individual Substance; *Nauka:* Moscow, 1981; Vol. 3, 472. [in Russian]
[77] Peletskiy, V.E; Chekhovskoy, V. Ya.; Belskaya, E.A. Thermophysical Properties of Titanium and its Alloys; *Metallurgiya:* Moscow, 1985; 103. [in Russian]
[78] Bobylev, A.V. Mechanical Properties and Processing Characteristics of Metals: Guide; *Metallurgiya:* Moscow, 1987; 208. [in Russian]
[79] Terekhova, O.G.; Lepakova, O.K.; Kostikova, V.A. *Chemistry for Sustainable Development.* 1998, 6, 195-198.
[80] Antsiferov, V.N.; Mazein, *Fizika I Khimiya obrabotki materialov.* 1996, 1, 105-109.

[81] Korchagin, M.A.; Grigorieva, T.F.; Bokhonov, B.B.; Barinova, A.P.; Lyakhov, N.Z. Proccedings of Russian Conference *"Combustion and Explosion Processes in Physics, Chemistry and Technology of Inorganic Materials"*, ISMAN: Chernogolovka, 2002, 200-204.

[82] Naiborodenko, Yu.S.; Kasatskiy, N.G.; Lepakova, O.K. Proceedings of Russian Conference *"Combustion and Explosion Processes in Physics, Chemistry and Technology of Inorganic Materials"*, ISMAN: Chernogolovka, 2002, 287-290.

[83] Leitsin, V.N.; Kolmakova, T.V.; Dmitrieva, M.A. (2007). Teplum-01: Program of Physical and Chemical Processes Simulation in Powder Materials Capable to Surface Luminescence of Reactive Compacts. Daily Computing teaching programs and innovation (Telegraph sectoral program fund and algorithm), 12, http://ofap.ru/portal/newspaper/2007/12_35.pdf

[84] Leitsin, V.N.; Dmitrieva, M.A.; Kobral, I.V. (2007). Shock_SHS_01: Program of Physical and Chemical Shock Synthesis Processes Simulation in Powder Materials Capable to Self-propagating High-temperature Synthesis. Daily Computing teaching programs and innovation (Telegraph sectoral program fund and algorithm, 12, http://ofap.ru/portal/newspaper/2007/12_35.pdf

In: Powder Metallurgy Research Trends
Editors: L. J. Smit and J. H. Van Dijk ISBN: 978-1-60456-852-3

Chapter 4

POWDER ADDITIVE PROCESSING WITH LASER ENGINEERED NET SHAPING (LENS®)

Baolong Zheng[1], Yuhong Xiong[1], Jonathan Nguyen[1], John E. Smugeresky[2], Yizhang Zhou[1], Enrique. J. Lavernia[1] and Julie. M. Schoenung[1]
[1] University of California, Davis, CA 95616, USA
[2] Sandia National Laboratories, Livermore, CA 94551-0969, USA

ABSTRACT

Powder metallurgy techniques have been extensively used to fabricate net shape components, and are popular in the current trends in materials processing. Laser Engineered Net Shaping (LENS®) is one of the fastest growing laser metal deposition processes, which has combined high power laser deposition and powder metallurgy technologies with advanced methodologies of rapid prototyping, to convert complex virtual objects (CAD solid models) into functional advanced structural components without the need for part-specific tooling. It is an additive versus a subtractive material forming process. With LENS® processing, the microstructure can be tailored by controlling both process parameters and composition. In addition to fabricating complex geometries of fully dense metals that require minimal finish machining, the mechanical properties have typically been found to be superior to those of components obtained by conventional processes. In this chapter, recent research and progress associated with the LENS® process are reviewed, such as laser materials processing and rapid manufacturing; metal alloys, composites, and cermets development with the LENS® process; the effect of process parameters on microstructure, properties, and build of part height; the application of *in-situ* molten pool sensor and Z-height control subsystems; thermal behavior measurement with thermal imaging and thermocouple methods, and numerical simulation; as well as a cost-benefit analysis of the LENS® process. Special emphasis is placed on LENS® deposited materials and process control, along with trends and challenges of laser direct fabrication of bulk metallic glasses and nanocrystalline materials, and analysis of residual stress and porosity in deposited materials.

1. INTRODUCTION

The consistency and precision of high-energy lasers have been widely used in the area of materials processing since being invented in the 1960s [1]. A variety of techniques, such as laser welding, laser cladding, laser cutting, and laser drilling, have been developed during the last two decades [2]. By examining the various laser based materials processing techniques, it becomes apparent that the high power density of the laser is commonly utilized as a heating source. The idea of incorporating lasers to melt metal powder and produce a bulk "near-net shape" rapidly solidified structure can be traced back to Breinan and Kear [2], who developed the Laserglaze™ technique. The concurrent development of microprocessors and Computer Aided Design (CAD) techniques aided in accurate control of the laser beam and building of complex three-dimensional (3D) solid structures, respectively. Advances in laser technology, computers, CAD software, motion control systems, and materials have revolutionized rapid prototyping capacity as we know it, and will probably continue to do so for years to come [2-12].

Over the past decades, a number of material/powder additive manufacturing technologies: Stereolithography (SL), Selective Laser Sintering (SLS), Laminated Object Manufacturing (LOM™), Three-Dimensional Printing (3D Printing), Directed Light Fabrication (DLF), Electron Beam Melting (EBM), and Laser Engineered Net Shaping (LENS®), have been developed. A number of these have been referred to as Solid Freeform Fabrication (SFF) methods and are some of the fastest growing manufacturing technologies. Unlike the traditional material subtractive or removal processes, additive manufacturing technology allows for the realization of near-net shapes as well as the potential of having intentional compositional variations as a function of position within the parts. Among these processes, SL, SLS, 3D Printing, DLF, and LENS® are classified as laser based additive processes. The early SFF processes converted CAD models into prototype parts using surrogate materials for rapid manufacturing purposes. Advances over the past decade have enabled SFF to surpass rapid prototyping (RP) groundwork, producing fully dense functional metallic parts. Focal points in today's competitive market place are reductions in both time and cost.

As one of the novel SFF technologies, LENS®, which incorporates features from both stereolithography and laser welding via a powder additive manufacturing process, was developed at Sandia National Laboratories in the 1990's [4]. Near-net shape metallic components with nearly full density can be fabricated by a laser beam melting metal or mixed powder line-by-line and layer-by-layer directly from a 3D CAD solid model. The concept of laser direct manufacturing is realized by the LENS® process, which uses materials of construction in place of surrogate materials, and eliminates the need for extra steps such as making patterns typically required for casting of metal parts and dies for powder metallurgy processes. In addition to fabricating complex geometries of fully dense metals that require minimal finish machining, the mechanical properties have typically been superior to those of components made from conventional processes [13]. The time span from a design model to the final product has been reduced significantly through the use of the LENS® manufacturing technique. Compared with other direct fabrication techniques, the LENS® system is more robust due to its characteristics of high structural integrity and dimensional tolerance.

Theoretically, any metal powder system that exhibits the required size distribution and morphology can be used as feedstock powder for LENS® manufacturing. Furthermore, ceramic powder such as SiC and Al_2O_3 can be incorporated in the metal powder system to tailor the mechanical and other properties of the components. Sandia National Laboratories has presented preliminary results with regard to multi-material processing by LENS® [4]. Investigations into the essential process mechanisms of LENS® have failed to fully define how the process variables correlate to features of the final parts [14].

This review focuses on the powder additive processing of LENS®. The potential for the development of novel materials along with an understanding of process controls and their effects on the resulting microstructure and properties are of particular interest. The current level of understanding of the thermal behavior is presented to facilitate accurate modeling of the *in-situ* deposition process. Special emphasis is also placed on potential benefits of the LENS® process when compared with other powder metallurgy techniques, along with trends and challenges associated with LENS®.

2. Laser Materials and Additive Processing

In 1917, Einstein laid the foundation for the laser by introducing the phenomenon of stimulated emission [15]. Maiman demonstrated the first laser by using a ruby crystal in 1960 [1]. The acronym LASER stands for "Light Amplification by Stimulated Emission of Radiation". There are many types of lasers, each type having a characteristic wavelength. In principle, lasers pump electrons from the ground state to a higher energy level and obtain an inversion in population between the ground state vs. excited state. Subsequently, stimulated emission takes place when incoming radiation causes the emission of radiation in phase and coherent with incident radiation. By using a resonant cavity, the radiation with this wavelength can be amplified, creating a beam of coherent, highly monochromatic and directional light, which can be transported from the cavity by using a semi-transparent mirror [16, 17].

For laser welding and deposition, both high power density and high total power are necessary, which restricts the type of lasers to be used. Currently, there are two types of high-power infrared lasers available on the market: CO_2 lasers and Nd:YAG lasers. The active medium for a CO_2 laser contains a gas mixture: N_2, He and CO_2. Vibration transitions in the CO_2 molecule give rise to the laser effect with a wavelength of 10.6 μm. The maximum power of continuous wave CO_2 lasers is in the range of kilowatts. The active medium of Nd:YAG lasers is comprised of a Nd doped $Y_3Al_5O_{12}$ crystal. Similarly, the output power is in the range of kilowatts. The active element, Nd^{3+}, gives rise to the laser effect with a wavelength of 1.06 μm, which provides high power densities as well as making it possible to transport the beam by means of optical fibers. Both CO_2 and Nd:YAG lasers are currently used in industry for cutting and welding applications. The flexible fiber optic beam delivery of Nd:YAG lasers plays a crucial role for three-dimensional applications. Furthermore, for metals, the absorption of the laser radiation is much higher in the case of the Nd:YAG laser, which compensates for its low efficiency. Recently, there has been a tendency to replace the high power CO_2 lasers with new generation Nd:YAG lasers [17, 18].

2.1. Laser-Matter Interactions

When a laser beam strikes a surface, the radiation is either reflected, absorbed, or transmitted [18]. The fraction of reflected, absorbed, and transmitted energy sums to unity. The total energy interacting with the target can be given by the equation below:

$$E = R \cdot E + A \cdot E + T \cdot E \tag{2.1}$$

where the coefficients *R, A* and *T* refer to the reflectivity, absorbtivity, and transmissivity of the material, respectively. The energy absorbed is typically very low, around 10-20% of the total energy, because of high reflectivity of metals, especially at larger laser wavelengths, such as for CO_2 and Nd:YAG lasers. The amount of absorbed laser energy can be enhanced by redirecting the reflected energy back onto the target [19].

The intensity of a laser propagating into the material decreases exponentially following the Beer-Lambert's expression:

$$I(z,t) = I_0(t)(1 - R - T)e^{-Az} \tag{2.2}$$

where A is absorbtivity, which depends on the medium, laser wavelength and intensity, as well as the inverse of absorption length; I_0 $(=AE)$ is the initial incident intensity at the surface of the material; and t is time. This equation quantifies the beam intensity I at a depth z. Laser energy penetration depth for a given material is defined as $p=1/A$, at which about 63% of the initial energy has been absorbed [18].

Lasers have the capability of melting, boiling, and forming plasma if sufficient power is applied. A vapor cavity or keyhole around the laser beam in the material occurs if there is sufficient energy density per unit area to form a plasma, and motion of the liquid metal in the molten pool is very turbulent [17]. A majority of laser processes are thermal in nature, and can be controlled very well for melting and vaporization processes. Moreover, the laser is controllable both spatially and temporally; meaning position and time of heat can be precisely controlled, however only the amount of energy absorbed by the material can contribute to processes like heating and melting.

In addition to material properties and laser type (wavelength), the intensity and distribution of the laser power are also important factors. The energy intensity of a laser beam can be determined by dividing the power of the beam by the spot area as below:

$$I = \frac{P}{vd} \tag{2.3}$$

where P is laser power, v is laser scanning speed, and d is laser spot size in diameter. For a given laser power, increasing the spot size and scanning speed will decrease the intensity.

For laser material processing, delivery of the laser beam is critical. The beam must be transported from the source to the target and properly focused to obtain appropriate energy density and spot size. Optical systems used to deliver the laser beam include reflective-transmissive and fiber optic beam delivery. Reflective-transmissive methods consist of a series of reflective mirrors aligned at specific angles to deliver the beam to the target. The fiber optic system, commonly used with Nd:YAG systems, incorporates a fiber optic cable to transmit the beam to the target. The reflective-transmissive method is used with the LENS® technology.

2.2. Laser Materials Processing

Recently, lasers have become acceptable and reliable materials processing tools for the fabrication industry with a wide variety of applications. From an applications point of view, laser material processing can be broadly divided into four major categories: deposition forming (manufacturing of near net-shape or finished products), joining (welding, brazing, etc.), machining (cutting, drilling, etc.), and surface engineering (cladding, heat treatment, etc.).

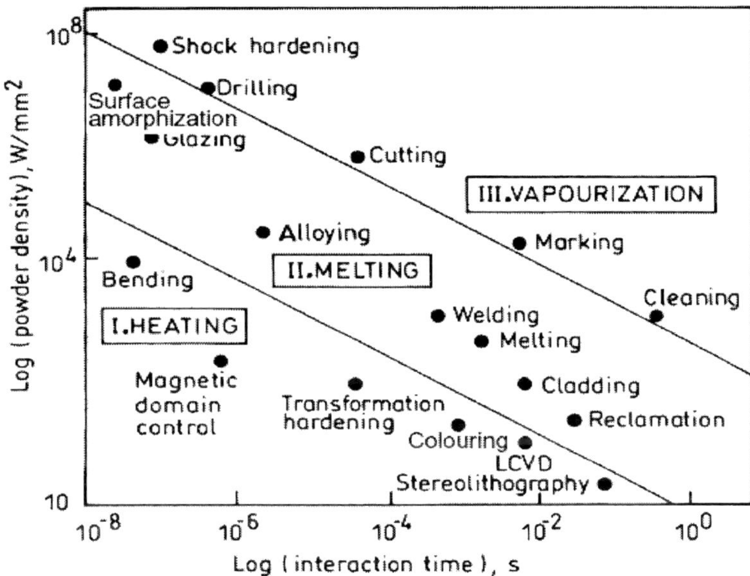

Figure 2.1. Schematic of process map in terms of laser power density as a function of interaction time for different laser material processes [18].

From the desired degree of heating and phase transition perspective, laser processes are divided into three major classes: heating (without melting/vaporizing), melting (no vaporizing), and vaporizing, as shown in figure 2.1 [18], which illustrates the domain for different laser material processing techniques as a function of laser power and interaction time. Low power density is required for transformation hardening, bending, and magnetic domain control, which rely on surface heating without surface melting. On the other hand, high power density is required for surface melting, glazing, cladding, welding, and cutting, which involve melting. Similarly, delivery of a substantially high power density within a very short interaction/pulse time is required for cutting, drilling, and machining.

One of the unique advantages of laser material processing is its flexibility. When combined with a multi-axis work piece positioning system or robot, the laser beam can be used for drilling, cutting, welding, deposition, and heat-treating processes, all with a single machine. Since energy transfer between the laser and material occurs through radiation, laser material processing is a non-contact process. A goal of materials processing is to produce products with correct shape, geometry, and dimensions. Laser materials processing offers a unique possibility for manufacturing finished products directly from raw powder materials without any elaborate intermediate operations [18, 20-23]. Direct one-step fabrication is a breakthrough and is very attractive because of the economy in time, cost, material, and manpower when compared with conventional fabrication.

2.3. Laser Rapid Manufacturing

A recently developed application of lasers in material processing is for use in rapid manufacturing (RM) and rapid prototyping (RP) technologies, where a computer aided designed (CAD) 3D model of an object can be converted into a physical component by coupling lasers with computer controlled positioning stages and special sintering, layering, or deposition techniques [18, 24-29]. These RM processes, sometimes referred to as Solid Freeform Fabrication (SFF) processes, have drawn much attention with several conferences, symposia [30], books [8, 31, 32], and journal papers devoted to related topics [33-39] in recent years. The development of RM implies that manufacturers are no longer constrained to shape metals by removing unwanted material. Instead, components can be formed into net shape components by additively building objects in lines and layers one after another.

There are four main steps in most laser RM/RP processes: (1) a three dimensional electronic solid model is drawn using a 3D CAD software, such as SolidWorks® or ProE®; (2) a software program electronically slices the 3D solid model into 2D layers to get a series of cross-sections; (3) the sliced model is then interpreted by process control software to determine appropriate processing conditions during the build; and (4) the physical functional prototype is built line-by-line and layer-by-layer.

This layer additive approach has been advanced through the use of laser power to either cure a liquid resin, or sinter, or melt solid particles to form a 3D component. Many different methods have been categorized and defined involving different target material systems, delivery methods, deposition methods, and power sources [8]. A brief description of the common techniques that are technologically relevant: Stereolithography (SL); Selective Laser Sintering (SLS); and Laser Engineered Net Shaping (LENS®), are provided below.

2.3.1. Stereolithography

Stereolithography (SL), which was patented in 1986 [40] and established the foundation for the rapid prototyping revolution, involves curing of a photosensitive liquid resin or polymer using a low power highly focused UV laser. Models are built upon a platform situated just below the surface in a vat of either the epoxy or acrylate resin. The laser scans the first layer, solidifying each cross section while leaving excess areas liquid. Next, the platform Z-direction is stepped down away from the focal point of the laser. A sweeper re-coats the solidified layer with liquid resin. Ensuing laser scans of the next slice atop the first and the processes are repeated until the complete prototype is built. Afterwards, the solid component is removed from the vat and rinsed clean of the excess resin before undergoing a complete curing treatment in an ultraviolet oven [8].

The Stereolithography Apparatus (SLA), developed in 1988 by 3D Systems [41], was the first commercially available 3D layer-additive process using CAD data. SLA is able to maintain a dimensional accuracy of up to ±0.1 mm and a layer thickness of 0.1-0.5 mm [40]. It is one of the most widely used rapid prototyping techniques for plastic models. SLA can produce integral parts with fine and complex structures. The disadvantages of the SLA process include the target material (resin, which is costly and toxic) and post processing requirements. Applications include master models for tools, medical models, and form-fit functions for assembly tests [41, 42].

2.3.2. Selective Laser Sintering

The Selective Laser Sintering (SLS) process was developed at the University of Texas and patented in 1989 [10, 43-46], and was commercialized by the DTM Corporation. SLS is similar to SL in most ways except that the material is in powder form and a laser beam (modulated CO_2 laser) is used to heat solid rather than liquid materials, such as nylons, thermoplastics, modeling wax, ceramics, and metals, to form a solid net-shaped object. Powder is fed into the machine a layer at a time of a specific height, and the laser scans the surface in a prescribed pattern to sinter the particles where material is desired. Each new layer of powder is leveled with a roller and scanned to form the desired sintered regions. The part is entirely surrounded by powder and the excess powder not sintered in each layer serves as a support structure for overhanging layers. Powder is fused together primarily by solid state bonding, but melting can also occur [30]. Sintering aids or polymer coatings are applied to ceramic or metallic powders to form a green (less-than-fully-dense) component, which is later sintered or infiltrated to eliminate porosity.

In addition to having a powder bed that precludes the need to create a structure to support overhanging features, the SLS process has the advantage of a wide variety of materials, as any powdered substance can be used. The disadvantages of the SLS process include the need for excess powder, the lack of achieving full theoretical density, and the roughness of the surface after fabrication [11, 47]. However, for applications such as visual representation and functional prototypes, the disadvantages are minimal.

It is interesting to note that the Electron Beam Melting (EBM) process, a competitive method of RM and similar to SLS, has recently attracted more attention. The EBM process uses an electron beam generated by an electron beam (EB) gun, rather than a laser beam. The EB gun passes electrical current through a tungsten filament, heating it up and causing it to emit electrons. Electrons are accelerated to a velocity between 0.1 and 0.4 times the speed of light. Since the EB gun is stationary and mechanical parts are not needed to deflect the beam, the result is high scanning speeds (up to 1000 m/s) and high positioning accuracy [48]. Figure 2.2 shows the parts of the EB gun. Kinetic energy of high-speed electrons is transformed into thermal energy when they impact the powder bed. This energy transformation immediately heats a predetermined layer thickness of the metal powder to above its melting point commensurate with the energy from the EB gun. The EBM process takes place in a vacuum chamber made up of a building tank with a vertically adjustable process platform, two powder hoppers and a rake system for powder application [49]. The building process starts by filling the build tank with powder, forming a bed on which the start plate can lie. A computer controlled electron beam scans the powder surface and melts loose powder based on the desired shape as specified by a CAD model. When scanning is complete, the process platform is lowered by one layer thickness, and computer controlled powder raking and scanning is repeated until each layer of the CAD model has been built.

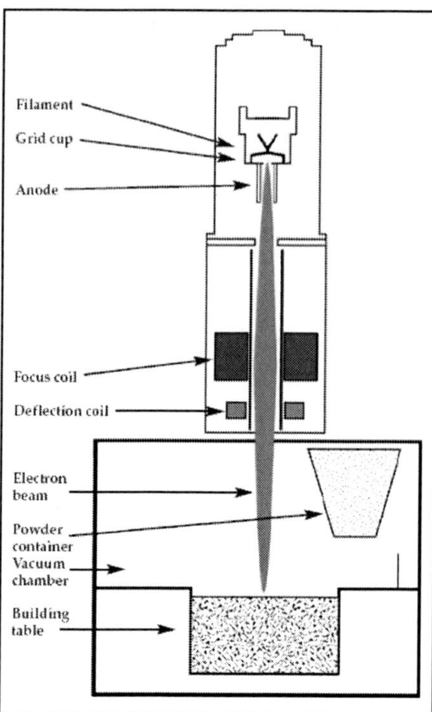

Figure 2.2. Schematic diagram of Electron Beam Melting process [48].

Advantages of EBM technology include the ability to process reflective metal powders such as Al alloys, to create complex parts with overhangs, and to generate biocompatible components. The disadvantages of the EBM process include high surface roughness, frequent failure of the tungsten filament, and formation of an additional "lip" along the edges of the fabricated parts. Potential applications are in biocompatible reconstruction of bones, functional prototypes, and production of large components.

2.3.3. Laser Engineered Net Shaping

The Laser Engineered Net Shaping (LENS®) process was originally developed at Sandia National Laboratories and commercialized by Optomec Company, Albuquerque, NM [50]. The LENS® process, as shown in figure 2.3, uses a high power laser (750–2200W Nd:YAG) to melt injected metal powders [4, 36, 51], supplied through four equally spaced nozzles of a deposition head to the focal point of the laser beam, where a molten pool is formed by the focused laser beam and increases in size resulting from injection and melting of the powder particles. The substrate, on which the powder is being deposited, can move in a plane in the X and Y directions, and as the laser power is switched on and off to a CAD controlled pattern, the desired cross-section of the component is fabricated. The focused laser beam and deposition head are moved up in the Z direction as each layer is completed, and a fully dense metal component can be deposited. There is no pre-placed powder for support. Powder is delivered and distributed through the head and the four nozzles by using a pressurized inert carrier gas. The deposition is done in a glove box, where the inert gas is used to shield the molten pool from atmospheric oxygen for better control of properties, and to promote layer-to-layer adhesion by providing better surface wetting.

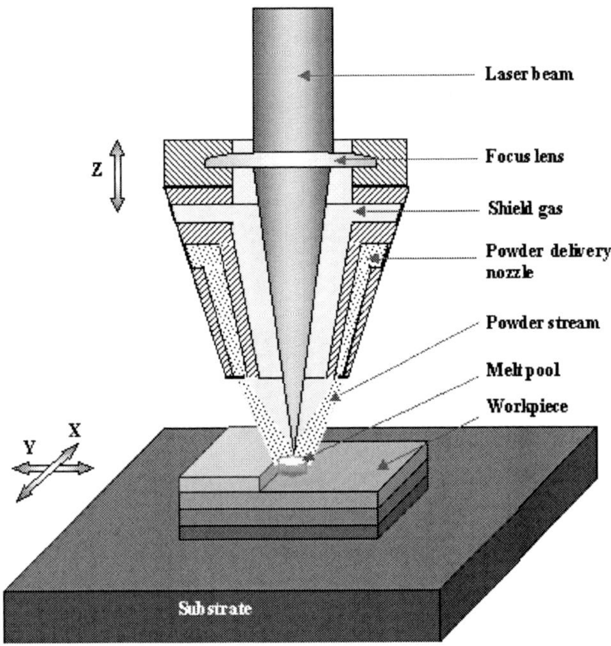

Figure 2.3. Schematic diagram of laser deposition with Laser Engineered Net Shaping (LENS®) [52].

Different properties at different locations of the component can be produced by in-situ alloying or grading of the material composition with independently controlled powder delivery feeders and/or by controlling deposition parameters. The LENS® process is capable of producing net shape components within a tolerance of ± 0.07 mm at a deposition rate of approximately 2500 mm^3/hr (laser speed ≈ 7 mm/s, layer thickness ≈ 0.25 mm, layer width ≈ 0.4 mm) [53] or about 0.045 kg/hr [54]. Figure 2.4 shows a LENS® 750 workstation equipped with *in-situ* molten pool size sensor control and Z-height sensor control subsystems [52], and figure 2.5 gives an example of a laser direct deposited part made with LENS® [55].

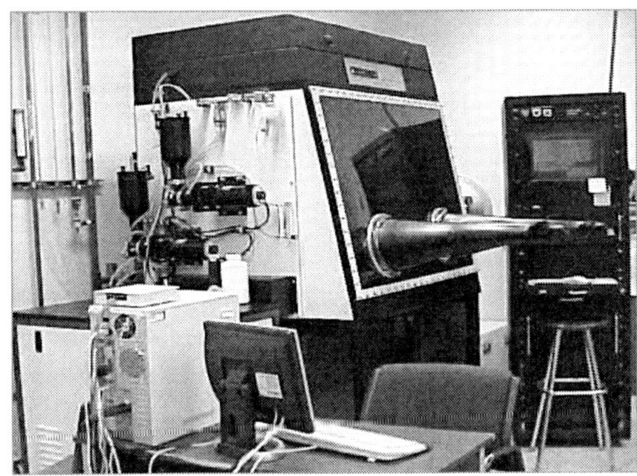

Figure 2.4. LENS® 750 workstation equipped with *in-situ* molten pool size sensor control and Z-height sensor control subsystems [55].

Figure 2.5. Photos showing: (a) LENS® processing of blade and (b) final built blade [54].

In comparison to other RM/RP technologies, the high power laser used in LENS® currently generates a surface roughness inferior to that of SLA and SLS [56-60]. Fabrication by means of SLA typically consists of photopolymer-based material on a liquid monomer synthesized by a laser line-by-line, although researchers have been able to fabricate components made from ceramics such as alumina and silicon nitride by SLS [61]. Normally, SLS uses materials with a stainless steel or bronze matrix mixed with a plastic binder. Interestingly, SLS components have been produced from ultra-high molecular weight polyethylene allowing for special applications such as the production of patient-specific prostheses. LENS® has fewer restrictions than SLA or SLS. The wide range of metals that can be potentially used allows for fabrication of functional parts in the as-deposited state. Furthermore, material composition can be altered dynamically by changing powder flow rate(s), unlike for SLA and SLS which use a liquid or a pre-deposited powder of a single composition. Both SLA and SLS require post-curing processes like furnace firing to consolidate the metal powders held together by the sintered plastic binder, while LENS® produces fully dense metallic parts.

The main advantages associated with the LENS® process are: 1) small heat affected zone (HAZ) with high cooling rate resulting in refined microstructures, 2) easy gradient deposition of multiple materials within a single part by precision deposition, and 3) deposition of fully dense metallic components.

The value of the LENS® process lies in its ability to fabricate fully dense net shaped metal components with good metallurgical properties, even though finish machining is generally required. The deposited materials have fine grain structure with properties equivalent or superior to forged materials, and do not require secondary firing operations compared to other RP processes. The disadvantages derive from the powder delivery system, which uses only spherical powders in the size range of 36-150 μm in diameter. In addition, deposition of overhanging geometries requires special designs such as by using a structure for support or by using a rotary stage.

The LENS® process has additional applications, including fundamental materials research, functional prototyping, addition of material and features to existing components such as thin wall components, and service and repair applications [4, 6, 50, 54, 62-64]. An example of a current focus for a LENS® application is the production of low-volume tooling [54]. Other researchers have focused on component repair [8, 50, 54, 65]. The LENS® system delivers significant and functional advantages when repairing high-performance metal components. LENS® is a highly targeted solution that can precisely add material to worn or damaged areas with minimal heat effect, enabling repair of the most sensitive thin wall components such as those found in gas turbine engines [4, 50]. LENS® is an inherently efficient approach that reduces production costs and shortens time to market for high-value components. The process enables fabrication of novel shapes, hollow structures, and material gradients that are not otherwise feasible. LENS® can also be used to add features to cast or forged parts. Optomec has received government grants to develop aircraft engine components and is pursuing research projects on gradient materials and metallic foams [50], while Tensegra used a NASA grant to create biomimetic, functionally adapted structures using LENS® [66]. LENS® is also an emerging solution for medical device manufacturing, including the development, prototyping, and production of specialty surgical instruments and prosthetic implants [50, 67, 68], such as hip, knee, and spinal prosthetics. Other applications envisioned include compositionally graded materials and components, custom implantable devices, and producing single crystal metals [69-74].

3. Materials Development with LENS®

Development of novel materials with LENS® powder additive technology is continuing at a steady pace. Many traditional components have been improved upon or replaced by LENS® deposited components, which offer enhanced performance, cost effectiveness, and environmental benefit. Advances in the development of components for structural applications with LENS® technology is more mature than for other evolving areas, such as functional components.

A literature survey revealed that the LENS® technology has been successfully applied to the development of materials since 1996 [51, 53, 72, 73, 75-78], with emphasis on exploitation of specific materials and their mechanical and physical behavior. It is impossible to cover all of the recent advances in materials development relevant to the LENS® process in detail here. This section will feature several developments that are representative of advances in materials deposition with LENS® including metallic alloys, metallic matrix composites (MMCs), graded materials and cermets.

3.1. Metallic Materials

Various Fe-, Ti-, Ni- and Co-based alloys have been used to fabricate components by LENS® [35, 36, 51, 53, 74, 76, 79]. Table 3.1 provides a comparison of the room temperature mechanical properties for typical alloys fabricated by both LENS® and wrought processes. In most cases, the properties of the LENS® deposited alloys are equivalent to or better than those of materials fabricated via traditional manufacturing processes. A wide variety of microstructures, some of which are unique, can be achieved with LENS®, depending on processing controls and parameters used.

Table 3.1. Mechanical properties for various alloys fabricated by LENS®

Properties	YTS, MPa		UTS, MPa		Elongation, %		Reference
Processing	LENS®	WRT	LENS®	WRT	LENS®	WRT	
316 SS	434	234	758	586	45-70	50	[51, 76, 80]
304 SS	324	276	655	-	70	55	[51]
PH13-8Mo	634	621	1324	931	13	14	[76]
H13	1462	1448	1703	1724	3	12	[51]
Ti-6Al-4V	931	855	965	931	16	10	[51]
IN 600	428	-	731	-	40	-	[51]
IN 625	614	400	931	834	38	37	[51]
IN 718	1117	1158	1400	1379	16	20	[51]

YTS - Yield Tensile Strength; UTS – Ultimate Tensile Strength; SS – Stainless Steel; IN - INCONEL; WRT – Wrought.

3.1.1. Stainless Steel

Gas atomized stainless steel (SS) (316L, 304, and 17-4 PH) powder with mesh size of -100/+325 (44 to 150 μm) has been widely used for LENS® processing and testing [35]. Excellent material properties are routinely achieved such as yield strengths twice that of wrought annealed bars yet with no sacrifice in ductility. Research on 316L SS demonstrated mechanisms as simple as grain size refinement leading to Hall-Petch strengthening. Typical grain sizes of LENS® processed 316L SS range from 1-10 μm [4, 35, 51] as shown in figure 3.1, whereas the grain size for traditional wrought material is approximately 40 μm [81], and for annealed materials is roughly 100 μm. Differences in grain size is believed to be the primary source for improved strengths of laser-based direct-fabricated structures and is consistent with the Hall-Petch relationship between yield strength and grain size. Consequently, full size components can be made with a wide range of strengths. Hardness values were also reported from 180 to 232 on the Knoop scale, equivalent to Rockwell B values of 85 to 96.

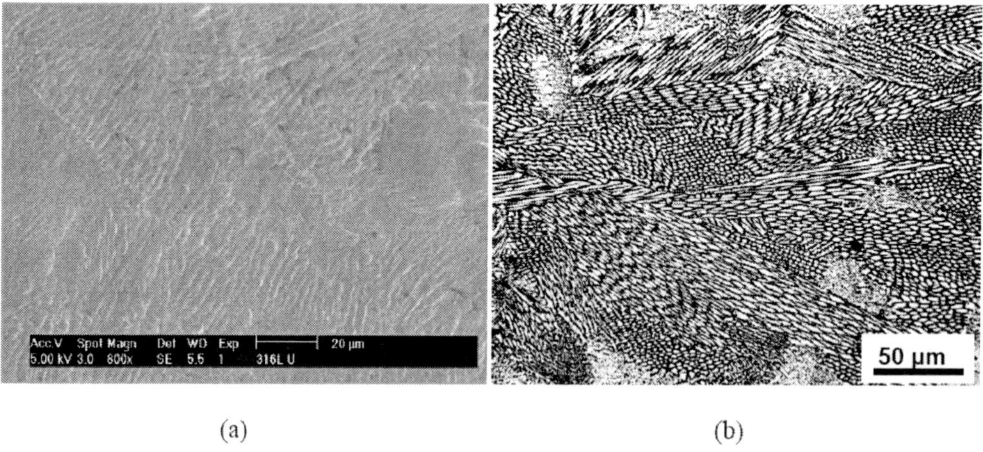

Figure 3.1. (a) SEM micrograph showing columnar morphology near the interface of the molten pool [82], and (b) optical micrograph showing austenite solidification cellular microstructure of 316L SS [80].

The solidification microstructures of LENS® deposited 316L SS are complex and varied as one might expect for the case of a molten pool that experiences rapid solidification. In general, a dendritic microstructure with columnar growth morphology predominates near the interface of the molten pool [36], as shown in figure 3.1(a) [82]. Epitaxial growth off the prior solid interface can be observed with each layer. The thickness of each layer is uniform, determined by the selected increment in Z-height from layer-to-layer, and re-melting of the top portion of the previous layer occurs, the amount determined by the height of the molten pool in relation to the increment in the Z direction for each successive layer. The resulting interface between each layer consists of a thin columnar region about 12 microns thick, as shown in figure 3.1(a). Figure 3.1 (b) is a magnified view of the top surface showing the microstructure of the cross-section parallel to the substrate surface [80], and reveals a typical fine solidification cellular microstructure. The morphology is primary austenite cells with intercellular ferrite presented at the cell boundary triple points and cell walls.

PH13-8Mo (Fe-13Cr-8Ni-2Mo-1Al wt.%) is a low carbon (<0.03), martensitic, precipitation hardening (PH) stainless steel, having high strength and hardness combined with good corrosion resistance in addition to exhibiting good ductility and toughness [76, 83]. For LENS® deposited PH13-8Mo, Smugeresky et al. reported that the ultimate tensile strength is higher than that of conventional heat treated and aged samples; yield strength and elongation at fracture are equivalent to that of conventional heat treated and aged PH13-8Mo materials [76, 84]. A sample observed in TEM indicated a morphology of nearly parallel martensite laths with retained austenite. These martensite laths contained a very high density of dislocations, which could have been generated either by austenite-martensite transformation and left by the advancing boundary or by thermal stress/strain during the LENS® process. This high level of dislocation density coupled with the fine lath size (0.2 μm) is responsible for the high strength achieved in this alloy. PH13-8Mo SS is a martensitic precipitation hardening (PH) steel, generally having NiAl precipitates. However, no precipitated NiAl particles were found in the TEM observation, which could be attributed to the low volume fractions of the precipitates involved [76].

3.1.2. Tool Steel

The fabrication and maintenance of tooling contributes significantly to the cost of many manufacturing operations. Laser-based LENS® direct fabrication of tooling has the potential for significant industrial and economical importance. Utilizing the unique metallurgical qualities of the LENS® processed materials could lead to increasing the useful life of tools and may also reduce costs. H13 tool steel is a commercially available secondary hardening alloy that exhibits a martensitic structure tempered with the formation of alloy carbides. H13 tool steels processed by LENS® exhibit varying properties attributed to both coarse and fine microstructures, depending on the position and processing conditions. Due to the layer additive nature, thermal cycles associated with the LENS® process result in numerous reheating cycles. Microstructural changes in deposited materials are observed in response to these cycles, with noticeable changes in hardness observed between top and bottom regions, as shown in figure 3.2, with a consequential decrease in hardness from Rc57 to Rc49 with increasing distance from the substrate, leveling off for distances 3 mm or larger. It is apparent that process parameters directly affect the properties, and that there can be designed variation of same with location.

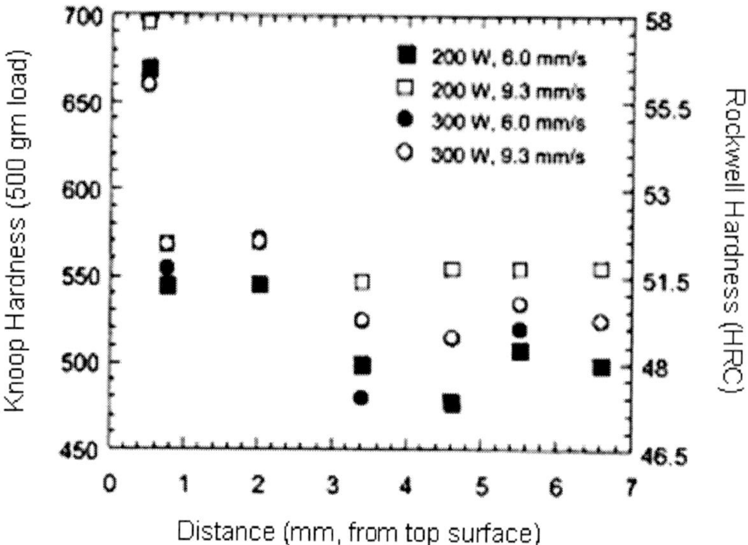

Figure 3.2. Hardness decreases with increasing distance from the top of H13 tool steel sidewall for four deposition conditions [33].

At the top of the build, the material solidifies as a high carbon austenite with intercellular ferrite. Some austenite is retained at the top of the build. The retained austenite transforms to martensite and is tempered on subsequent laser passes. Although the intercellular ferrite is partially transformed to martensite, a significant amount of ferrite remains [35]. The final microstructure of the lower area consists of tempered martensite with a bimodal distribution of fine vanadium- (A areas) and larger chromium- (B areas) containing carbides, as shown in figure 3.3. A kinetic model, developed to relate thermal history to hardness for H13, was combined with LENS® thermal data to estimate build hardness [53]. The steep thermal gradients and rapid softening kinetics in H13 require very precise knowledge of the thermal history to accurately predict properties.

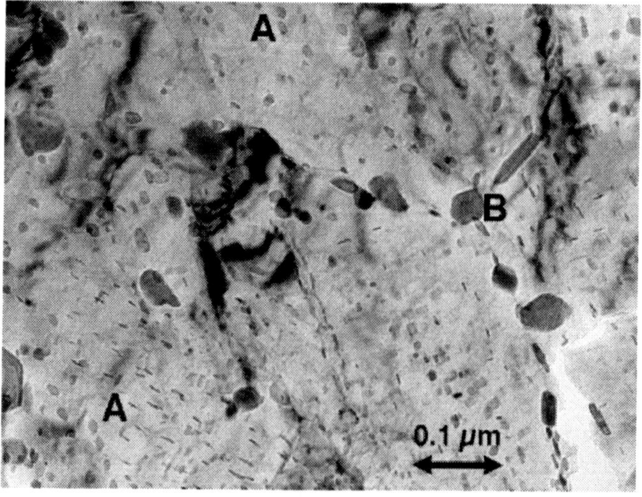

Figure 3.3. TEM image of LENS® deposited H13 showing bimodal distribution of fine vanadium- (A), and coarser chromium- (B) carbides [53].

3.1.3. Titanium Alloys

Ti-6Al-4V is the most common α-β titanium alloy, which contains various two phase microstructures at room temperature, depending upon thermal history. The extent of β to α transformation and size and morphology of α phase that forms depend primarily upon the cooling rate experienced by the β phase during solidification [85]. Heat treatments can be used to alter the distribution, size, and morphology of the α phase, but generally do not have a large effect on the prior-β grains.

The macrograph of the LENS® deposited Ti-6Al-4V (figure 3.4 (a)) shows the presence of large columnar prior-β grains in the build direction, and the presence of macroscopic layer bands, which appear in every layer except for the last 2-4 layers of the deposit. The microstructure of LENS® deposited Ti-6Al-4V consists of α-Ti and β-Ti phases, and exhibits fine acicular α grains about 1 μm in width and of different orientations, as shown in figure 3.4 (b). The acicular HCP α phase in a Widmanstatten pattern is formed from BCC β, and outlined in retained, yet small volume fraction, of very fine BCC β laths, which is the normal Ti-6Al-4V microstructure produced by laser deposition [79, 85-87].

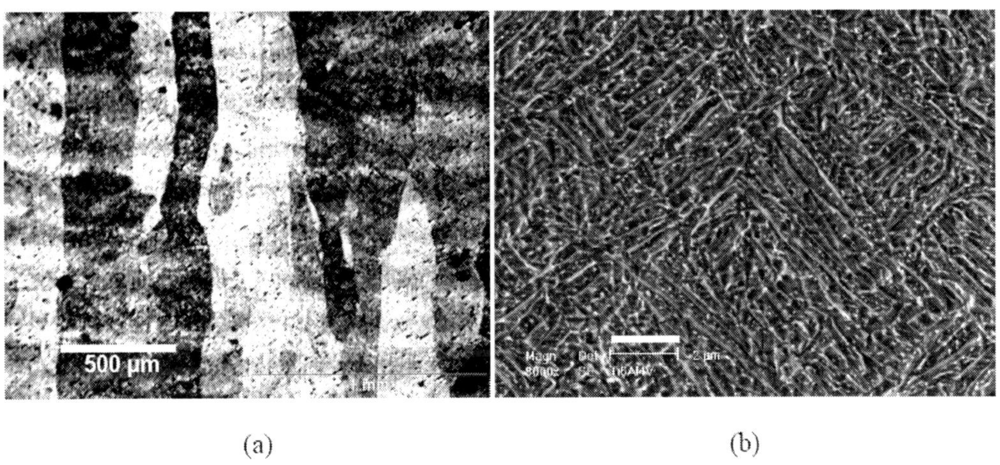

(a) (b)

Figure 3.4. Morphologies of LENS® deposited Ti-6Al-4V showing (a) prior β columnar grains, and (b) microstructure consisting of colony α grains outlined in retained β phase.

It is well known that titanium alloys are subject to embrittlement by oxygen and nitrogen, impurities commonly introduced during conventional processing. Comparison of the composition of LENS® deposited Ti-6Al-4V and the starting powder composition has shown that the amount of oxygen and nitrogen contained can be held very low under properly controlled conditions. Tensile and hardness properties of laser-based direct-fabricated Ti-6Al-4V material are also within ASTM B 367 Grade C5. Fatigue testing has shown that laser-based direct-fabricated Ti-6Al-4V is equivalent to both cast and hot isostatically pressed materials [88].

Fully dense TiAl alloys were also deposited using LENS® technology [89]. Depending on operating parameters, either equiaxed γ-TiAl/α_2-Ti_3Al or single phase α_2-Ti_3Al microstructures can be obtained. The LENS® method was also used to fabricate porous Ti implants [68, 90]. Porous pure Ti structures with controlled porosity in the range of 17–58 vol.% and pore size <800 μm were produced by controlling the LENS® process parameters.

The deposited porous Ti showed a broad range in mechanical strength of 24–463 MPa and a low Young's modulus of 2.6–44 GPa [90].

3.1.4. Nickel-Base Superalloys

INCONEL 625 is an austenitic nickel based material with high temperature as well as corrosion and oxidation resistant properties. It is generally regarded as the most widely used nickel-base alloy in the world. The properties of LENS® processed INCONEL 625 exceed those of annealed bars [51]. INCONEL 718 is a widely used γ'-strengthened nickel based superalloy. Both strength and ductility of laser deposited INCONEL 718 are higher than typical values in cast material [51, 91].

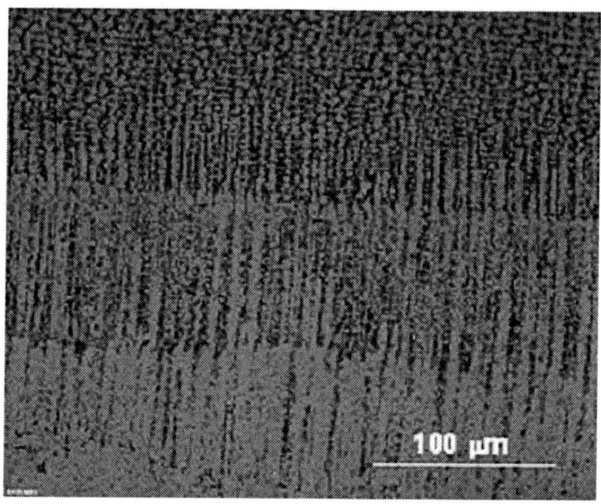

Figure 3.5. Microstructure of LENS® deposited Ni-base superalloy [74], showing variation in the microstructure with distance away from the substrate.

LENS® technology has been used as a general repair method for single-crystal nickel-base superalloy turbine blades [92, 93]. For a variety of deposition conditions, it was possible to produce deposits that had the same crystallographic orientation as the substrate. The primary difficulty encountered was the development of cracks in the deposits. The use of the LENS® process to deposit Rene′ 80 powder as a filler material onto GTD-111 substrates has also been reported [74]. Microstructural analysis showed that metallurgical bonding has been achieved between the deposit and the substrate. A multi-layered and directionally solidified microstructure can be achieved with optimized process parameters. The microstructure of deposited materials was considerably finer than the substrate. Figure 3.5 shows the microstructure of the deposited Rene′ 80 superalloy [74]. The first layer appears to be of cellular dendritic morphology; subsequent layers become progressively more columnar dendritic. Defects in the deposits, including porosity, microfissuring, and formation of stray grains, were observed.

Ni–25at.% Mo alloys have been successfully deposited from a blend of 75% Ni powder and 25% Mo powder using the LENS® process [75]. The laser deposited Ni-Mo alloys exhibited a rapidly solidified microstructure consisting of a single FCC phase. The highly exothermic enthalpy of mixing of elemental Ni and Mo appears to play a significant role in the homogeneous mixing of the laser deposited alloy as well as in determining the rate of solidification.

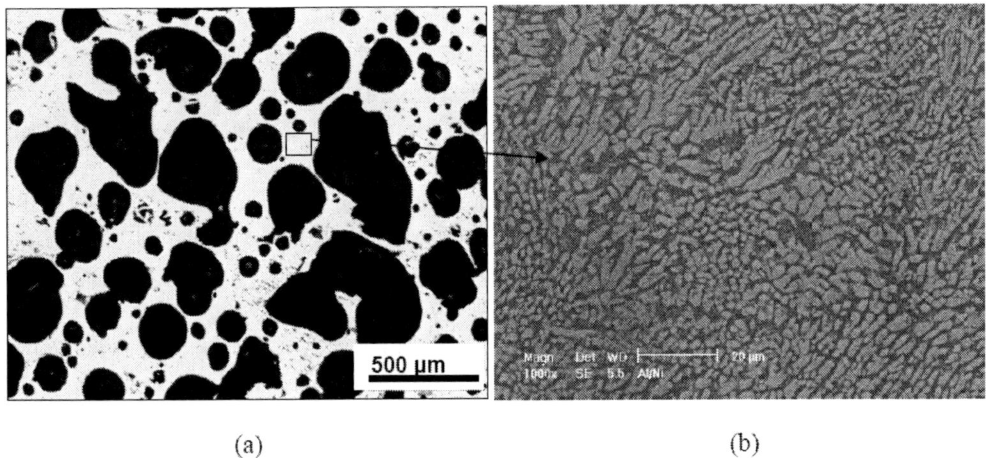

(a) (b)

Figure 3.6. Micrograph of LENS® deposited Al/Al_3Ni foam materials showing (a) high volume of pores, and (b) intermetallic compound (gray) in Al matrix (dark) [94].

Deposition of Al and Cu alloys is still a challenge for CO_2 or Nd:YAG laser-based materials processing due to the lower absorption characteristics of Al and Cu alloys. The laser used in the LENS® process is a continuous Nd:YAG laser. It has been reported that hybrid Al/Al_3Ni foam materials have been successfully synthesized by Zheng et al. via LENS® deposition of Ni-coated Al powder [94]. During the LENS® process, Ni reacted with the Al, leading to the formation of the intermetallic compound Al_3Ni, which can act as a reinforcement phase, while a high volume of pores (up to 50%) were formed *in-situ* for weight reduction, as shown in figure 3.6.

3.2. Metallic Matrix Composites and Graded Materials

One of the unique features of the LENS® process is that the feed material is delivered to the molten pool via a particle stream. This function allows for real time adjustments of the powder mixture entering the molten pool, thus altering the deposited material's composition. As a precision line-by-line and layer-by-layer fabrication process, LENS® has a deposition capability for composites and functionally graded structures. Therefore, the designer can tailor properties critical to component performance. One primary advantage associated with the LENS® process is the ability to provide both precise control of the reinforcement particle distribution and gradient precision deposition of multiple materials within a single component.

3.2.1. Metallic Matrix Composites

Particulate-reinforced metal matrix composites (MMCs), with high specific strength and stiffness, and low coefficient of thermal expansion (CTE), can be tailored to attain low density and dimensionally stable structures [95]. MMCs can be fabricated using a wide variety of approaches, and these can be grouped into two generalized categories: 1) casting and 2) powder metallurgical processing. In the case of casting, interfacial reactions are an issue because ceramic particles spend considerable time in contact with the molten metal matrix. Ti alloys are singularly susceptible to this problem, which is partly responsible for the limited application of Ti MMCs [96]. Moreover, particle segregation can occur during casting due to density differences between ceramics and metals. In contrast, conventional powder metallurgical methods can be used to attain elevated volume fractions of reinforcement, with limited or no interfacial reactions. Considering that low temperatures can be maintained and exposure controlled, the major disadvantage of conventional powder metallurgical routes is that they are relatively complex and limited in terms of product geometry [97]. LENS® provides a novel pathway to produce net shaped, lightweight MMC components [72, 98, 99].

Carbide-particle reinforced TiAl-matrix composites have been deposited using pre-mixed TiC (20vol%) and gas atomized Ti-48Al-2Cr-2Nb powders as feedstock materials for the LENS® process [98]. The deposited composites exhibit a hardness of more than twice that of Ti-6Al-4V alloy. The microstructure of deposited composites consists of un-melted TiC particles with re-solidified TiC and Ti_2AlC carbides distributed in γ-TiAl and α_2-Ti_3Al matrix phases. Both TiAl monolithic and composite materials exhibited susceptibility to solid-state cracking due to the high thermal stresses generated during the LENS® process. These cracks could be prevented by preheating the substrate during the LENS® process.

Ti–6Al–4V–TiB MMC composites have been *in-situ* deposited by Fraser et al. from a powder feedstock consisting of a blend of Ti–6Al–4V alloy and elemental boron powders using the LENS® process [72, 99-101]. As shown in figure 3.7, a homogeneous dispersion consisting of micron to sub-micron as well as nano-scale TiB precipitates formed within the Ti–6Al–4V α/β matrix. Nano-scale TiB precipitates can enhance the mechanical properties of these MMCs. The microstructure was substantially refined compared with those produced with other powder processing techniques. Heat treatments after deposition suggest that LENS® fabricated composites are thermodynamically stable, exhibiting limited TiB coarsening [72, 100].

It was reported by Zheng et al. that the LENS® process was also successfully used to fabricate Ti-6Al-4V MMC components with Ni-coated TiC particles as the reinforcement phase for enhancing the interface bonding between the metallic matrix and the ceramic reinforcement particles [71, 102, 103]. Figure 3.8 shows the presence of discrete and uniformly distributed fine re-solidified TiC particles (approximately 1-5 μm) (RSC) and the un-melted or partially melted coarse TiC particles (approximately 10-40 μm) (UMC). The melting of coarse TiC and subsequent re-solidification of fine TiC particles result in a fine microstructure as well as a stronger interfacial bonding between TiC particles and matrix. The LENS® deposited Ti-6Al-4V MMCs showed much higher strength compared to that of the monolithic matrix alloy, as well as isotropic properties. Ti-Ni intermetallic phases, formed in the deposited Ti6Al4V+TiC/Ni composites, can contribute to the strength, but are detrimental to ductility. The significant strength improvement can be attributed to the reinforcing effect of the TiC particles, the formation of Ti-Ni intermetallic compounds, and grain size refinement of the Ti-6Al-4V matrix produced during the LENS® process. Other types of MMCs deposited by LENS® with coated ceramic particles include: Ti-6Al-4V+TiN/Ni, 316L+B_4C/Ni, 316L+TiC/Ni, and INCONEL 625+TiC/Ni [65, 103].

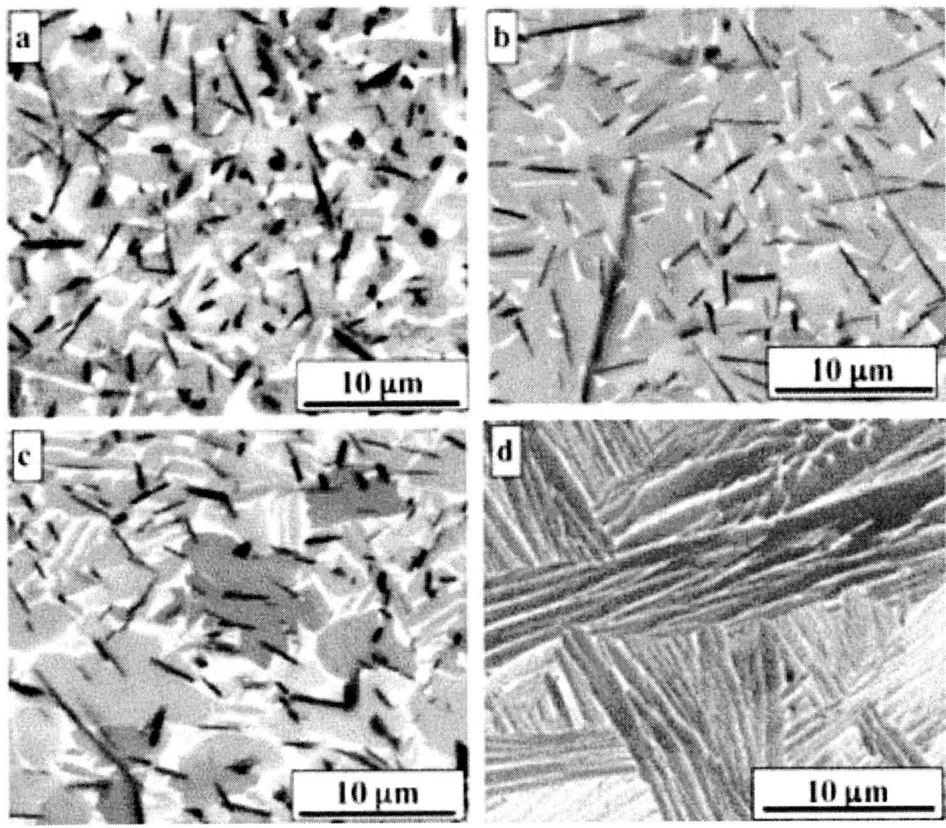

Figure 3.7. (a) SEM (BSE) micrograph of LENS® deposited Ti–6Al–4V–TiB composite; (b) microstructure after heat treatment at 700 °C for 100 h; (c) microstructure after heat treatment at 1100 °C for 10 h; (d) micrograph of deposited Ti–6Al–4V alloy [72].

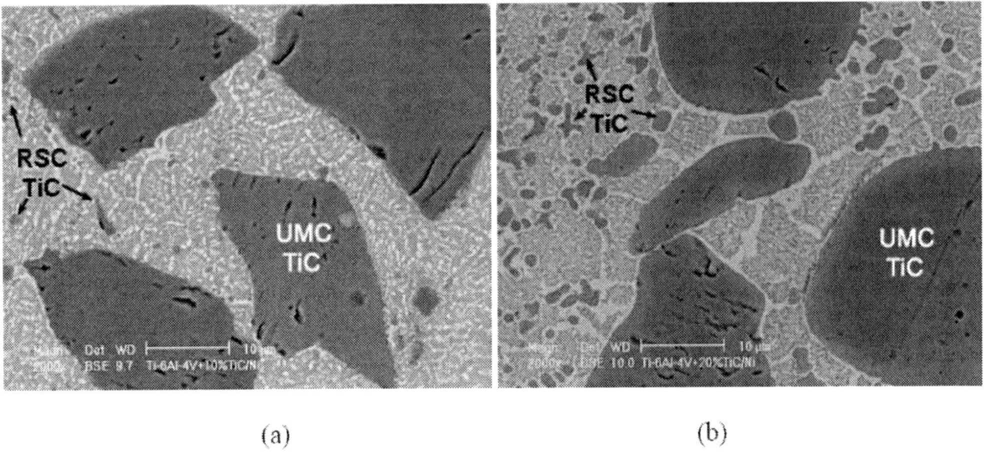

(a) (b)

Figure 3.8. SEM (BSE) micrographs of the LENS® deposited Ti6Al4V+TiC/Ni MMCs, showing re-solidified fine carbides (RSC), and unmelted coarse carbides (UMC): (a) Ti6Al4V+10wt.% TiC/Ni, and (b) Ti6Al4V+20wt.% TiC/Ni [102].

3.2.2. Functionally Graded Materials

Functionally graded materials (FGM) are a relatively new generation of engineered materials wherein the microstructural details are spatially varied through non-uniform distribution of the reinforcements, by using reinforcements with different properties, sizes and shapes, as well as by interchanging the roles of reinforcement and matrix phases in a continuous manner. The ability to pick-and-place material compositions, which has been a major challenge to successfully fabricating metallurgically sound, graded compositions, is a key benefit of the LENS® process, which allows designers to tailor localized material characteristics within a component [73, 77]. The ability to selectively apply different material compositions at user-defined regions within a component requires a thorough understanding of the dynamics of the process [77] and precise powder feeding capabilities.

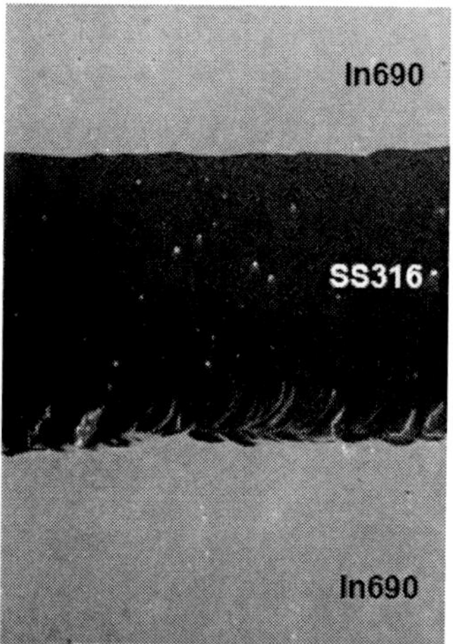

Figure 3.9. Graded structure of 316L SS (Dark) and IN 690 (Light) [73].

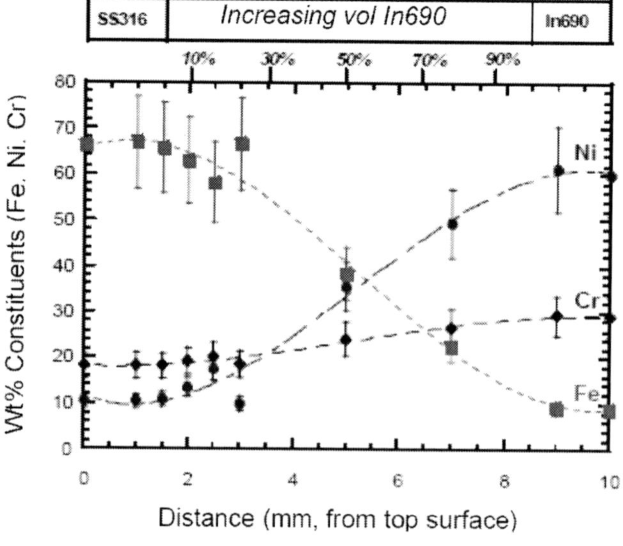

Figure 3.10. Alloyed constituent results for blending IN 690 into 316L SS [73].

Previous work has been done on graded structures. The microstructure and hardness was tailored when blending INCONEL alloy 690 (IN 690), a nominally 59wt.% Ni–30wt.% Cr alloy, into 316L SS [73]. Dense structures were fabricated in a 4-layer composite for two materials, 316L SS and IN 690, and 316L SS and MM10, a carbide strengthened powder metallurgy tool steel. Figure 3.9 shows a SEM micrograph of three layers in the alternating 316L SS/IN 690 part, where the darker shade material is 316L SS. During LENS® fabrication of each layer, a small fraction of the previous layer is re-melted to create the molten pool into which the new material powder is injected. Figure 3.10 shows the quantitative elemental constituent results for blending IN 690 into 316L SS. As expected, the Fe content decreases as more IN 690 is added, resulting in the increase in Ni content.

A graded ternary Ti–8Al–xV alloy has also been deposited using the LENS® process. The feedstock powder mixture used for depositing the graded alloy consisted of elemental Ti, Al, and V powders [104]. The LENS® process has also been used by Fraser et al. to successfully deposit functionally graded Ti–xV and Ti–xMo alloys [105]. A compositional gradient, from elemental Ti to Ti–25at.% V or Ti–25at.% Mo, has been achieved within a length of about 25 mm. The feedstock used for depositing the graded alloy consists of elemental Ti and V (or Mo) powders. While the as-deposited microstructure offers interesting possibilities in terms of microstructure gradients across the compositionally graded alloys, subsequent thermo-mechanical treatments can be introduced in order to tailor the microstructure for specific applications, such as aerospace structures.

It has also been reported that a functionally graded, hard and wear-resistant Co–Cr–Mo alloy was deposited on a Ti–6Al–4V alloy with a metallurgically sound interface using LENS® technology [106]. The coating of Co–Cr–Mo alloy is biocompatible and significantly increases the surface hardness without any intermetallic phases formed in the transition regions. Surface hardness increased from 333 to 947 HV with Co–Cr–Mo concentration increasing from the bottom to the top of the graded layers.

3.3. Cermets

Cermets are composites composed of ceramic and metallic materials. They have unique combinations of high temperature resistance and hardness from the ceramic, as well as high toughness contributions from the metallic material. In this section, the application of the LENS® process to hard materials, such as tungsten carbide (WC) and titanium carbide (TiC), is reviewed. Cobalt (Co), Ni, and Ti are usually the metallic components used to toughen these types of hard materials. Cutting tools, abrasives, wear-resistant coatings, and heat-resistant coatings are their representative applications.

The first use of the LENS® process to deposit WC-Co cermets was reported by Wei and co-workers [78]. A sub-micron sized WC-18 wt.% Co powder was used to build thin wall parts. A microstructure with alternating coarse and fine WC layers was observed. However, the LENS® processed parts were not dense enough at this first attempt. Further investigation of the LENS® process on WC-Co cermets were demonstrated by Xiong et al. [31]. In this work, nanocrystalline WC-10 wt.% Co powder without any grain growth inhibitors was used. Dense WC-Co specimens (97% of the theoretical density) without any cracks were produced. The WC-Co thin wall samples were deposited at a laser power of 200 W and a traverse speed of 3 mm/s with dimensions of 20 mm long by 13 mm high containing four parallel lines for each layer. Similar to Wei et al.'s work, SEM images indicated the presence of an alternating layer microstructure with large WC particles in the light layers and fine WC particles in the dark layers, as shown in figure 3.11. Figure 3.11(a) is a side view image showing alternating layers; and figure 3.11(b) shows a boundary area between the alternating layers. Coalescence and random touching along faceted WC grains contribute to the formation of larger WC particles when the top portion of the previous layer is re-melted. It was suggested that a dense feedstock powder and appropriate process parameters are critical to density improvement.

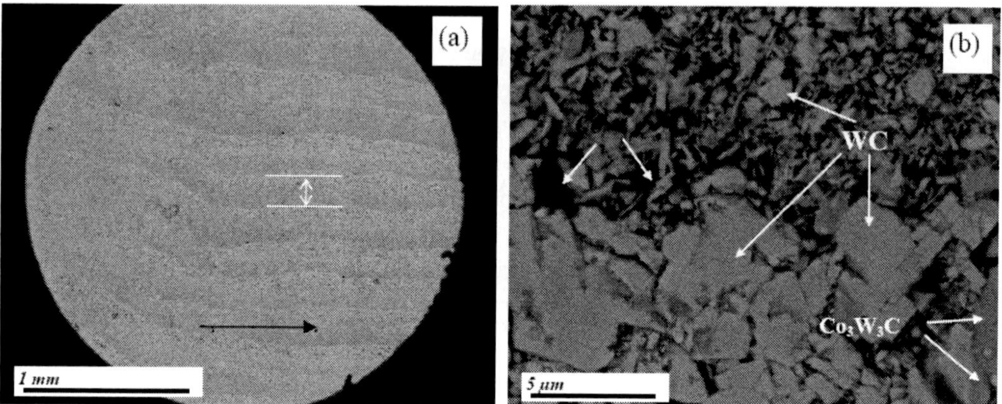

Figure 3.11. SEM (BSE) images of a WC-Co specimen: (a) an image from side view showing alternating layers (the black arrow indicates the horizontal deposition direction, and the white arrows show the layer thickness of one deposition layer); (b) enlarged image of (a) showing a boundary area [34].

In addition, it appears that a correlation exists between hardness and cooling rate during solidification. It has been reported by Hofmeister et al. [35] that cooling rate decreases with increasing part height. With the cermet samples, the hardness values decreased with increased height of the deposited specimen, which suggests that the decrease in cooling rate results in lower hardness.

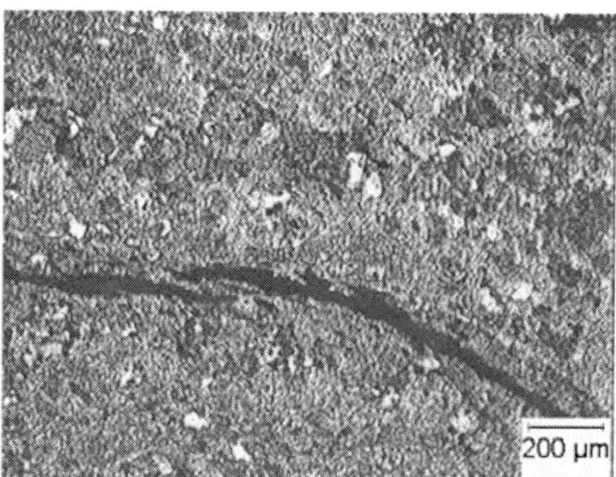

Figure 3.12. Image of a TiC-Ti specimen showing cracks originating near the interface of the specimen and the substrate [107].

Experimental results were reported by Liu and DuPont for LENS® deposits made from micron-sized TiC-20 vol.% Ti powder [107]. The square column TiC-Ti specimens were deposited at a laser power of 420 W and traverse speed of 8.5 mm/s with 12.7 mm sides and 8 deposited layers. Both the layer thickness and hatch spacing were set at 0.38 mm. A transverse crack was observed between the specimen and the Ti-6Al-4V substrate, as shown in figure 3.12. It was reported that processing-generated thermal stresses and interfacial mismatch stresses were the primary cause.

Crack-free FGM TiC-Ti was also fabricated via LENS® process in this work. The composition changed from pure Ti for the bottom layer up to 95 vol% TiC for the 20th layer. Figure 3.13 demonstrates the microstructure of a functionally graded TiC-Ti sample deposited at a laser power of 420 W and traverse speed of 8.5 mm/s. Re-solidified TiC (RSC) and un-melted TiC (UMC) were observed in the Ti-rich bottom layers and TiC-rich top layers, respectively. It is believed that thermal stresses are reduced and ductility is improved by gradual compositional variations, resulting in crack-free microstructures. Hardness results show a gradual increase from the bottom layers to the top layers due to the increasing concentration of the TiC component in this FGM.

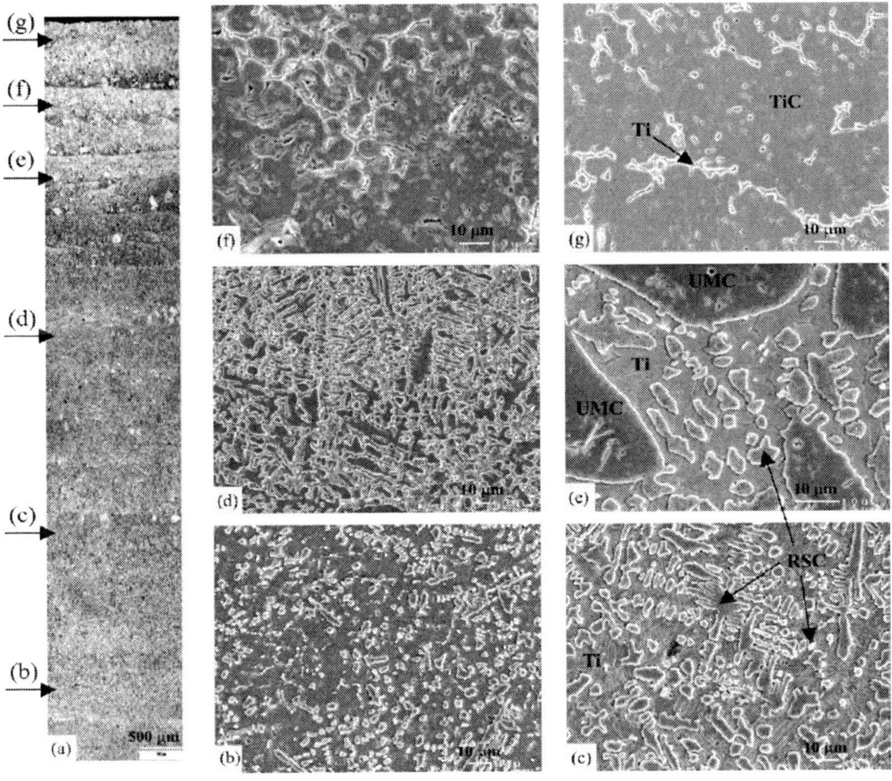

Figure 3.13. Microstructures of the FGM TiC-Ti: (a) optical microscopy photomicrograph of the deposit; (b)–(g) SEM photomicrographs with increasing TiC contents in different locations, showing re-solidified TiC (RSC) and un-melted TiC (UMC) [107].

Although reports on the application of the LENS® process to cermets are limited, WC-based cermets have been widely studied using other laser-related processes, such as laser sintering, laser cladding, Automated Laser Fabrication (ALFa), and Laser-based Direct Metal

Deposition (LBDMD) [108-111]. Glaeser et al. processed both WC-Co and WC-Ni cermets with a laser sintering machine equipped with a 200 W CO_2 laser. However, the porosity of the sintered cermets was reported to be 30%-40% resulting from incomplete liquid phase sintering [108]. Sears and Costello reported good cladding results with NiTung60® (a WC-Ni cermet) on H-13 tools. However, due to porosity and cracks, poor cladding of a WC-Co cermet on the same substrate were also reported [109]. Dense WC-Co bulk cermets produced by the ALFa process (equipped with a pulsed Nd:YAG laser) with proper process conditions were reported by Paul and Khajepour using conventional WC-Co powder [110]. Further investigation into the application of the LENS® process to cermets is clearly needed.

4. LENS® Process Control

In order to understand the development of microstructure and the associated properties of the LENS® deposited components, it is important to know the thermal gradient and cooling rates in and around the molten pool because these control the scale and morphology of the first solid formed in the liquid-solid interface [36, 51]. The thermal gradient and cooling rates, which control solidification behavior of the LENS® process, are controlled by process parameters such as laser power density, X- and Y-traverse speed, heat transfer conditions, etc. So, the purpose of this section is to review the solidification behavior and the process control of the LENS® process.

4.1. Solidification Behavior

LENS® materials processing is, in essence, a thermal process, which involves heating, melting, possible vaporization, and re-solidification of deposited materials. Due to the laser scanning and layer additive nature of the LENS® fabrication process, a complex thermal history is experienced in different regions of the build. The thermal histories associated with the LENS® process involve partial re-melting of the top layer, and numerous reheating cycles for all previously deposited layers at successively lower temperatures. The complex thermal behavior during the LENS® process can result in complex microstructure evolution, especially in precipitation hardened alloys. In addition, the use of a finely focused laser beam to form a rapidly traversing pool can result in relatively high solidification velocities and cooling rates, so it is important to understand and control the thermal and solidification behavior in order to control microstructure evolution and achieve desired quality of processing for fabrication of structural components.

The thermal history of a deposited component is primarily determined by the heat transfer mechanisms that take place during LENS® fabrication. The size of the molten pool is very small (~1 mm in diameter), so we can assume that heat transfer via radiation and convection can be ignored. The heat transfer in the LENS® process is primarily dominated by the heat conduction between the substrate and the part, and the top layer and the layers immediately below it. Therefore, the governing equation of heat transfer in the LENS® process can be expressed as:

$$\rho c \frac{\partial T}{\partial t} = K \nabla^2 T \tag{4.1}$$

where ρ is the density of the material (g/cm^3), C is the specific heat (J/g °C), K is the thermal conductivity (W/cm °C). With the temperature distribution in and around the pool attained, the thermal gradient ($G = dT/dx$) and cooling rate ($\mu = dT/dt$) can be determined accordingly.

Thermal images of the molten pool have been obtained, and thermal gradients and cooling rates along the trailing edge of the molten pool have been evaluated by Hofmeister et al. [36, 51] to have cooling rates at the solid-liquid interface ranging from 10^2 to 10^3 Ks^{-1} [35]. Quench rates for the LENS® process are sufficiently fast to reduce segregation and provide a microstructural scale consistent with rapid solidification microstructures. In solidification processes, the freezing rate (the inverse of cooling rate) is proportional to the length scale of the molten zone. Figure 4.1 illustrates the effect of different processes and corresponding cooling rates for metallic material processes plotted against length scales: laser pulse heating of copper [112], splat quenching of iron on a copper substrate [113], drop-tube splats [114], and chill-cast slab [115] use the length scale of thickness; atomized droplets [116] use the droplet diameter; arc welding [117] and large castings [114] use volume to surface area; the LENS® processing scale is the length of the molten pool. The microstructural scale of LENS® deposited materials falls between arc-welding and atomization processes. One advantage of the LENS® process is that parts of considerable size can be fabricated with microstructures having rapid solidification characteristics.

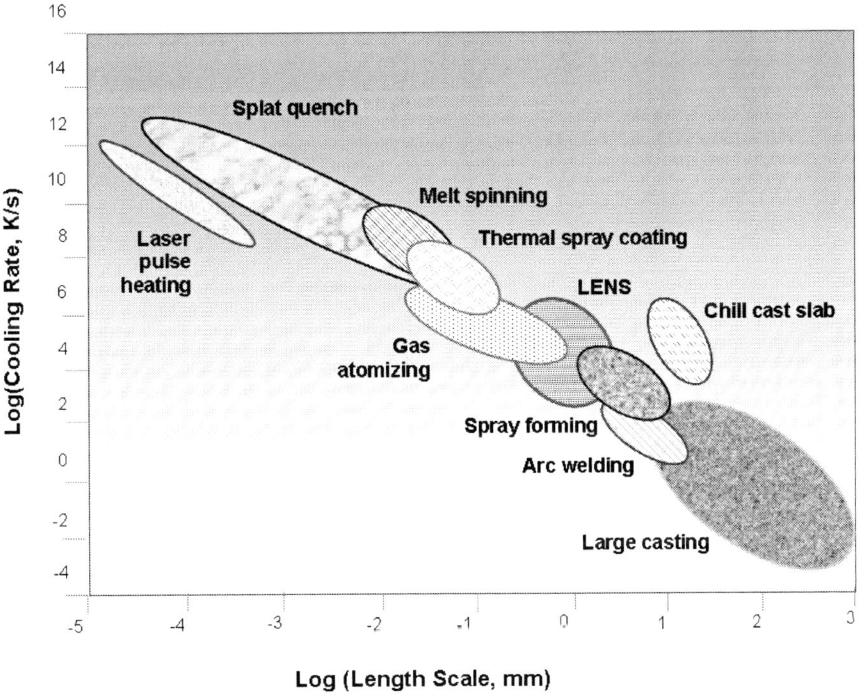

Figure 4.1. Cooling rates for metallic material processes plotted against length scales [55].

The microstructures of LENS® deposited materials are complex yet consistent with rapid solidification. In general, a dendritic microstructure with columnar growth morphology predominates near the interface of the molten pool, as is clearly shown in figure 4.2, for microstructures of deposited 316L SS [80]. The presence of a directional columnar microstructure demonstrates that the heat flux is highly unidirectional. Crystals grow in relation to the direction of heat flow, and the fine columnar dendrites grow almost parallel to the direction of growth.

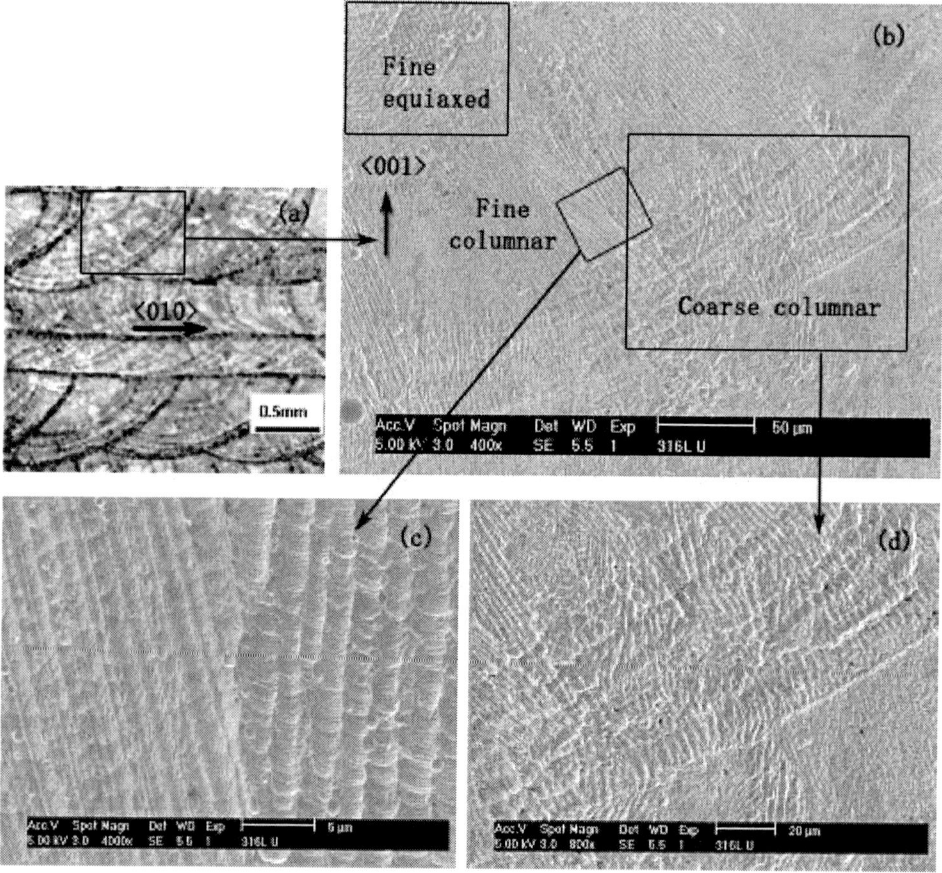

Figure 4.2. Laser track cross-section microstructures of deposited 316L SS [80].

Figure 4.2(a) shows the typical cross-section microstructure perpendicular to the laser travel direction and substrate, obtained with optical microscopy, revealing inhomogeneity of the molten pool. The characteristic rows and layers of the LENS® material reflect solidification from the continuously moving laser molten pool tracks. In figure 4.2(b), evidence of epitaxial growth off the prior solid interface can be observed within each layer. Figure 4.2(c) is a magnified view of a fine columnar structured region, where columns are about 1.5 μm in size, and figure 4.2(d) is a magnified view of a coarse columnar structured region, where the columns are about 4 μm in size. Figure 4.2(b) displays the typical columnar dendrite structure grown toward the top center of the molten pool. Around the center, there is also a small fine equiaxed structure zone, which is the cross-section of columnar dendrites parallel to the laser travel direction.

LENS® processing can be further considered as a dynamic thermal process because the heat source is continuously moving. As the laser beam moves away, the maximum temperature gradients are constantly changing directions. The growing columnar crystals are thus trying to follow the maximum temperature gradients, which can furthermore result in changes in the crystal growth direction at the molten pool center, as shown in figure 4.3, due to the solidification front trying to keep pace with the moving laser beam. Given that the crystal growth velocity attempts to keep pace with laser travel velocity, the velocity of the solidification front can be given by:

$$R = v \cdot cos\theta \tag{4.2}$$

where θ is the angle between the scan direction and the growth direction, and v is the laser travel speed. The vector of R representing crystal growth rate needs to continuously adjust itself as growth proceeds toward the laser track centerline. From Eq. 4.2, the solidification rate is largest when θ is about zero. The initial crystal growth rate is low and associated with a planar solidification front, and the front changes to cellular as the growth rate increases. Another reason for the local change in columnar growth direction is that re-nucleation occurs from the fragments of dendrite arms, which are fractured and dispersed by the turbulent convection in the molten pool. Variations in the laser travel speed can also influence the columnar direction since the growing crystals tend to follow the highest temperature gradients. Therefore, the microstructure of deposited components can be controlled by managing the process parameters during LENS® deposition, as described below.

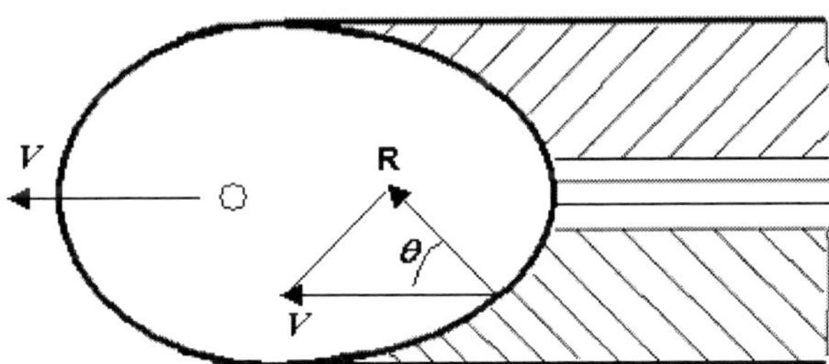

Figure 4.3. Illustration of columnar crystal growth of laser deposition track [80].

4.2. Effects of Process Parameters

To explore the application of the LENS® technique, an understanding of the relationship between processing conditions, microstructure and properties is of special interest. However, reports on this topic are mainly limited to the effects of laser power and traverse speed [34, 51]. In this section, a comprehensive analysis of the effects of processing conditions, including laser power, traverse speed, powder feed rate, temperature of substrate, and working distance, is provided.

4.2.1. Laser Power

Laser power is one of the key process parameters determining microstructure and properties of the resulting specimens during the LENS® process. If the laser power is too low and the other process parameters are kept constant, lack-of-fusion occurs resulting in pores within layers or gaps between layer boundaries. Figure 4.4 displays the effect of low laser power on microstructure of a niobium deposit [118]. When laser power is sufficiently high, dense specimens can be built. However, molten pool height varies with laser power. With an increase in laser power, more powder is included in each molten pool. The width and depth of the molten pool increase resulting in a higher layer thickness. The molten pool may reach a superheated condition if there is extra energy to heat the melted material.

The layer thickness is also controlled by the increment in Z-height set by the operator. If the height of the molten pool is close to but greater than the set Z-height increment, the part builds with layer thickness equal to the Z-height increment. Only the top layer will have the height of the molten pool resulting in a slight increase in final height. If the height of the molten pool is much greater than the set Z-height increment, the part builds initially with a layer thickness higher than the Z-height increment, which could later result in a defocus situation where the focal point of the laser beam ends up below the top surface of the part being built. Laser power achieved in the active region would then be reduced and less powder would be deposited onto the previous layer. Consequently, layer thickness may become much smaller in the upper portion of the specimen. This condition will continue until the distance between the top of the specimen and the focal point is close to zero. Then a layer higher than the increment in Z-height will build again.

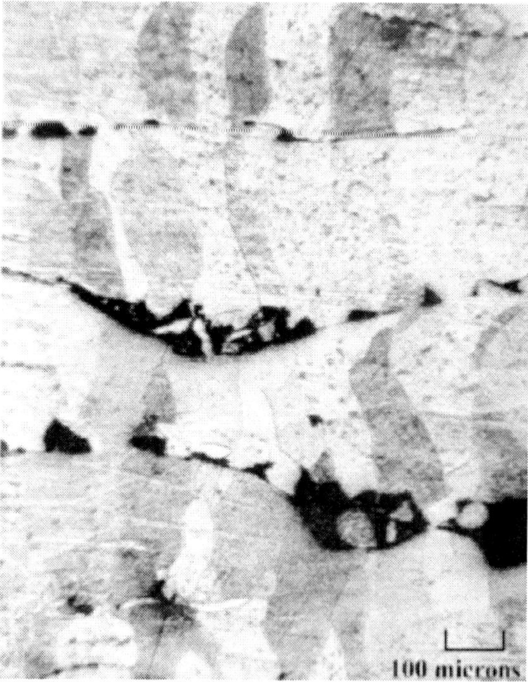

Figure 4.4. Pores and gaps observed between layer boundaries in a niobium deposit because of lack-of-fusion [118].

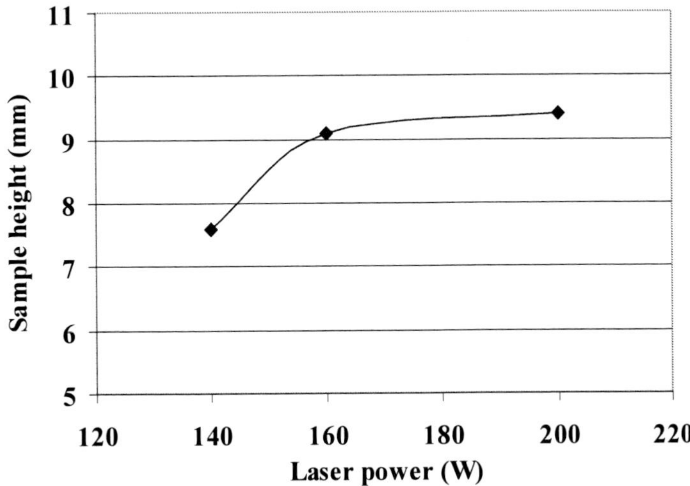

Figure 4.5. The influence of laser power on the height of LENS® deposited WC-Co thin wall samples with other parameters remaining constant [70].

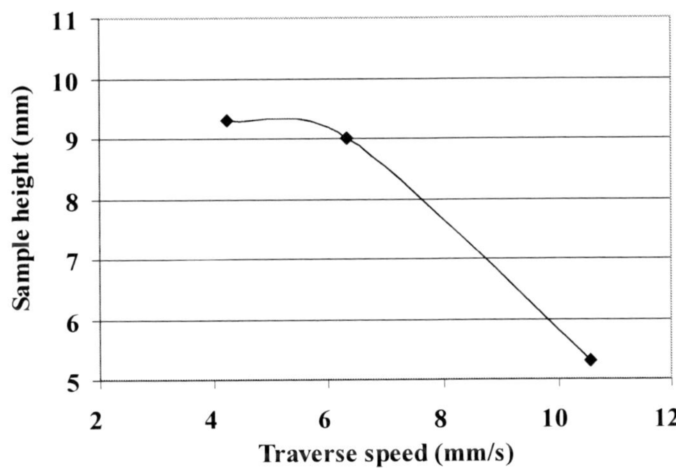

Figure 4.6. The influence of traverse speed on the height of LENS® deposited WC-Co thin wall samples [70].

Xiong et al. reported the effect of laser power on sample height (figure 4.5) [70]. WC-Co thin wall samples were deposited using variable levels of laser power. It was observed that specimen height increased rapidly with laser power in the lower power range. In the higher power range, specimen height increased slightly with laser power because a defocus condition of the laser beam occurred. In this way the height is self-correcting because the focal point is below the deposit metal surface. In addition, if the top layer is higher than the streams of powder particles, the molten pool will be material starved and no additional powder is injected into the molten pool.

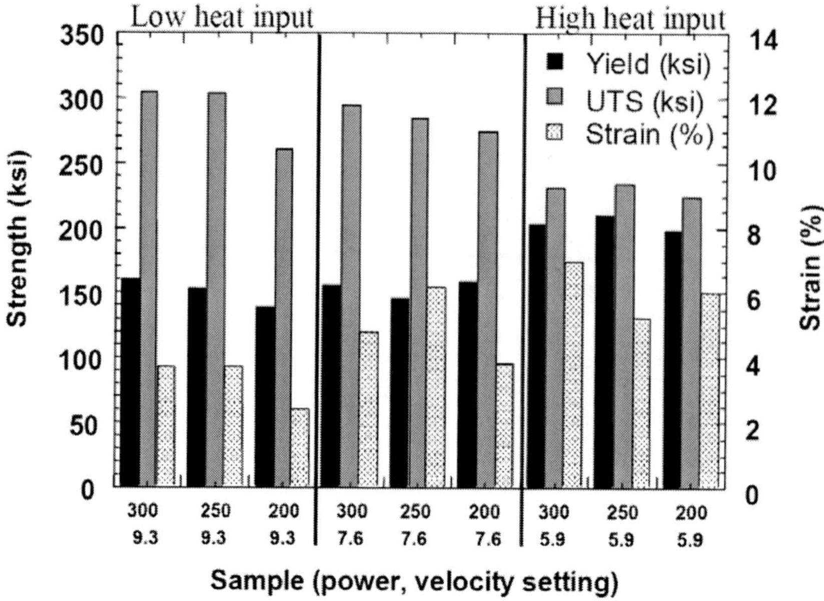

Figure 4.7. Comparison of tensile properties for LENS® deposited H13 tool steel fabricated at various power and velocity values [51].

4.2.2. Traverse Speed

Traverse speed, the speed of the moving substrate, is another critical process parameter that controls material height, microstructure and properties. High traverse speed can cause insufficient powder deposition per unit length. Thus, an insufficient amount of powder is melted by the laser beam resulting in a lower layer thickness compared to the desired value. Since the increment in Z-height for each layer is set equal to the desired layer thickness, an increase in working distance results. Consequently, the top of the previous layer will be too low to deposit any new powder on its surface. Thus, deposition stops. As a result, the final height is much smaller than expected. Figure 4.6 shows the effect of traverse speed on the height of a WC-Co thin wall specimen [34, 70]. Other process parameters were held constant. It can be seen that specimen height decreased with an increase in traverse speed. In this case 5 mm/s would be the maximum critical speed to sustain continuous void-free deposition.

In fact, the ratio of laser power to traverse speed, energy input (J/mm), is a more useful value to control the process. Cooling rate is a significant term in controlling microstructure and material properties. Kou describes the relationship between cooling rate and energy input using Rosenthal's three-dimensional solution [119]:

$$\partial T/\partial t \propto k(T - T_0)^2/(P/V) \tag{4.3}$$

where T is the temperature of the molten pool, t is time, k is the thermal conductivity of the material, T_0 is the substrate temperature, P is laser power, and V is traverse speed. The cooling rate is clearly inversely proportional to energy input. A relevant study by Griffith et al. [51] states that when energy input is lower (with low laser power or high traverse speed), cooling rate is high resulting in fast solidification of the molten pool. For these conditions, LENS®

processed H13 tool steel was found to have high yield strengths and ultimate tensile strengths. On the contrary, when energy input is higher (with high laser power or low traverse speed), high ductility or strain was observed for the lower cooling rate. Figure 4.7 demonstrates the effect of energy input (heat input) on mechanical properties. These observations may be because H13 tool steel is a heat treatable alloy, and the multiple excursions in the temperature as the materials are deposited result in microstructure variations with time and height of the build. It was also shown by Mazumder et al. that the microstructure of H13 varies with distance from the top surface [120, 121].

4.2.3. Powder Feed Rate

Powder feed rate is a process parameter that is used to control the amount of powder that is fed into the focal point of the laser beam. At too low of a powder feed rate, the molten pool can be smaller than required, and the actual layer thickness is smaller than the desired value. Consequently the working distance will increase with each successive layer until the deposition stops. The final height is then much smaller than expected, indicating that low powder feed rate produces the same effect as high traverse speed. In contrast, with a high powder feed rate, more powder is fed onto the surface of the previous layer, a larger and taller molten pool is generated and if it is much larger than the Z-height increment, the final specimen height will be higher. Defocusing of the laser beam can occur, having similar effects to high laser power, resulting in higher sample height. Paul et al. [122] claim that for single-track deposition with a given laser power and a given traverse speed, the deposition rate (corresponding to layer thickness) increases linearly with powder feed rate up to a critical value. When powder feed rate is too high, laser power is not adequate to heat or melt extra powder, resulting in imbedded un-melted particles and possible porosity within the microstructure.

4.2.4. Substrate Temperature

According to Eq. (4.3), cooling rate is proportional to the square of temperature difference in the molten pool and substrate. It is well known that high cooling rates result in high thermal stress, which can cause the formation of cracks when thermal stresses are larger than the material's fracture strength [124]. Liu and Dupont studied the effect of substrate temperature on the microstructure of LENS® deposited Ti alloys and composites [123]. It was observed that without heating the substrate, cracks usually originate from the edge or surface of the samples, and at the interface between the deposit and the substrate, as shown in figure 4.8. It was concluded that it would be a challenge to prevent crack formation only by manipulating the process parameters, such as laser power and traverse speed. However, preheating the substrate to a certain level can significantly reduce the cooling rate, which results in lower thermal stress. The influence of preheating the substrate can be clearly seen in figure 4.9 (white dots in the photographs are the un-melted TiC particles) when compared with that in figure 4.8.

Hofmeister et al. reported that more than 90% of the laser heat is conducted and dissipitated through the substrate [35]. Thus, the temperature of the substrate is also an important factor in controlling molten pool size in addition to controlling cracks in the deposited materials.

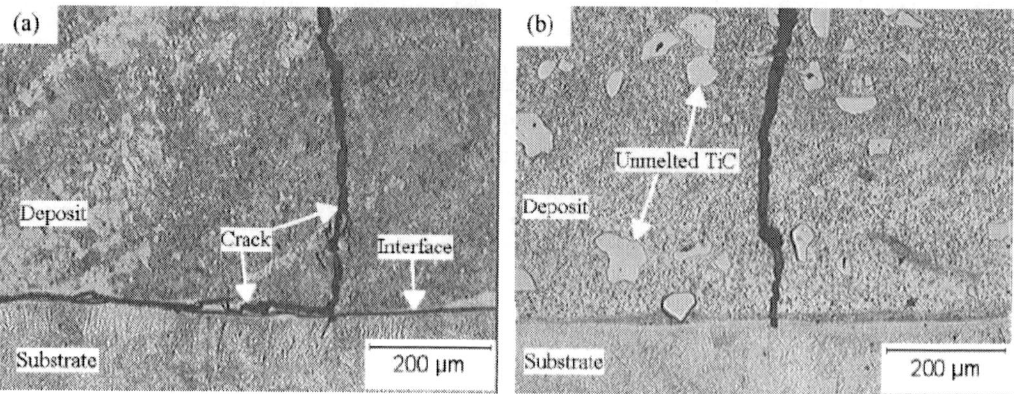

Figure 4.8. Cracks observed in (*a*) Ti-48Al-2Cr-2Nb (300 W, 8.5 mm/s) and (*b*) Ti-48Al-2Cr-2Nb1 20 vol%TiC (300 W, 16.9 mm/s) deposits [123].

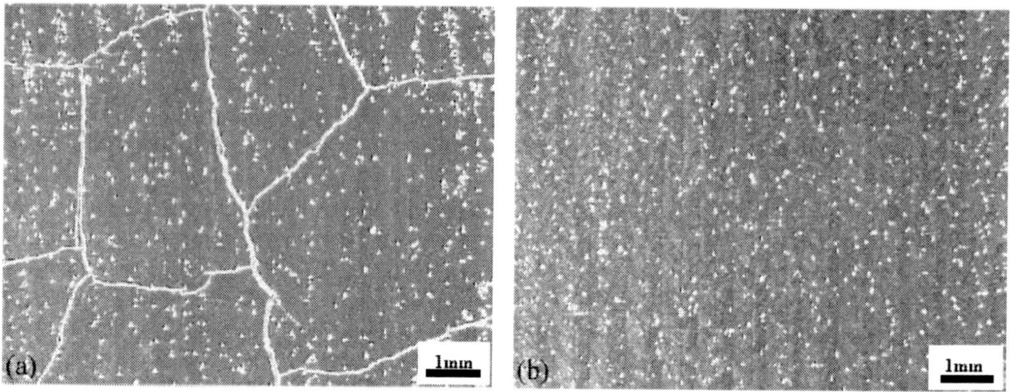

Figure 4.9. Effectiveness of preventing cracking in Ti-48Al-2Cr-2Nb1 TiC composite deposits: (*a*) without preheating, and (b) heating to 450 °C [123].

4.2.5. Working Distance

In contrast to processing parameters, such as laser power, traverse speed, powder feed rate and substrate temperature, the working distance is usually fixed before deposition begins, and is expected to stay constant during the LENS® process. However, in a recent study conducted by Xiong et al., it was found that variations in working distance can play an important role in microstructure improvement [125]. The effect of working distance and relative location of the focal plane to the deposition surface was studied for WC-Co cermet deposits made from nanostructured WC-Co powder, as discussed earlier in this chapter, and summarized below.

Dense thin wall specimens were produced using the LENS® technique by choosing proper deposition parameters. The working plane was 2 mm higher than the focal plane of the lens. An alternating layer microstructure was observed as shown in figure 3.11(a). Layers were deposited along the horizontal direction. One dark layer (consisting of fine WC particles) and one light layer (composed of large WC particles) form one deposition layer as indicated by the white arrows. Observations from figure 3.11(b) indicate grain growth is limited to a certain extent during laser deposition. Some grains remain in a size less than 100 nm, while some grow to a size in the submicron or micron regimes. Previous work reported that the large white particles are not WC single crystals, but polycrystals containing WC grains with a size ranging from tens of nanometers to one or two microns [34]. Figure 4.10 shows SEM images of a deposited specimen using similar process parameters (deposited at a laser power of 200 W and a traverse speed at 3 mm/s) to the one presented in figure 3.11. Only one change was made: the working plane was about 4 mm higher than the focal plane of the lens, which also means the working distance from the sample top to the nozzles is 2 mm shorter than the previous value used. Uniform microstructure without layer boundaries throughout the whole sample area was observed as shown in figure 4.10(a). Even more, particle coarsening was not significant, as shown in figure 4.10(b), which is believed to be related to the unique thermal behavior: a shorter working distance at a higher working plane (a defocus condition) results in lower laser intensity because of the increasing diameter of the laser beam.

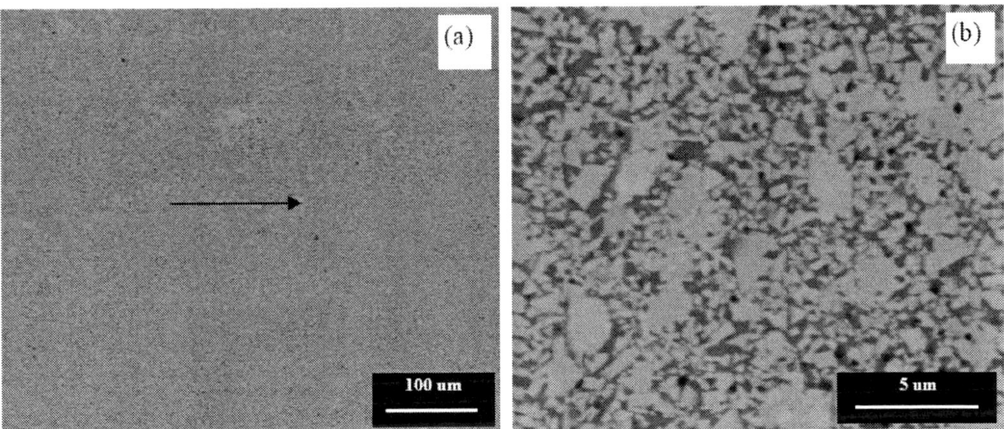

Figure 4.10. SEM images of a WC-Co specimen (a) at low magnification showing uniform microstructure (the black arrow indicates the horizontal deposition direction), and (b) at high magnification showing fine WC particles [125].

The effect of working distance on microstructure change can be explained as follows. When one layer is being deposited at the focal plane of the lens, the Co binder in the top portion of the previous layer will be re-melted resulting in coarsening of WC particles. However, the bottom portion of the previous layer will only be reheated, not re-melted, retaining the fine WC particles. As specimens are built up, microstructures with alternating layers are generated. With a shorter working distance at higher working plane, laser intensity is lower. So, less powder is deposited in the molten pool and the layer thickness is smaller. But, re-melting of the Co binder in the top portion of the previous layer can still take place when the working plane is not much higher than the focal plane. Thus, the microstructure with alternating layers will also appear as demonstrated in figure 3.11(a). To some extent, when the working plane is much higher than the focal plane, the layer thickness can be undetectably small. Due to the resulting low laser intensity, the effect of re-melting is not significant; reheating of fine particles for a few previous layers would occur instead.

Therefore, a homogeneous microstructure without much grain coarsening is produced, like the one shown in figure 4.10(a).

Other factors, such as the condition of the starting powder and possible chemical reactions during laser heating and melting, also affect the microstructure of the deposited material. Figure 4.11 shows a microstructure with spherical pores within a W-25Re alloy sample [118]. Residual gas or porosity in the feedstock powder or gaseous products from chemical reactions typically results in such spherical pores in the microstructure.

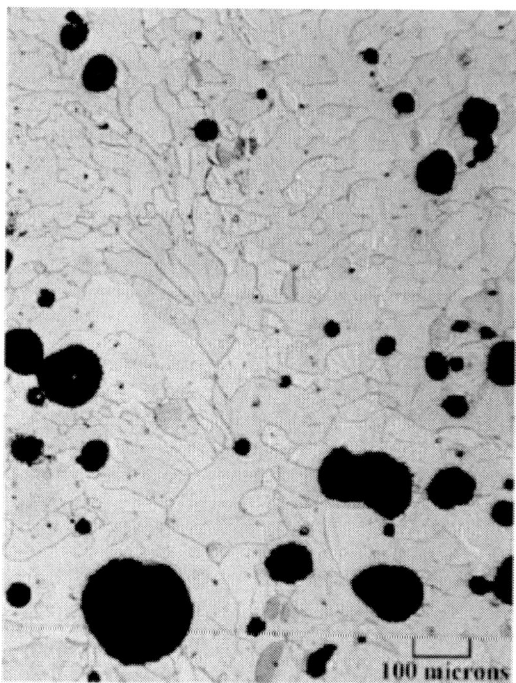

Figure 4.11. Porosity caused by residual gas in starting powders in blended W-25Re powder alloy [118].

4.3. Real Time Closed-Loop Control Systems

The challenge for LENS® process control is to simultaneously manage the dimensions and the properties. Close control of dimensions will result in substantial savings in post-machining for surface finish. The desired properties can be achieved and cost can be reduced through process control and the minimization of post-deposition heat treatment. During LENS® deposition, maintaining consistent molten pool size and deposition distance is critical, particularly for microstructural homogeneity and dimensional accuracy of the components. Accordingly, it is necessary to provide proper thermal management of the molten pool and deposited layer thickness during deposition. This can be achieved by two means: one is proper choice of process parameters of the traditional open-loop laser process, which relies heavily on an operator's expertise in order to establish a production process; another is via the implementation of external sensors and control strategies for closed-loop laser processing to insure that the molten pool size and deposited thickness are stable during the course of deposition [126]. This section reviews closed-loop control methods for laser direct deposition.

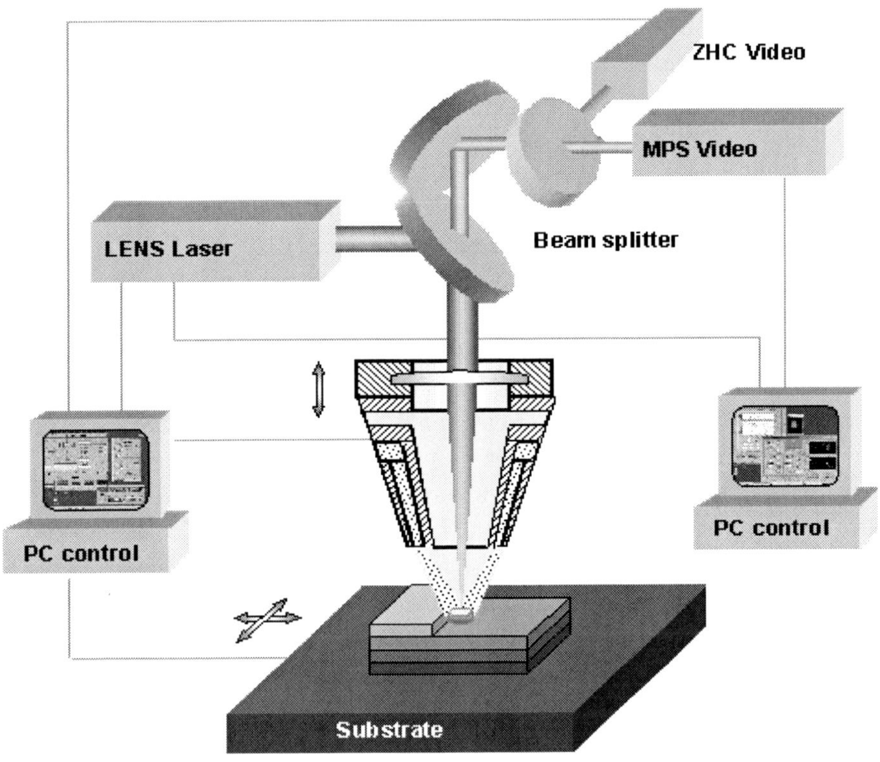

Figure 4.12. Sketch of *in-situ* molten pool sensor and Z-height control system applied to the LENS® process [55, 126].

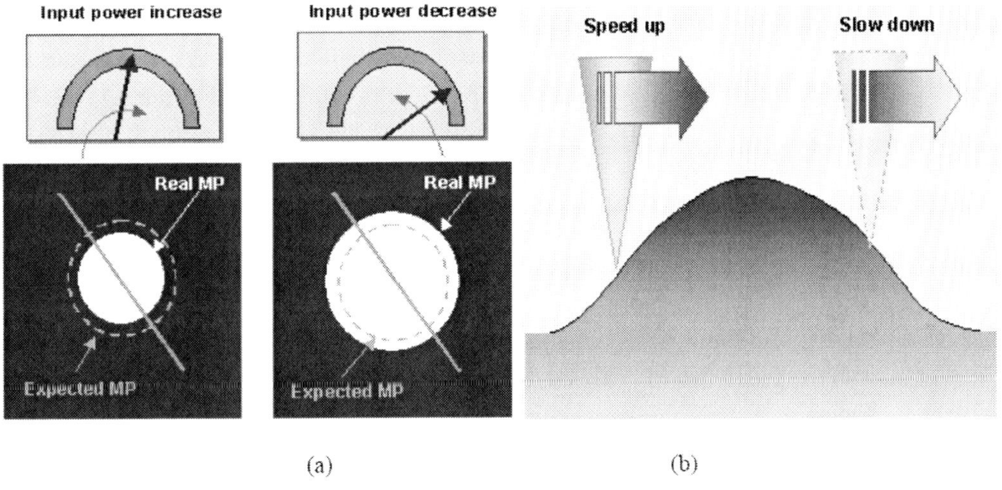

Figure 4.13. Sketch of *in-situ* (a) molten pool sensor control, and (b) Z-height control [126].

4.3.1. Molten Pool Size Control

In order to produce components with fine features, it is necessary to provide proper thermal management of the molten pool during deposition. Previous thermal modeling and experimental studies on LENS® deposition have shown that the temperature of a deposited component can increase during deposition due to heat accumulation [33, 51, 127, 128]. The molten pool size is proportional to the laser power input, and the overall deposition distance increases with decreasing traverse speed. Therefore, a real time molten pool sensor (MPS) and a Z-height control (ZHC) subsystems have been developed by Optomec and applied to the LENS® equipment to provide closed-loop control, as shown in figure 4.12 [55, 126].

The amount of heat already contained within the part plays an important role in the deposition, especially during a long build where the local temperature of the component can be significantly above room temperature [80]. The developed MPS subsystem is designed to automatically adjust the laser input power level to maintain a constant surface area of the molten pool, as shown in figure 4.13(a) [126], which is useful for the deposition of variable materials as well as for eliminating molten pool size variations that can occur due to geometrical changes in the component, thereby improving microstructure homogeneity. The MPS system consists of a camera for observing and detecting an infrared image of the molten pool, filters and attenuators, computer, thermal imaging software, and breakout board. It was found that laser power decreases as build height increases with MPS activated during the LENS® process. This fact suggests a constantly increasing temperature of deposited material as build height increases [126]. Control with the MPS closed-loop system gives consistent fusion zone areas with less variance, also improves dimensional accuracy of a deposited component relative to open loop process operation.

4.3.2. Z-height Control

In the LENS® process, the deposited dimensions in the X- and Y-directions can be precisely controlled, but variation in the process parameters (laser power, traverse speed, powder feed rate etc.) can cause height errors in the built part. Control of the deposit thickness or height is a critical issue since it impacts the quality of the product. Therefore, layer height control methods provide an important advancement to the process.

The Optomec ZHC subsystem is able to measure the height of the deposition surface coaxial to the laser beam delivery system to provide real-time control during LENS® deposition [55]. The light reflected from the deposition surface, detected through a series of optical filters and lenses, provides a relative distance from deposition surface to the final focusing lens. This signal can be used to control variables that ultimately control deposited height. The real-time ZHC subsystem is designed to automatically retain the same deposition distance by changing the laser travel speed, as shown in figure 4.13(b) [126]. The thickness of a deposit can be reduced by increasing the laser traverse speed, and vise versa. Controlling the deposited layer height with ZHC by changing the stage traverse speed consequently has minimal effects on the properties of deposited materials [80].

The basic function of the ZHC is to limit the height of metal deposition. Controlling Z-height by controlling laser shutters with multiple sensors had been developed and applied on a Direct Metal Deposition (DMD) system at the University of Michigan [120, 129], as shown in figure 4.14. The sample on the left in figure 4.14 was fabricated using the height controller. Visualization and measurement of the surface deformation of the molten pool is utilized by the reflective topography technique. The height controller has the ability to sense a difference in layer thickness during deposition and shuts off the laser until the area of excess build-up is passed. Multiple sensors were used for closed-loop feedback control of the deposit height. It was reported that the key benefit of applying multiple sensor height controllers is that the roughness of the surfaces can also be reduced on average 14–20%.

Figure 4.14. Example of fabrication with height controller (left) and no height controller (right) [120].

A non-feedback layer height controlling process for DLD has been developed at Liverpool University based on controlling the shape of the powder streams emitted from a four-port side feed nozzle [130]. This method limits the deposited layer height by causing a sharp reduction in catchment efficiency in the vertical plane at a fixed distance from the powder feed nozzles, and, as a result, is capable of depositing a consistent layer height in spite of power, powder flow, or process velocity variations.

General Electric Global Research Center (GE) has also been developing a molten pool sensor and Z-height control system for its Laser Net Shape Manufacturing (LNSM) technique [131, 132]. The Z-height closed-loop control developed at GE is based on dual vision feedback of deposition height. It was reported a side-view high speed digital charge-coupled device (CCD) camera was installed to detect both the substrate height in front of the molten pool and the clad height behind the molten pool in real-time [132]. The former is fed positively, while the latter is fed negatively into the process controller. Based on the dual height feedback, the laser scan velocity is adjusted in real-time to keep the clad height within the desired range.

5. Thermal Behavior during the LENS® Process

Although the LENS® process has been used to fabricate a broad range of materials successfully, to better explore the potential of the LENS® technique, there is a need to understand its thermal behavior and the relevant effects on material microstructure and properties. Basically, there are two experimental methods to implement this investigation: contact temperature sensor measurements, such as with thermocouples, and non-contact measurements, such as with a radiation pyrometer. In this section, experimental methods and the related numerical simulation results are introduced.

5.1. Thermal Imaging Method

Non-contact thermal imaging measurement, such as visible and near infrared (IR) radiation pyrometry, is required to obtain the temperature information of interest. Studies on the thermal behavior of the LENS® process using thermal imaging methods have been investigated within the past decade [33, 35, 36, 51, 133]. Two reports published by Hofmeister et al. [36] and Wei et al. [133] are summarized here.

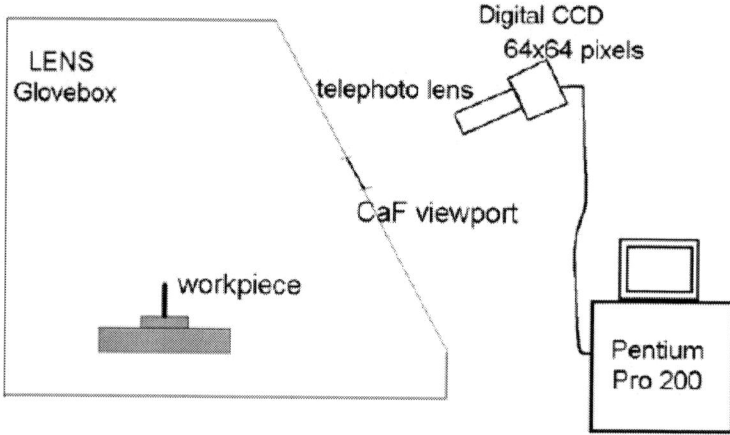

Figure 5.1. Experimental setup of the thermal imaging system [36].

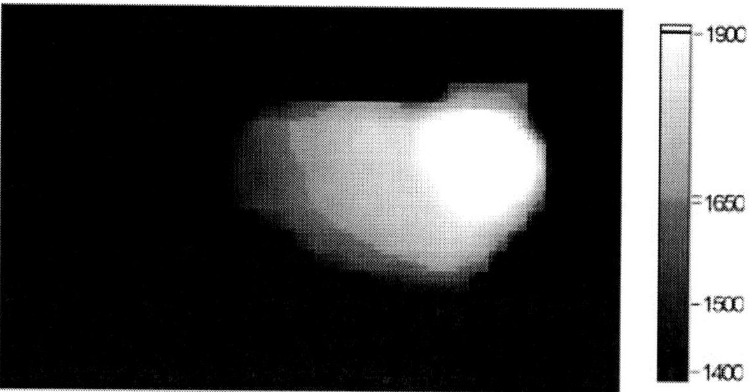

Figure 5.2. A typical thermal image of a 316L SS line build looking from the side [36].

A high-speed digital charge-coupled device (CCD) video camera was used to monitor the thermal images of a 316L SS line build during the LENS® process by Hofmeister et al. [36]. A telephoto lens and broad band pass filter centered at 650 nm were used in the image path. Figure 5.1 shows the experimental setup of the thermal imaging system. The image has been colorized to show temperature in degrees Kelvin. A CaF viewport placed in front of the

LENS® glove box was used for the thermal imaging camera to record the process. The whole system was calibrated for temperature measurements with a tungsten strip lamp radiant source. Figure 5.2 is a side view of a typical thermal image of a line build. Temperature distribution along the traverse direction for different laser power settings is demonstrated in figure 5.3. According to the solidification interface temperature of 316L SS at 1650 K, it can be found that the molten pool size increases with power up to 275 W. Higher power only increases the molten pool temperature. Figure 5.4 shows the calculated cooling rates as a function of laser power at the solid-liquid interface. These results reveal that the cooling rate is on the order of 10^3 K/s and the highest cooling rate can be obtained at the lowest power. In contrast, the cooling rate at higher power is lower resulting in a more coarsened microstructure due to grain growth.

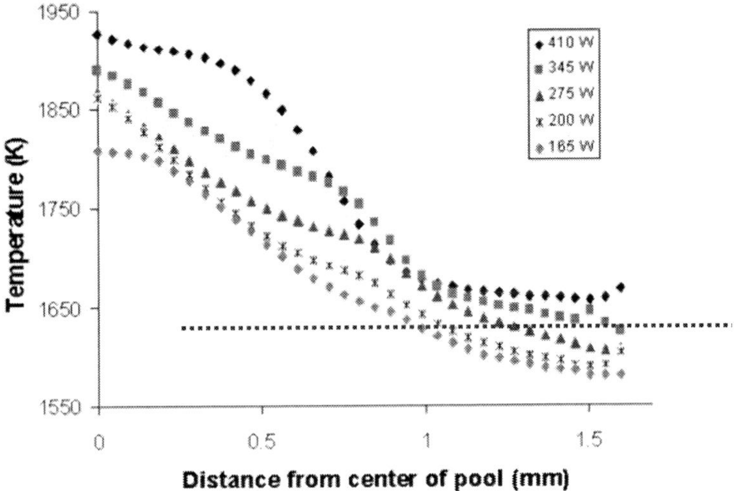

Figure 5.3. Thermal profiles from the center of the molten pool of 316L SS builds along the traverse direction for different laser power settings [36].

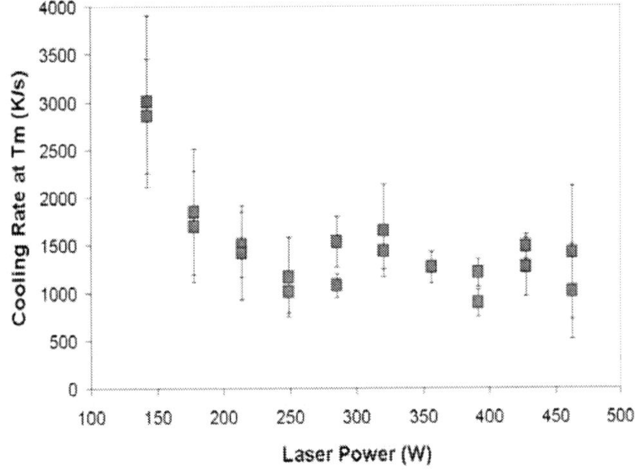

Figure 5.4. Cooling rates as a function of laser power at the solid-liquid interface of 316L SS builds [36].

Another type of experimental setup using a high-speed digital CCD video camera is shown in figure 5.5, with a top view of the molten pool [35]. Compared to the setup shown in figure 5.1, the camera in this experimental setup is stationary and has the same focal point as the laser. The camera is always in focus and views the molten pool regardless of the X, Y, and Z positions

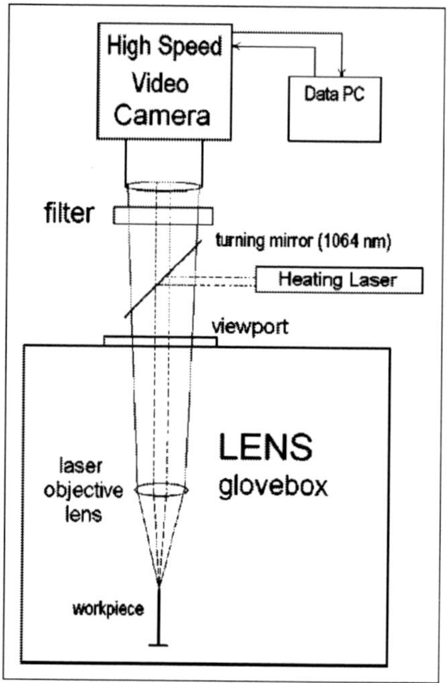

Figure 5.5. Experimental setup showing thermal measurement system positioned outside the LENS® system [35].

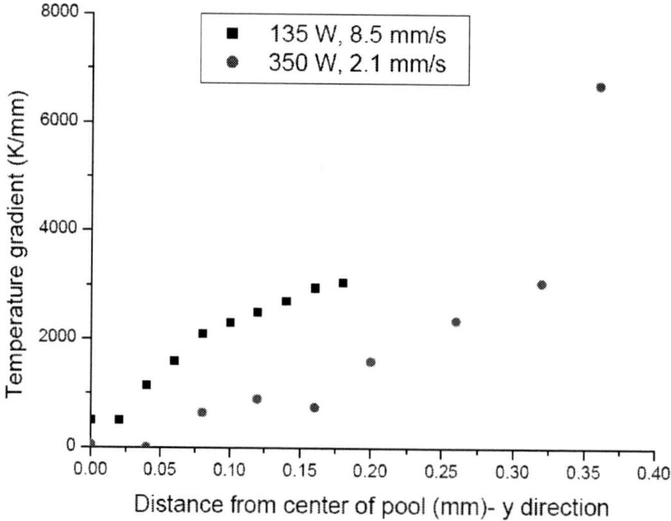

Figure 5.6. Temperature gradient from the center of the molten pool of 316L SS builds for different processing conditions [133].

Another thermal imaging system, a two-wavelength pyrometer has received considerable attention because it relies on the ratio of the relative intensities of radiation at two different wavelengths. Hence, this approach is independent of emissivity, thus providing the potential for a more accurate temperature measurement [133]. A Thermaviz™ two wavelength imaging pyrometer system was used to study the thermal behavior of LENS® processed 316L SS by Wei et al. [133]. The experimental setup was similar to the one shown in figure 5.1, except that there was not a CaF viewport between the laser and the pyrometer. Optical magnification provides a spatial resolution of 20 μm/pixel. The long band wavelength selected was from 650 to 750 nm, and the short band wavelength was from 600 to 650 nm, both of which fall into the transparent range of the laser safety shield. The imaging pyrometer records a dynamic range from 1500 K to 2500 K.

A standard tungsten filament source was used to calibrate the Thermaviz™ system. The temperature profile of a LENS® deposited 316L SS thin wall sample (11 mm long x 6.5 mm high x 0.25 mm wide) was recorded using this system. Laser power ranging from 135 W to 455 W was applied with various traverse speeds (2.1 to 8.5 mm/s). Temperature gradient distribution and cooling rate in the molten pool and the surrounding area were derived from the temperature profile, showing that the temperature gradient from the center of the pool is on the order of 10^2~10^3 K/mm, and the cooling rate is on the order of 10^2 ~10^4 K/s in the LENS® processed zone, as shown in figure 5.6 and figure 5.7 (the Y-direction is perpendicular to the traverse direction of the molten pool).

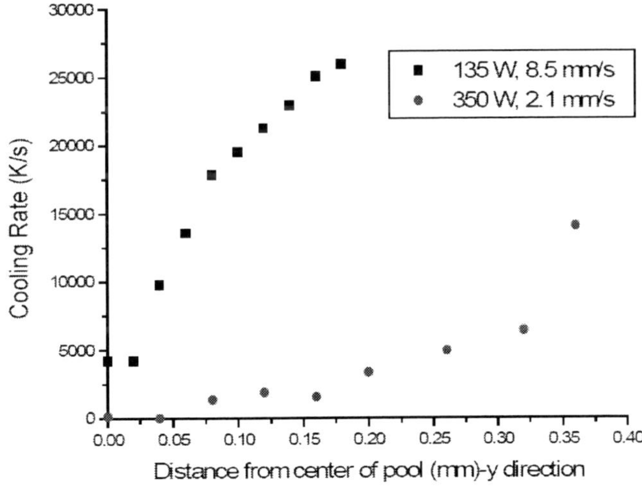

Figure 5.7. Cooling rate from the center of the molten pool of 316L SS builds for different processing conditions [133].

Both the high-speed digital CCD video camera and the two wavelength imaging pyrometer have been used to investigate the thermal behavior of a LENS® processed material in terms of temperature profile and cooling rate. Each offer unique advantages. The high-speed digital CCD video camera has a faster frame rate per second, which can show more detailed information. The two wavelength imaging pyrometer can provide more accurate temperature values, with a margin of error between ± 6 °C, due to its ability to avoid the effect of material emissivity. One limitation of thermal imaging methods is the inability to acquire the entire thermal history of deposited components, especially the temperature variation in the solidified materials.

5.2. Thermocouple Method

Thermocouple (TC) measurements of temperature provide a direct and efficient means of measuring temperature. Inserting TCs into the sample during LENS® fabrication has been used to obtain the thermal history [33, 51, 53]. Previous work by both Griffith et al. [33] and Brooks et al. [53] investigated the deposition of H13 tool steels using similar approaches with comparable results. Figure 5.8 shows the results of TC temperature readings obtained by Griffith et al. using a single TC and process parameters of 200 W laser power and 6 mm/s traverse speed depositing single width shell boxes with equal side lengths of 6.35 cm. Fine type-C thermocouple wire of 10 μm diameter was used for the measurements to ensure no reaction during deposition. Each peak represents the TC response as the laser passed over the TC from original insertion to successive layer depositions, with approximately 42 seconds between passes. After the initial temperature peak, heat is quickly dissipated to a nominal temperature of 150°C. This solidification process would characteristically produce a martensitic microstructure with high hardness. Nevertheless, successive passes of the laser reheats the material, which directly affects the properties of deposited materials, including residual stresses and mechanical properties due to tempering or aging effects [33].

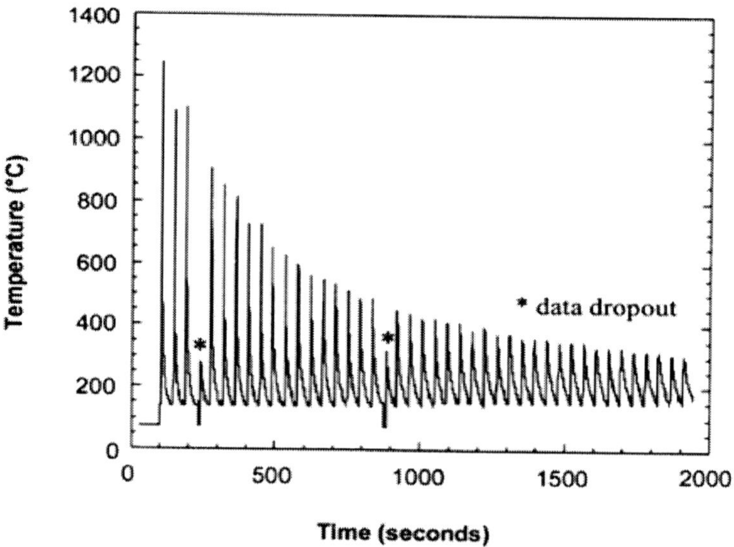

Figure 5.8. Thermocouple temperature readings vs. time during fabrication of H13 tool steel [33].

On the other hand, Brooks et al. inserted a series of 0.13 mm diameter TCs, collecting data at 100 Hz, with varying laser power and traverse velocity during LENS® processing. As in the Griffith et al. work, 6.35 cm, single line, square shells were built. Figures 5.9(a) and 5.9(b) show the experimental set-up and TC temperature readings, respectively [53]. Due to the layer additive nature of the LENS® fabrication process, a complex thermal history was experienced in different regions of the deposited material. The thermal histories associated with the LENS® process involves at least partial re-melting of the previous layer, and numerous lower temperature reheating cycles with each successive layer deposited. The complex thermal behavior during the LENS® process results in complex microstructure evolution. Both Griffith et al. and Brooks et al. correlated their TC data to changes in microstructure that derive from the cyclic thermal excursions above 600°C resulting in improved mechanical properties. In the case of H13 tool steel, it was seen that two distinct microstructure features were observed: one in the near surface layers; the other in the heart of the deposited samples. While no detailed report of the microstructure variation as a function of layer location was done, the thermal cycles suggest a continuously varying amount of second phases exists in the layers between the ones reheated to above 600 °C and those that have their most recent reheating to the steady state temperature of 150 °C.

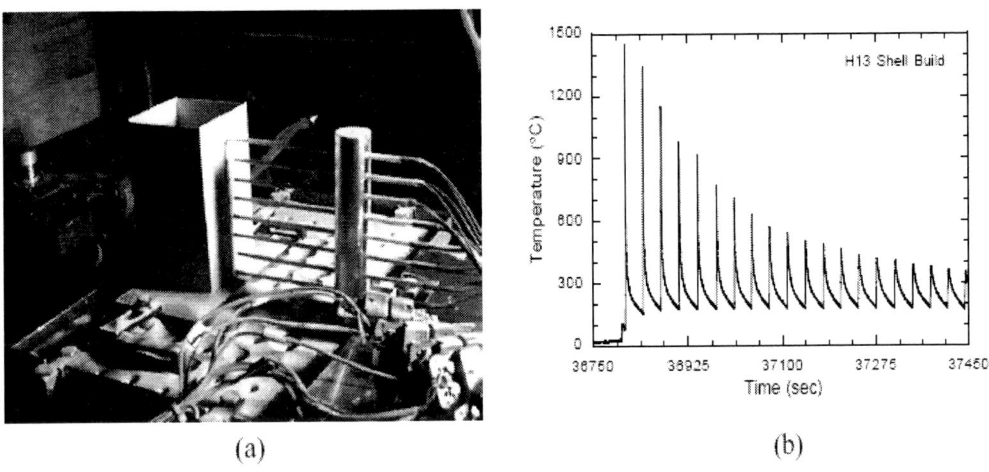

Figure 5.9. (a) Experimental setup with a series of thermocouples (TCs), and (b) TC temperature reading vs. time during fabrication of H13 tool steel [53].

5.3. Numerical Simulation

Because deposition during the LENS® process is a rapid solidification process, it occurs under non-equilibrium conditions and is completed in a few milliseconds. Recording temperature variation over such short times is a challenge with the thermocouple method. Another limitation of the thermocouple method is the inability to acquire the temperature information of the molten pool and its surrounding area during LENS® processing. The thermal imaging method has a much faster response rate. However, the thermal area that a thermal imaging system can record has a limited range, mainly focusing on the area in and around the molten pool. Thermal behavior outside the molten pool is also critical to understand the development of the microstructure, properties, and residual stress of the entire part. Accordingly, numerical simulation, such as finite element method (FEM) and finite difference method (FDM), can be implemented as an effective approach for investigating the thermal history during LENS® deposition, and can give information on the entire thermal behavior. A number of studies carried out by several authors have simulated thermal behavior during the LENS® process [36, 52, 127, 134-136].

An initial attempt with FEM was done by Hofmeister et al. on a 2D 316L SS thin wall part [36]. An eight-node solid cubic element with 1.1 mm on each side was modeled. The modeled thin wall had a size of 25.4 mm wide × 76.2 mm high with the element size as the thickness. In this study, the substrate temperature was fixed at 573 K, the initial temperatures of the eight nodes were set at 1650 K to 1900 K, and only thermal conduction was considered. No detailed results were shown in this study. However, it briefly concluded that the thermal gradients are steep near the molten pool, and reach a steady-state condition in areas far away from the molten pool.

A detailed 2D finite element thermal model was developed by Wang and Felicelli to calculate the temperature distribution in 316L SS plates during LENS® deposition [135]. Numerical simulations were performed on the upper region of a plate, which agree well with measured thermal profiles done by Hofmeister et al. [35, 36]. It was reported that the molten pool temperature contour could reach a steady-state condition when the laser beam moves to about 5 mm away from the part edges in each layer. The parameters of laser output power and thermal conductivity of deposited materials as well as its variation with temperature play an important role in the thermal profiles and the molten pool size.

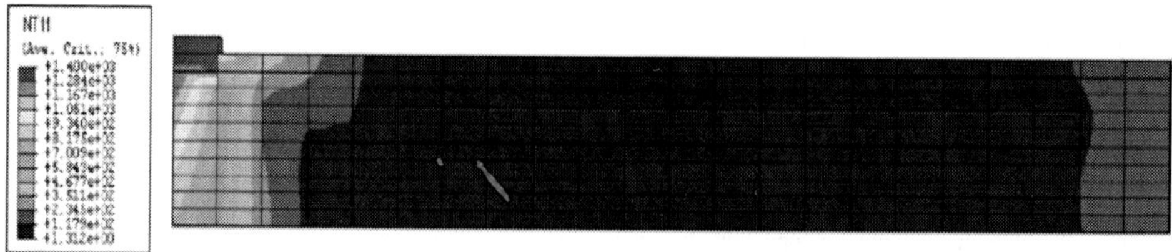

Figure 5.10. Finite element results for temperature distribution, in degrees Celsius, near the molten pool (when the wall is formed to 11 mm long (X-direction) and 1.3 mm tall (Z-direction)) [136].

The FEM software, ABAQUS, was employed by Ye et al. to calculate the temperature field for a 316L SS thin wall component [136]. A 3D eight-node solid element was chosen to discretize the geometric model. The element had a size of 0.5 mm long × 0.13 mm high × 0.25 mm wide. The height of the element was equal to the layer thickness of the thin wall (11 mm wide × 6.5 mm high) being built. They assumed the temperatures of the element nodes where the laser beam is focused are the same as the melting temperature of the 316L SS. Figure 5.10 shows the numerical results of the temperature distribution for a side view image in the thin wall when the wall is formed to 1.3 mm high (temperature is in units of Celsius). The model predicts that the temperatures drop sharply in the area just outside the molten pool. As the distance from the molten pool increases, the temperature decreases gradually. The calculated numerical results reveal that the highest temperature gradient around the molten pool is 588 K/mm, which is within the range of experimental measurements of 400 to 750 K/mm, taken at the center and obtained by using thermal imaging methods [35, 133]. It was concluded that numerical simulations have the potential to provide information on the entire thermal behavior, especially to provide insight into the temperature in the area outside the field of view for a pyrometer.

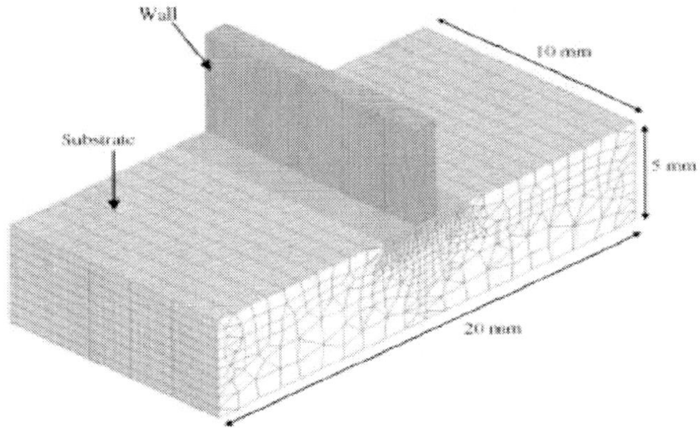

Figure 5.11. Finite element mesh and geometry to simulate the LENS® process for a 10-layer plate [134].

Another 3D FEM was performed by Wang et al. on a 410L SS thin wall part with 10 deposition layers to simulate the LENS® process using the SYSWELD software tool, which takes into account temperature dependent material properties and phase transformations [134]. Dense meshes were applied to areas where higher thermal gradients are expected, such as the area between substrate and initial deposited material, as seen in figure 5.11.

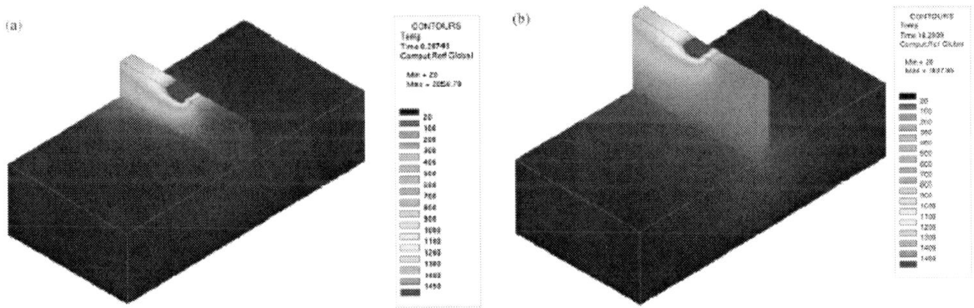

Figure 5.12. Three dimensional temperature distribution: (a) 5th layer, (b) 10th layer [134].

In this model, a fixed mesh for the wall and substrate with elements initially inactive and then activated during deposition was applied. The substrate was also 410L SS having an initial temperature of 293 K. This study also employed a dummy material method, using three different types of materials for activation. Laser energy transfer efficiency of the Gaussian profile of heat flux produced by a moving heat source was considered in this model, while phase transition and Marangoni convection were not. Results from the model were compared to the experimental results. Figures 5.12(a) and 5.12(b) show the three dimensional temperature distributions as the laser beam moves to the center of the fifth layer, and tenth layer, respectively.

Figure 5.13 shows the calculated thermal cycles during deposition, which are comparable to experimental results on H13 tool steel by Griffith et al. [33]. Although, the maximum temperature was approximately 2373 K after deposition of the first layer, cooling was down to a nominal 373 K between depositions, and melting occurred to a depth of about one and a half layers for each pass. Further inferences from the model are described, such as cooling of parts after deposition of each layer and cyclic reheat cycles affecting the residual stresses and mechanical properties.

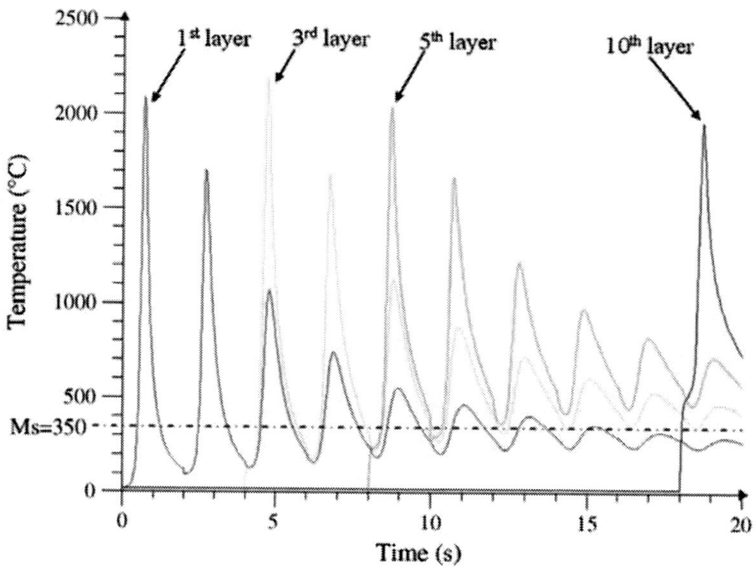

Figure 5.13. Thermal cycles for the midpoints of layers 1, 3, 5, and 10 of the deposition profile as a function of time [134].

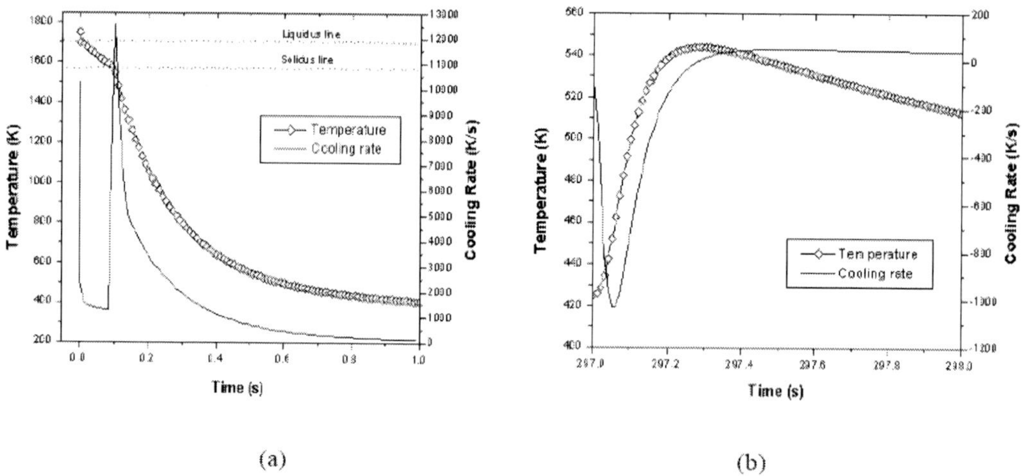

Figure 5.14. Variation in temperature and cooling rate during the early stage of depositing (a) the 1st layer and (b) the 10th layer [127].

The alternate-direction explicit (ADE) finite difference method (FDM) was used by Zheng et al. to simulate the thermal behavior during the LENS® process [52, 127]. Figure 5.14 shows the variations in temperature and cooling rate versus time during the early stages of the first layer deposition and the tenth layer deposition, respectively. Four different cooling stages can be seen in the early thermal cycles, as shown in figure 5.14(a): liquid phase cooling prior to the liquidus line; two phases of liquid and solid cooling while the temperature is in between the liquidus temperature and the solidus temperature; rapid solid state cooling below the solidus temperature, and a slow solid-state cooling stage. The temperature decreases slowly within the liquid–solid zone due to the release of latent heat during solidification. Once the deposited layer becomes fully solid, the temperature decreases rapidly. As a result, a sharp peak appears on the cooling rate–time curve near the solidus temperature. Therefore, the maximum values are related to the complete solidification of the material at a local level. As more layers are deposited, the temperature changes become less pronounced, and the quench effects decrease or even disappear, as shown in figure 5.14(b). During laser deposition, the thermal excursions dampen out when either the thermal energy source moves away from the substrate during deposition of a layer or subsequent layers are deposited. The high cooling rate results from a very small molten pool in the substrate or deposited material, which behaves as a very large heat sink.

Numerical simulations for 3D parts, such as thick wall parts or parts with complex 3D geometries, have not been reported thus far. Modeling of 3D shaped parts has the potential to provide more information, such as thermal management, dimensional accuracy and accumulated thermal stresses. This is an area of special interest and needs further investigation.

6. Benefit Analysis for LENS® Processing

Since its development in the 1990's, the LENS® technique has proven to have the ability to produce engineering parts into near net shape with little or no post fabrication operations. This process has been used to fabricate a broad range of materials with improved physical and mechanical material properties, as described in the previous sections of this chapter. The rapid prototyping features and accurate shape control capability make LENS® a unique technology compared to other powder metallurgy (P/M) technologies [137], which basically include pressing, sintering and post machining multiple steps with specific equipment and tools for each step. This single step process has the potential to provide benefits such as increased manufacturing productivity and reduced fabrication costs when compared to P/M methods. Quantitative studies on this topic are very limited, and yet very important. In this section, WC-Co cermets have been chosen as a reference to compare the environmental impacts and costs during the fabrication stage between the LENS® process and the conventional P/M process. The environmental impacts of the two processes are compared by using Life Cycle Assessment (LCA). The fabrication costs are evaluated with a Technical Cost Modeling (TCM) approach.

LCA is a useful tool to estimate the environmental impacts during the life cycle of a product. It has been successfully applied to compare environmental assessment of products/processes and to improve both pollution prevention and quality of goods [139-143]. Xiong et al. conducted an LCA on WC-Co cermet fabrication [138]. In this study, an 11 cm^3 WC-12 wt% Co die was selected as the functional unit, comparing the LENS® process followed by fine grinding with the conventional P/M process followed by rough/fine grinding. Specific process parameters for these two processes were assumed for this study (table 6.1). The relevant impact categories were limited to renewable resource use, nonrenewable resource use, energy use, global warming, acidification, photochemical smog, particulate matter, and landfill use. Table 6.2 shows the characterization scores for each impact category. It can be seen that, with the current process parameters and functional unit, the LENS® process has more benefits in landfill use saving because of its low material scrap rate. Benefits from nonrenewable resource (WC-Co) savings and one renewable resource (argon) saving are also significant. The P/M process consumes less energy and therefore, results in another renewable resource (water) saving and less air pollution impacts. However, it is clear that the benefits of the P/M process relative to these environmental impacts are not significant.

Table 6.1. Process parameters for the LENS® and the P/M processes according to the specific functional unit [138]

LENS®		P/M	
Parameters	Setting	Parameters	Setting
Equipment power	5.5 kW	CIP power	1.0 kW
Flow of argon gas [a]	1.0 l/min	CIP time	0.2 hr
Deposition time	3.3 hr	Sintering power	3.0 kW
Material scrap rate [b]	0.10	Flow of argon gas [a]	2.0 l/min
		Sintering time	4.5 hr
		Grinding power	2.0 kW
		Grinding time	1.4 hr
		Material scrap rate [b]	0.35

[a] Gas condition is 20 °C and 1 atm.
[b] Material scrap rate represents the ratio of waste to input resource.

Table 6.2. Characterization scores for the LENS® and the P/M processes (low values preferred) [138]

Inventory	Impact categories	Unit	Characterization score	
			LENS®	P/M
Water	Renewable resource use 1	l	52	47
Argon gas	Renewable resource use 2	l	198	540
WC and Co	Nonrenewable resource use	g	175	242
Electricity	Energy use	kWh	18	16
CO_2	Global warming	g	5256	4756
NO_X and SO_2	Acidification	g	4.75	4.30
VOCs	Photochemical smog	g	0.34	0.31
Particle	Particulate matter	g	0.24	0.21
Solid waste	Landfill use	cm^3	1.22	5.92

After applying weighting factors, final evaluation scores were determined, as shown in Figure 6.1, indicating a lower score of 0.84 for the LENS® process and a higher score of 0.93 for the P/M process (lower scores are preferred). The evaluation results demonstrate that, given the assumptions that have been made, the P/M process has, on a relative basis, 11% more impact on the environment than the LENS® process. However, due to the uncertainty in process parameters and functional unit, sensitivity analyses were also conducted. It was concluded that the LENS® process is more preferable if powder recycling for the LENS® process is applied, part size is small, or part shape is complex (such as shapes with hollow and internal structures). Conversely, the P/M process is more suitable to produce large parts of simple shape.

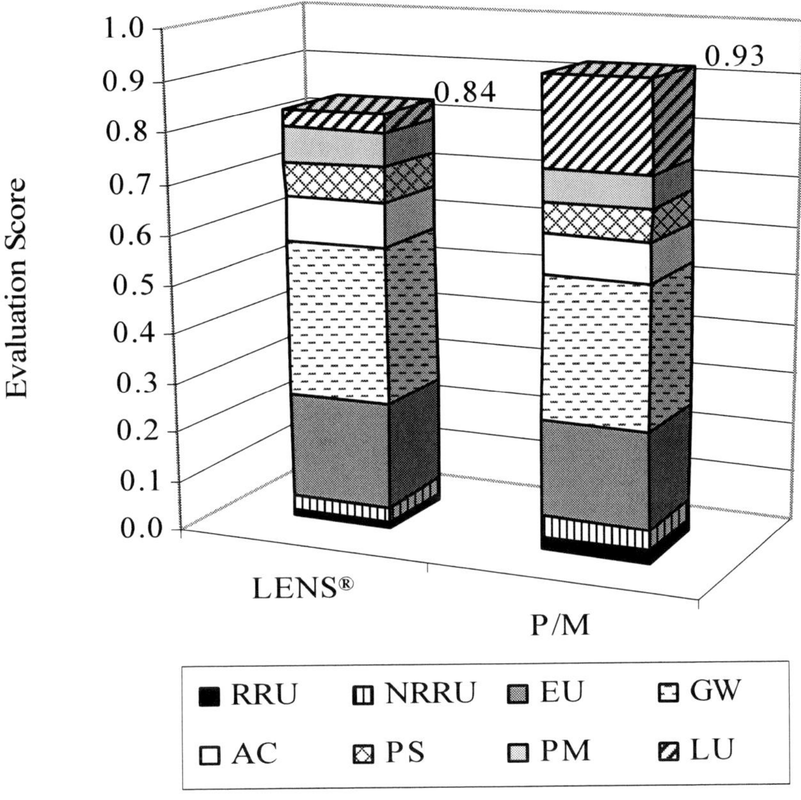

Figure 6.1. Evaluation score for the LENS® and the P/M processes: low values indicate less environmental impact. Legend: RRU (renewable resource use), NRRU (nonrenewable resource use), EU (energy use), GW (global warming), AC (acidification), PS (photochemical smog), PM (particulate matter), LU (landfill use) [138].

Schoenung et al. [144] applied technical cost modeling to quantitatively estimate the manufacturing costs. The same functional unit as for the LCA study, an 11 cm^3 WC-12 wt% Co die, was selected. Two fabrication processes were chosen as the alternatives: the LENS® process followed by fine grinding and the conventional P/M process followed by rough/fine grinding.

Figures 6.2 and 6.3 show the key results of this study. The total cost for an 11 cm^3 WC-12 wt.% Co die is calculated as $152 for the LENS® process followed by fine grinding. It can be seen from figure 6.2 that the main machine (machine used for each step) and material costs are the cost drivers among the cost elements. The cost of laser deposition is about 13 times the cost of fine grinding revealing that the LENS® processed sample requires only a short grinding time. This is because the LENS® process has the advantage of producing near-net shape products, thus reducing grinding costs. Seen from figure 6.3, for the P/M process, material and labor costs are the cost drivers among the cost elements. Costs for the three sequential steps are relatively equal: pressing, sintering and grinding. The total cost for the same product made by the P/M process is estimated to be \$145, which represents approximately a 5% savings compared to the LENS® process. However, the difference in total costs is within the uncertainty range for this cost modeling effort. It was concluded that the manufacturing costs of the specific WC-Co product made by the two processes are essentially equivalent. However, the total costs are sensitive to uncertainties, such as process parameters and functional unit. Similar to the LCA study, sensitivity studies on these uncertainties indicate that the LENS® process is more preferable if powder recycling for the LENS® process is applied, part size is small, or part shape is complex. Equipment type and costs also affect the final manufacturing cost to a certain extent.

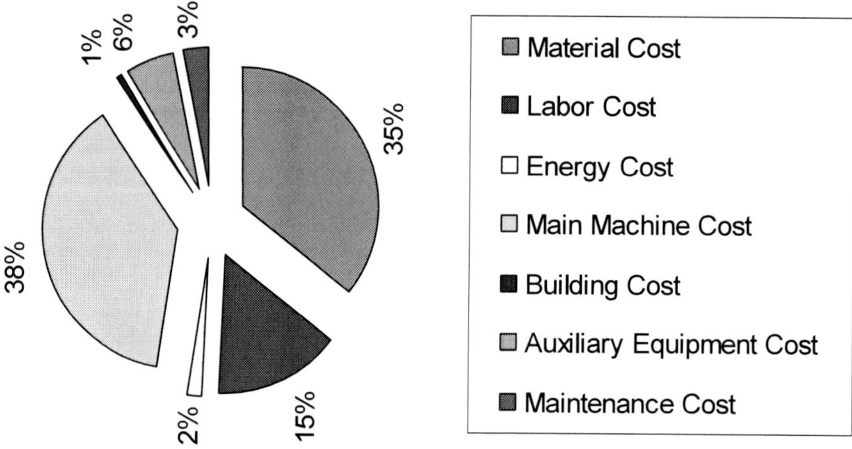

(a) Distribution of cost by cost element.

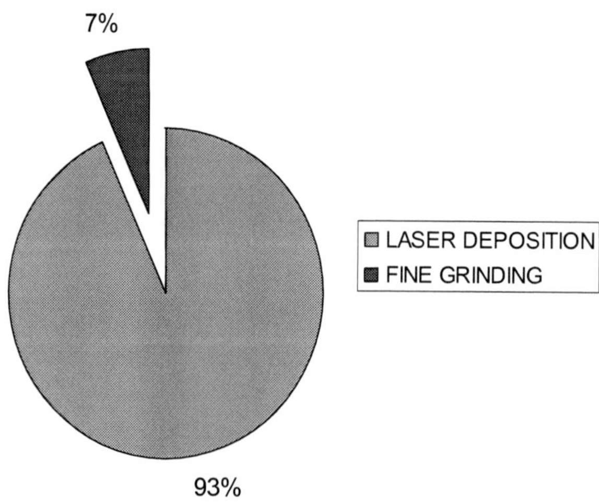

(b) Distribution of cost by process step.

Figure 6.2. Distribution of manufacturing costs for an 11 cm^3 WC-Co die made by the LENS® process followed by fine grinding; total cost is \$152 [144].

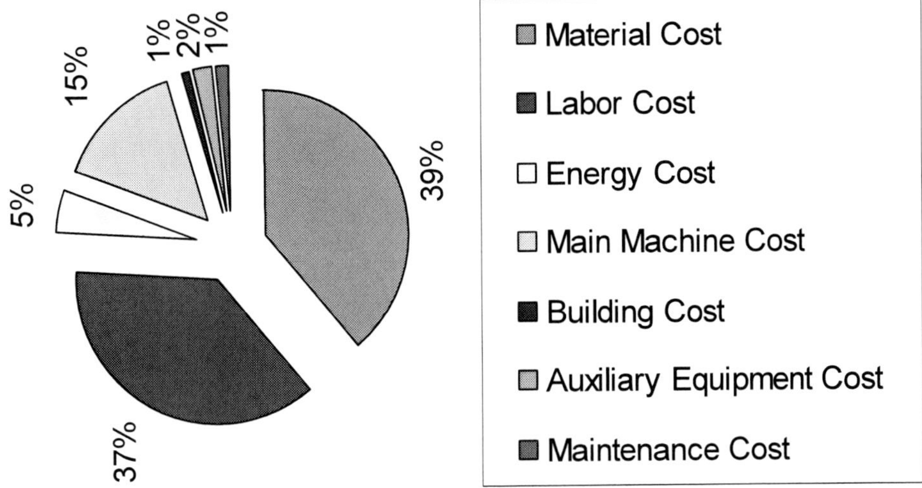

(a) Distribution of cost by cost element.

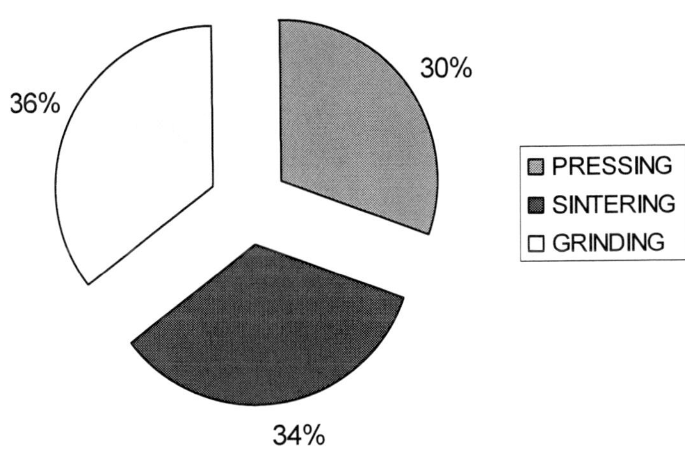

(b) Distribution of cost by process step.

Figure 6.3. Distribution of manufacturing costs for an 11 cm^3 WC-Co die made by the P/M process followed by rough/fine grinding; total cost is $145 [144].

The LENS® manufacturer, Optomec, also compared the costs between the LENS® process and forging process for medical devices, such as hip stems [50]. A standard hip stem usually costs $90 when forged in volume. A hip stem for custom versions are machined at $200. However, the estimated cost for the same part produced in volume by the LENS® process is $60. The LENS® technique is more cost-effective in the application of repairing high-performance metal components instead of making new parts. It can precisely add material to worn or damaged areas with minimal heat effect, and repair sensitive thin wall shaped components, such as gas turbine engine parts. Cost saving benefits will be further enhanced with further improvements and continuous development of the LENS® technology.

7. TRENDS AND CHALLENGES

Some of the trends and challenges associated with continued innovation in LENS® deposition can be classified in the following distinct areas: metallic glasses, nanocrystalline materials (i.e., with grain sizes in the 10-100's of nm), particle-pool interactions, and residual stress and porosity.

7.1. Metallic Glasses

Development of a series of new bulk metallic glasses (MGs) with multicomponent chemistry and high glass forming ability (GFA) have been developed in Zr-, Mg-, La-, Pd-, Ti-, Al- and Fe-based alloy systems [145-149] using various rapid solidification techniques. In turn, the properties of these bulk MGs, which are usually superior to the properties of their crystalline counterparts, have stimulated researchers to explore techniques to fabricate net shaped bulk MG components. Compared with most other MGs, such as Zr-based and Pd-based MGs, Fe-based MGs cost less in conjuction with having excellent mechanical and physical properties [150]. The major obstacle to form Fe-based MGs has typically been their limited GFA, although some progress has been documented in the literature [150-153].

Gas atomization (GA) is a practical and effective approach to producing MG powder. However, the gas-atomized MG powder cannot be used directly as a structural component. Accordingly, gas-atomized powder is generally processed through a series of conventional powder metallurgy steps, such as degassing, cold isostatic pressing (CIP) or hot isostatic pressing (HIP), and extrusion to form dense bulk materials, which are then machined into engineering components.

As discussed above, the LENS® process can produce relatively high localized cooling rates due to the very small size of the molten pool and the conduction of thermal energy into the substrate. The material that is initially deposited on the cold substrate during LENS® experiences a high quenching effect, and it is in this region that an amorphous microstructure is likely to be observed.

In related studies it was reported that gas-atomized $Fe_{58}Cr_{15}Mn_2B_{16}C_4Mo_2Si_1W_1Zr_1$ (at.%) powder was used for LENS® deposition of bulk MG components [55, 154]. Figure 7.1 shows the microstructure evolution in a laser deposited Fe-based MG shell component, processed with a laser output power of 280 W and a traverse speed of 12.7 mm/s. The microstructure of the initial two layers of the deposit, as shown in figure 7.1(d) is featureless, indicating that the cooling rate experienced by the deposited materials was high enough to form an amorphous microstructure. As the thickness of the deposited material increased, as shown in figures 7.1(c), (b) and (a) respectively, the microstructure became coarser and precipitation became increasingly evident, particularly at the boundary of the molten pool region. These results confirm that the cooling rate decreases as the number of layers deposited increases. In addition, the heat generated by subsequently deposited layers can promote crystallization of the preceding layers.

Nanocrystalline α-Fe with a size of 10-100 nm precipitated and embedded in an amorphous matrix were observed via TEM as shown in figure 7.2 [55]. With increasing deposit thickness, new crystal phases precipitated, and the microstructure coarsened due to the imposition of reheating cycles in combination with a decreasing cooling rate. The microhardness of the laser deposited Fe-based MG materials was approximately 900 HV (9.52 GPa), which is much higher than other conventional steels [154].

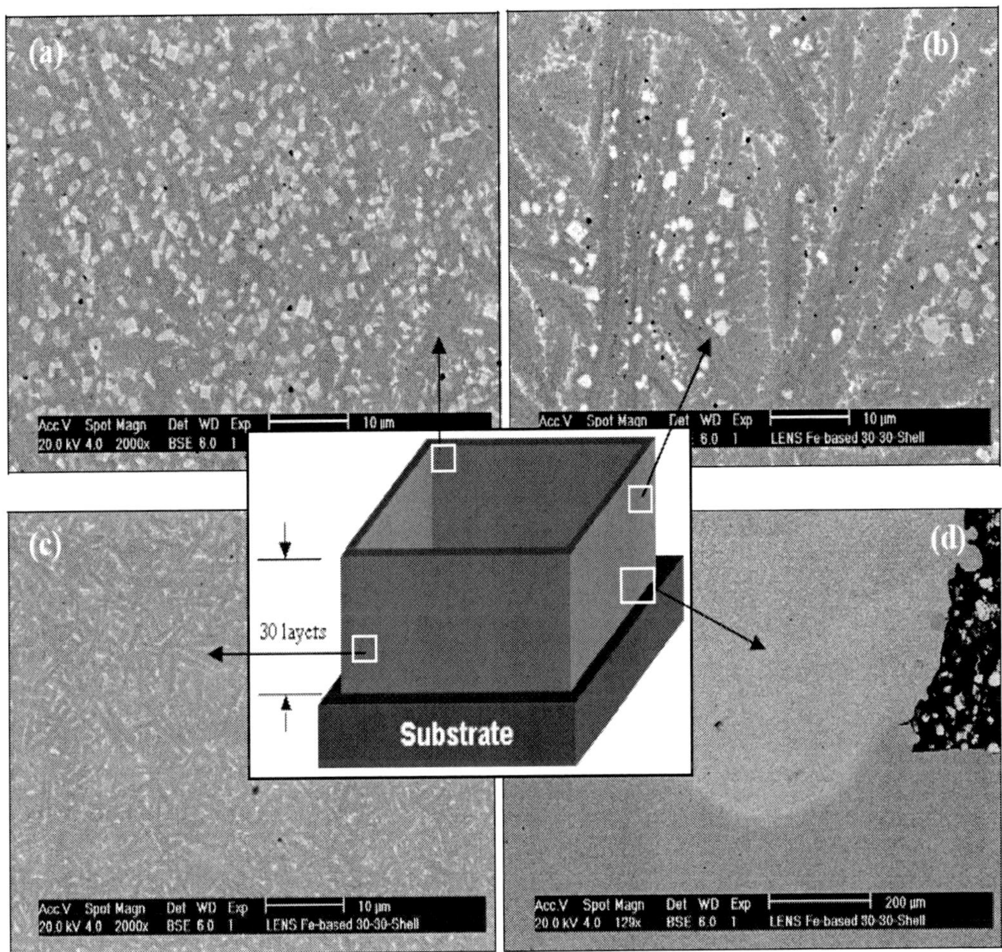

Figure 7.1. SEM (BSE) micrographs of different position of LENS® deposited Fe-based metallic glass shell component [154].

These preliminary research results on the LENS® deposited MG components indicate the flexibility of LENS® in processing novel materials, even those in which the thermal and solidification conditions must meet certain conditions. In order for the LENS® process to mature into a prominent manufacturing technology, particularly for large-sized MG components, the following difficulties and hurdles are likely to be encountered: epitaxial nucleation and growth of crystalline phases from the compatible interface with the crystal substrate; recrystallization or nucleation of crystalline phases at the overlap regions between adjacent laser tracks due to thermal activation; and the possibility of diffusion-controlled partial crystallization from the melt or adjacent amorphous regions due to a relatively longer interaction time necessary for a continuous wave (CW) laser compared to a pulsed laser.

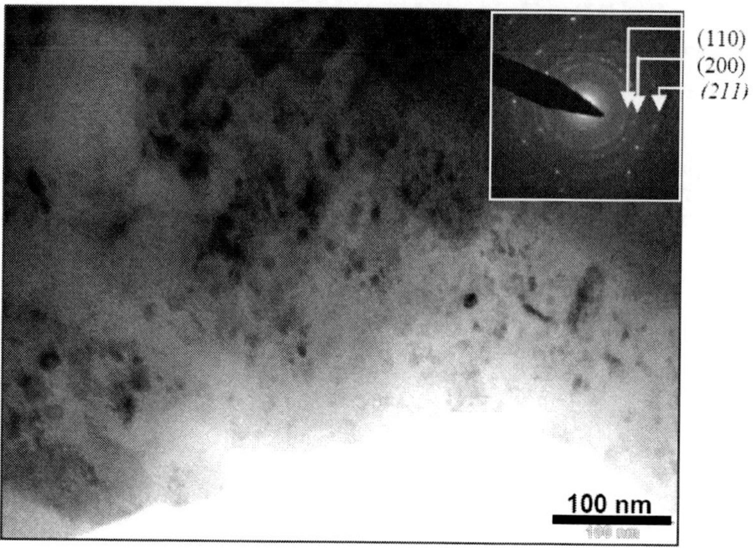

Figure 7.2. TEM micrograph and its diffraction pattern of the initial layers of laser deposited Fe-based metallic glass [55].

7.2. Nanocrystalline Materials

Nanocrystalline or nanostructured materials, with grain size less than 100 nm, are of particular technological interest because they exhibit different properties compared to those with microcrystalline features, such as optical, electrical, thermodynamic and mechanical properties. Nanotechnology has been widely applied in optical, electronic, and biomedical fields. As mentioned before, the LENS® technique has advantages in rapid prototyping and manufacturing, proper shape control, and high cooling rate. High cooling rate is important for maintaining fine grain size or controlling grain growth during laser melting and heating. Metallic alloys and composites with grains in the nanoscale have superior mechanical properties, such as high strength and high hardness. Therefore, there is a promising trend for the LENS® technique to produce nanostructured metallic alloys and composites.

To date, however, very few studies have been reported on the application of the LENS® process to nanocrystalline alloys and composites. Cryomilled nanostructured 316L SS powder was used to fabricate thin wall parts by the LENS® process [155]. The results of this study show that a sub-micron grain structure was ultimately achieved (figure 7.3). Apparently, grain growth occurred during deposition; however, this grain structure is still much finer compared to those made by other conventional methods. It is believed that fast cooling rate and rapid solidification during the LENS® process is responsible for the frequently reported fine grain structure. Nanostructured WC-Co powder was applied to produce thick-wall shape cermets using the LENS® process [34]. The SEM images shown in Section 4.2.5, figure 4.10, reveal that some WC grains remain within the nanoscale, and some grow to a size in the submicron or micron regimes after laser deposition. Particle coarsening was not significant, which results from proper selection of process parameters. Bulk parts consisting of nanocrystals throughout the bulk part, made by the LENS® process, have not yet been reported. Challenges associated with this might be: maintaining nanocrystals in bulk samples requires high cooling rate and short heating time during laser deposition; however, high cooling rate may generate potential cracks along sample edges, and between the sample and its substrate; although deposition of one layer only takes a few seconds or a couple of minutes depending on the part dimensions, the layer additive feature of the LENS® process causes re-melting or reheating of the deposited layers during the deposition of subsequent layers [33, 53], which results in coarsening of grains. Therefore, dealing with these challenges requires further investigation, such as careful control of process parameters, selection of material composition, and the design of an optimal motion path.

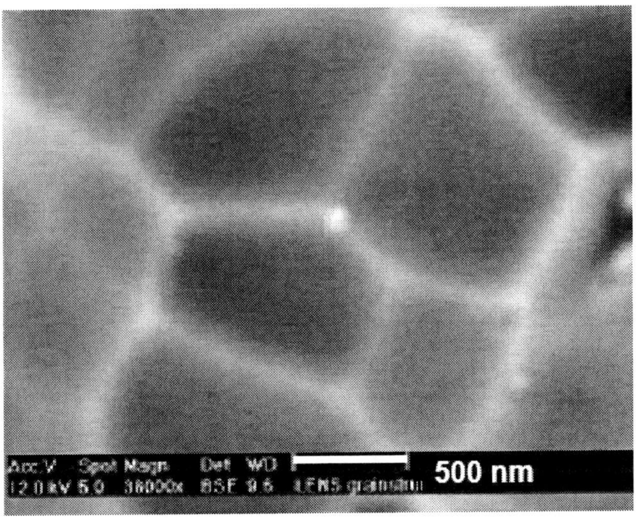

Figure 7.3. SEM image of the LENS® deposited thin wall parts from cryomilled nanostructured 316L SS powder [155].

7.3. Powder and Particle-Pool Interactions

Ensuring good flowability of powder particles represents a key element of the LENS® process. During LENS® deposition, powder is delivered from containers, which utilize a perforated disc mechanism rotating at a controlled speed to meter powder into a stream of inert gas, which passes the powder through four nozzles focused on the molten pool. The powder streams diverge somewhat when they leave the nozzles, forming conical streams of powder. These powder cones intersect in the vicinity of the molten pool, making inter-particle collisions inevitable.

The morphology and size distribution of the feedstock powder is critical to the dimensional accuracy and properties of the near net shape components. The surface finish of the deposited parts is a strong function of the powder particle size, with small powder particle size giving a better surface finish [4]. According to reference [39], the surface roughness of LENS® deposited INCONEL alloy 690 alloy is approximately 12 μm. The suitable particle size for LENS® deposition is in the range of 40-150 μm [35, 156]. Inherent issues associated with particles that are too small include agglomeration and the associated decrease in flowability, whereas oversized particles are unsuitable for a gas-carried-particle process such as LENS®. Furthermore, the stability of the laser-generated molten pool requires a narrow particle size distribution [4]. Inspection of the published literature shows that no systematic work has been carried out addressing particle-pool interaction behavior and how this will affect the formation of the microstructure of the LENS® deposited component. Considering that the thermal history during the LENS® process is significantly different than for other thermal/mechanical processes, detailed study of the particle-pool interactions remains a formidable challenge, given the small size of the pool, the presence of large thermal gradients, and the displacement of the solid/liquid interface. For example, if nanostructured feedstock powders experience complete melting prior to impingement onto the molten pool, the formation of a nanocrystalline microstructure will be strongly influenced by the cooling rate. If the feedstock powder experiences a high temperature but no melting prior to impingement onto the molten pool, the formation of a nanocrystalline microstructure will depend on the thermal stability of the feedstock powder. A high-speed camera could be used to characterize the particle-droplet melting and particle-pool interaction behavior during the LENS® process. Also, high-speed photography could provide the possibility to characterize the morphology and structure of the feedstock powder in different stages before entering the molten pool.

7.4. Residual Stresses and Porosity

The production of geometrically complex components with enhanced physical and mechanical properties via LENS® will require careful control of residual stresses and porosity [51]. This requirement will become increasingly difficult in the case of complex materials systems, such as functionally graded composites, non-equilibrium microstructures, and directionally solidified components, which are bound to challenge our ability to exercise careful control of numerous, and often interrelated process parameters.

7.4.1. Residual Stresses

Large differences in the coefficients of thermal expansion will undoubtedly contribute to the formation of thermal strains during the cyclic heating and cooling that are characteristics of the LENS® process. The primary factors that control morphology and microstructure of the initial deposited layer are thermal gradients and cooling rates [33]. Residual stress development is an intrinsic outcome of successive deposition processes such as LENS® [14]. The associated thermal stress behavior in the deposited structure is an important phenomenon that necessitates a more in-depth understanding. Some preliminary work has been done by Maziasz et al. [157] and Griffith et al. [33] regarding the residual stresses in a LENS® deposited H13 tool steel structure. The determination of residual stress after deposition is carried out by using existing methods such as X-ray diffraction [158, 159] for near surface stresses and neutron diffraction [157] for bulk stresses; a holographic-hole drilling technique [33] can also be used. Currently, challenges arise when trying to predict the level of stress in the LENS® deposited structure according to its thermal history and in developing numerical simulation models of the residual stress.

Both thin walls and towers of 316L SS and INCONEL 718 have been built via LENS® for subsequent residual stress measurements using neutron diffraction [160]. Measurements on thin walls indicate that stress increases in the build direction (Z) and decreases as the free end is approached. Likewise, tower components demonstrate compressive axial stress strongly decreases approaching the free end. Stresses in the Z-direction are seen 2 mm away from the free end, while stresses in the X- and Y-directions are compressive at the center and tensile towards the edges, hence balancing the stress concentrations. From a conventional manufacturing approach, tensile stresses at the edges are highly undesirable due to the increased tendency of being predisposed to surface crack growth. Figure 7.4 shows the stresses present along the center of the tower [160].

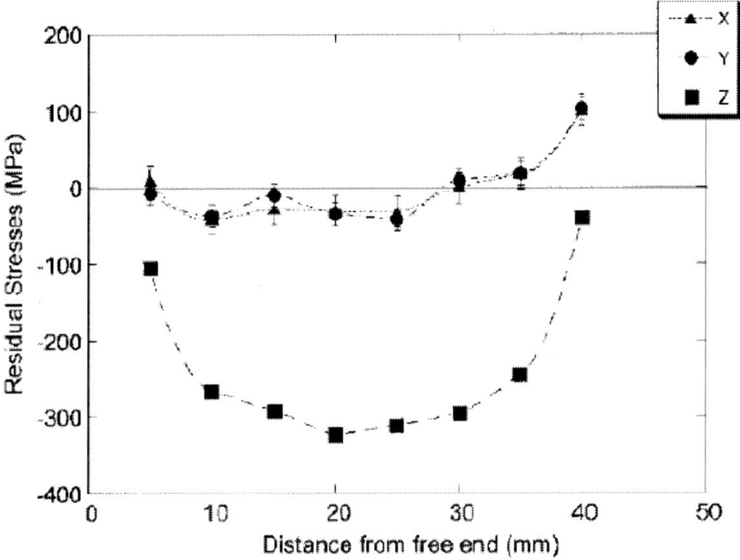

Figure 7.4. X, Y, Z components of residual stress in a LENS® deposited 316L SS tower along the centerline [160].

Compressive stress at the interior and tensile stress concentrations at the edges may be caused by different cooling rates, if the edges remain hotter than the interior. In this case, the shrinkage of the outside is constrained by the cooler interior. Varying the process parameters to maintain the same molten pool temperature throughout the part build and reducing the travel time along the edges may lead to a reduction in residual stress in the bulk material [160, 161].

7.4.2. Porosity

Generally, there are two sources of porosity observed in parts built with LENS®: lack-of-fusion and gas porosity. Figure 7.5(a) illustrates the interlayer lack-of-fusion porosity from LENS® deposited 304L SS [37], and figure 7.5(b) illustrates intralayer porosity in LENS® deposited PH13-8Mo SS [76]. The lack-of-fusion porosity always occurs along layer boundaries with irregular shape, as shown in figure 7.5(a), and the cause is probably from an insufficient deposit thickness due to insufficient laser power in relation to the focal point of the laser. This occurs when there is a mismatch between the molten pool or layer height and the Z-height increment set by the operator. Lack-of-fusion porosity is not frequently observed and is readily avoided with proper selection of process parameters. The general shape of gas porosity is spherical, although there is no specific location at which it may occur, as shown in figure 7.5(b), which is an image of an etched cross-sectional surface of LENS® deposited PH13-8Mo SS [76].

Similar to laser welding and cladding processes, the primary cause of gas porosity is that gas dissolved or entrapped in the melt may not have sufficient time to escape to the top of the molten pool due to rapid solidification rates during LENS® deposition, therefore forming gas porosity in deposited components. There are also complicated sources of gas porosity, such as porosity caused by the collapse of unstable keyholes [162, 163], porosity due to entrapment of gases by surface turbulence [30, 47, 48], gas entrainment during turbulent impact of particles into the molten pool, contamination by powder-feed gases, gases contained within the powder [30], moisture absorbed on powder surfaces [164], vaporization of high vapor pressure alloy constituents [163], and shrinkage porosity [165, 166], which hinder a straightforward explanation in many cases [37, 38, 70, 76, 167]. The amount of gas porosity in deposits is generally less than 0.05 pct. by area [38].

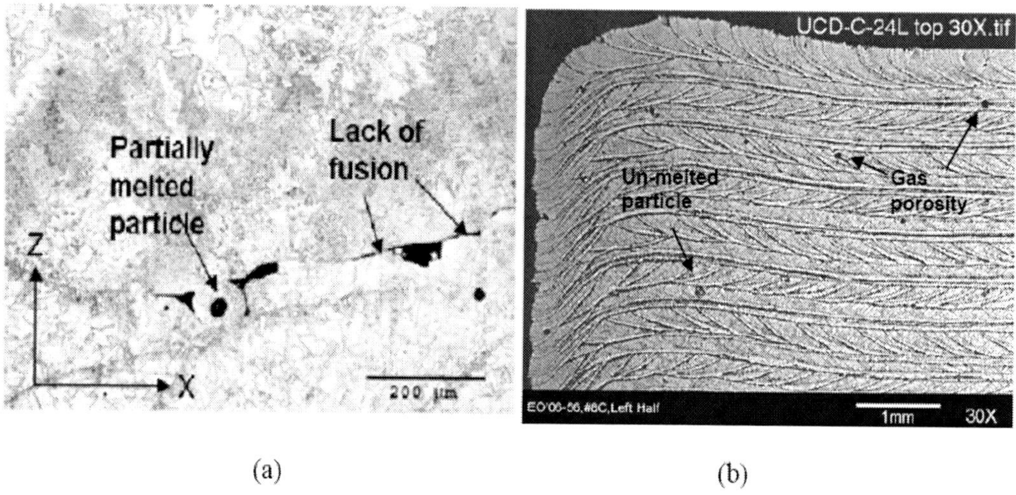

Figure 7.5. Photomicrograph of (a) interlayer lack-of-fusion porosity in 304L blocks [37], and (b) spherical intralayer porosity in PH13-8Mo SS blocks [76].

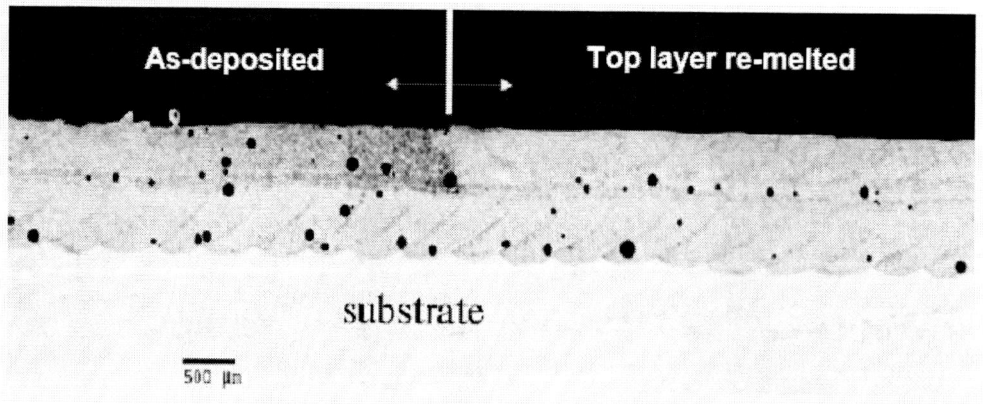

Figure 7.6. Photomicrograph of two layers of LENS® deposited 17-4PH SS. The top layer on the right has been re-melted to reduce the porosity content [37].

The work performed by Susan et al. suggests a direct relationship between deposit porosity and gas atomized powders used for deposition [37]. The gas entrapped in gas atomized powder can be retained in a LENS® deposited component in the form of gas porosity. Well-controlled re-melting of a deposited layer by a second run of the laser beam can significantly reduce porosity by allowing some pores formed during the first run to escape. Figure 7.6 shows a promising result for LENS® deposited 17-4PH SS, in which the top layer is re-melted to promote porosity removal [37]. Kobryn et al. reported that porosity in LENS® deposited Ti-6Al-4V decreases with increasing traverse speed and power level, and lack-of-fusion porosity is more prevalent when thin substrates are used [38].

The formation of gas porosity is greatly influenced by deposition processing parameters [38, 168]. At slow laser traverse speeds, the interaction time is long, and gas porosity tends to nucleate, grow and escape from the molten pool, mostly as a result of buoyancy or other convective flow processes [17]. The porosity volume fraction in the molten pool decreases with increasing laser traverse speed due to insufficient time for gas pores to nucleate and grow. The density of pores increases with increasing laser traverse speed and decreasing laser power due to the reduced time for pores to coalesce. On the other hand, high laser traverse speed or laser power could result in lack-of-fusion porosity. The porosity in LENS® deposited parts can be controlled to a certain extent by properly manipulating process parameters, such as laser power and laser traverse speed, which control the size and temperature of the molten pool. The scope of future work will require further investigation into the influence of interactions among laser beam, injected powder, and molten pool on porosity formation along with identification of the mechanisms that are active.

8. Concluding Remarks

As one of a number of competing direct laser deposition processes, the LENS® technique was developed with the goal of automatically fabricating complex shaped or featured metallic components based on 3D CAD solid models of components, especially for development work for parts or new materials, and low volume production applications. As a powder additive process, LENS® provides the capability for a wide range of materials to be fabricated including metallic alloys, metal matrix composites and cermets. The major hurdles that restrict wider use of LENS® in routine material processing are limitations of the beam size with respect to the deposited component size, cost, residual stress and porosity, as well as the need for skilled manpower. Production speed also needs improvement in order to broaden production opportunities compared to high volume production processes such as injection molding or die casting. The above limitations can be overcome if more areas are identified where LENS® offers unmatched advantages in terms of end properties and product quality in comparison to those achieved by other conventional powder metallurgy techniques.

The successful application of the LENS® process to novel materials systems such as metallic glasses and nanocrystalline materials will require further research and development. Moreover, the underlying scientific principles that control the microstructure evolution during the LENS® process require further study before complete microstructure control can be achieved.

In this chapter, we have reviewed the major LENS® material processing routes that are either routinely used in commercial applications or are slated for future exploitation. The processes discussed have broadly been categorized into LENS® deposited materials, processing control, thermal behavior, and environmental and cost analysis. The focus of this review centers on the basic principles, scope, and mechanisms associated with the LENS® process. Conscious effort has been made to outline the key issues and define the future scope of research and development in some selected areas. It is hoped that the present review will provide an impetus for further development and applications of powder additive processing with Laser Engineered Net Shaping.

REFERENCES

[1] T. H. Maiman, *Nature,* 187 (1960) 493.

[2] E. M. Breinan and B. H. Kear, in: Laser Mater. Proc., (*Ed. M. Bass*), *North-Holland*, Amsterdam, 3 (1983) 236.

[3] P. L. Blackwell and A. Wisbey, *J. Mater. Proc. Tech.,* 170 (2005) 268.

[4] C. L. Atwood, M. L. Griffith, L. D. Harwell, D. L. Greene, D. E. Reckaway, M. T. Ensz, D. M. Keicher, M. E. Schlienger, J. A. Romero, M. S. Oliver, F. P. Jeantette and J. E. Smugeresky, *Sandia Report SAND99-0739*, Sandia National Laboratories, USA, (1999), 1-30.

[5] C. L. Atwood, M. L. Griffith, M. E. Schlienger, L. D. Harwell, M. T. Ensz, D. M. Keicher, J. A. Romero and J. E. Smugeresky, *Proc. of ICALEO '98,* Orlando, FL, USA, 11/16-19/1998, (1998) E1.

[6] K. P. Cooper, *JOM*, 53, 9, (2001) 29.

[7] M. L. Griffith, E. Schlienger, C. L. Atwood, J. A. Romero, J. E. Smugeresky, L. D. Harwell and D. L. Greene, *Proc. of the Rapid Prototyping and Manufacturing Conference '97*, Dearborn, MI, USA, 04/22-24/1997 (1997) 260.

[8] D. T. Pham and S. S. Dimov, Rapid Manufacturing: The Tech. and Appl. of Rapid Prototyping and Rapid Tooling, Springer, London, (2001).

[9] A. Vasinonta, J. L. Beuth and J. L. Griffith, *J. Manuf. Sci. Eng.,* 123 (2001) 615.

[10] F. Klocke, T. Celiker and Y-A Song, *Rapid Prototyping J.,* 1 (1995) 32.

[11] S. Das, N. Harlan, G. Lee, J. J. Beaman, D. L. Bourell, J. W. Barlow, T. Fuesting, L. Brown and K. Sargent, *Mater. and Manuf. Proc.*, 13 (1998) 241.

[12] M. Y. Zhou, J. T. Xi and J. Q. Yan, *J. Mater. Proc. Tech.,* 146 (2004) 396.

[13] D. M. Keicher, J. E. Smugeresky, J. A. Romero, M. L. Griffith and L. D. Harwell, *Proc. of SPIE, the Int. Society for Optical Engineering,* 2993 (1997) 91.

[14] J. Beuth and N. Klingbeil, *JOM, J. the Minerals*, Metals and Materials Society, USA, 53 (2001) 36.

[15] A. Einstein, *Phys. Z,* 18 (1917) 121.

[16] W. V. Smith and P. P. Sorokin, *The Laser*, McGraw-Hill (1966).

[17] J. F. Ready and D. F. Farson, *LIA Handbook of Laser Mater. Proc.*, Laser Institute America Magnolia Publishing Inc. (2001).

[18] W. M. Steen, *Laser Mater. Proc.*, Springer-Verlag (2003).

[19] V. M. Weerasinghe and W. M. Steen, *Appl. Laser Tooling,* (1987) 183.

[20] W. W. Duley, *Laser Surface Treatment of Metals: NATO-ASI Series (E)*, 115, C. W. Draper, P. Mazzoldi, Boston: Martinus Nijhoff (1986) 3.

[21] J. Laeng, J. G. Stewart and F. W. Liou, *Int. J. Prod. Res.,* 38 (2000) 3973.

[22] B. L. Mordike, *Mater. Sci. Tech.*, (Eds. R.W. Cahn, P. Haasen and E.J. Kramer), Weinheim, VCH (1993).

[23] C. W. Draper, *Laser and Electron Beam Proc. of Mater.*, C.W. White, P.S. Peercy eds. NewYork, Academic Press (1980) 721.

[24] W. Wang, M. R. Holl and D. T. Schwartz, *J. Electrochem. Soc.,* 148, (2001) C363.

[25] A. Greco, A. Licciulli and A. Maffezzoli, *J. Mater. Sci.*, 36 (2001) 99.

[26] L. Lu, JY H. Fuh, Z. D. Chen, C. C. Leong and Y. S. Wong, *Mater. Res. Bull.,* 35 (2000) 1555.

[27] K. Daneshvar, M. Raissi and S. M. Bobbio, *J. Appl. Phys.,* 88 (2000) 2205.

[28] M. C. Wanke, O. Lehmann, K. Muller, Q. Wen and M. Stuke, *Sci. of the Total Environment,* 275 (1997) 1284.

[29] O. Lehmann and M. Stuke, *Science,* 270 (1995) 1644.

[30] *Proc. of Solid Freeform Fabrication Symposia (SFF)*, Univ. of Texas at Austin TX, USA, (1990-2002).

[31] J. J. Beaman, J. W. Barlow, D. L. Bourell, R. H. Craford, H. L. Marcus and K. P. McAlea, *SFF*, Kluwer, Boston, USA, (1997).

[32] P. F. Jacobs, *Stereolithography and other RP and M. Tech.*, SME, Dearborn MI, USA, (1996).

[33] M. L. Griffith, M. E. Schlienger, L. D. Harwell, M. S. Oliver, M. D. Baldwin, M. T. Ensz, M. Essien, J. Brooks, C. V. Robino, J. E. Smugeresky, W. H. Hofmeister, M. J. Wert and D. V. Nelson, *Mater. Design*, 20 (1999) 107.

[34] Y. Xiong, J. E. Smugeresky, L. Ajdelsztajn and J. M. Schoenung, *Mater. Sci. Eng.: A*, In Press, Corrected Proof (2008).

[35] W. Hofmeister, M. Griffith, M. Ensz and J. Smugeresky, *JOM,* (2001) 30.

[36] W. Hofmeister, M. Wert, J. E. Smugeresky, J. A. Philliber, M. Griffith and M. Ensz, *JOM-e* (http://www.tms.org/pubs/journals/JOM/9907/Hofmeister/Hofmeister-9907.html), 51 (1999).

[37] D. F. Susan, J. D. Puskar, J. A. Brooks and C. V. Robino, *Mater. Characterization,* 57 (2006) 36.

[38] P. A. Kobryn, E. H. Moore and S. L. Semiatin, *Scripta Materialia,* 43 (2000) 299.

[39] G. K. Lewisa and E. Schlienger, *Mater. Design,* 21 (2000) 417.

[40] C. Hull, Apparatus for production of three-dimensional objects by stereolithography, (1986).

[41] P. F. Jacobs, *Rapid Prototyping and Manuf.: Fundamentals of Stereolithography*, McGraw-Hill, New York, (1992).

[42] S. Jayanthi, B. Hokuf and Lawton, *Mech. Prop. of Stereolithographic*, Materials Group, 08 (1995).

[43] G. Zong, Y. Wu, N. Tran, I. Lee, D. L. Bourell and J. J. Beaman, *Proc. Solid Freeform Fabrication Symposium*, University of Texas, Austin, TX, USA, (1992) 72.

[44] R. Knight, J. Beaman, D. Freiteg, *Proc. Solid Freeform Fabrication Symposium*, University of Texas, Austin, TX, USA, (1996) 349.

[45] J. J. Beaman and S. Das, *Direct selective laser sintering of metals*, Board of Regents, University Texas System, US Patent 6676892, (2004).

[46] C. R. Deckard, Method and apparatus for producing parts by selective sintering, US Patent 5017753, (1991).
[47] S. Das, T. P. Fuesting, G. Danyo, L. E. Brown, J.J. Beaman and D.L. Bourell, *Mater. Design,* 21 (2000) 63.
[48] J. Hiemenz, *Ad. Mater. Processes*, (2007) 45.
[49] P. Heinl, A. Rottmair, C. Körner and R. F. Singer, *Adv. Eng. Mater.,* 9 (2007) 360.
[50] Optomec, http://optomec.com/ (2008).
[51] M. L. Griffith, M. T. Ensz, J. D. Puskar, C. V. Robino, J. A. Brooks, J. A. Philliber, J. E. Smugeresky and W. H. Hofmeister, *Symposium Y Proc.*, Materials Research Society, USA, V625, 04 (2000).
[52] B. Zheng, Y. Zhou, J. E. Smugeresky, J. M. Schoenung and E. J. Lavernia, Part I Numerical Calculations, *Metall. Mater. Trans. A*, 39A (2008) 2228.
[53] J. A. Brooks, C. V. Robino, T. Headley, S. Goods, R. C. Dykhuizen and M. L. Griffith, *Proc. of the Solid Freeform Fabrication Symposium*, Austin, TX, USA, (1999) 375.
[54] Sandia Technology, *the LENS success story,* http://www.optomec.com/downloads/ Optomec-LENS-Sandia-Success-Story.pdf (2003).
[55] B. Zheng, *Synthesis and Behavior of Metallic Glasses via Gas Atomization and Laser Deposition*, University of California, Davis Ph.D. Dissertation (2006).
[56] M. Agarwala, D. Bourell, J. Beaman, H. Marcus and J. Barlow, *Rapid Prototyping J.,* 1 (1995) 26.
[57] E. Berry, J. M. Brown, M. Connell, C. M. Craven, N. D. Efford, A. Radjenovic and M. A. Smith, *Medical Eng. Phys.,* 19 (1997) 90.
[58] Y. P. Kathuria, *Surface and Coatings Tech.,* 116-119 (1999) 643.
[59] H. J. Niu and I. T. H. Chang, *Scripta Materialia,* 41 (1999) 25.
[60] W. Cheng, J. Y. H. Fuh, A. Y. C. Nee, Y. S. Wong, H. T. Loh and T. Miyazawa, *Rapid Prototyping J.*, 1 (1995) 12
[61] M. L. Griffith and J. W. Halloran, *Journal of the American Ceramic Society,* 79 (1996) 2601.
[62] J. E. Smugeresky, R. Grylls, D. M. Keicher and and C. Robino, *Powder Mater.:* Current Research and Industrial Practices III as held at the Materials Science and Technology, (2003).
[63] D. D. Gill, C. J. Atwood and J. E. Smugeresky, *Performer: Sandia National Labs.*, Albuquerque, NM. Sponsor: Department of Energy, Washington, DC, USA, 1 (2006) SAND2006~6551.
[64] D. D. Gill, J. E. Smugeresky, C. J. Atwood, M. D. Jew and S. Scheffel, *Performer: Sandia National Labs.*, Albuquerque, NM. Sponsor: Department of Energy, Washington, DC, USA, 1 (2006) SAND2006~6431.
[65] B. Zheng, J. Nguyen, J. E. Smugeresky, Y. Zhou, D. Baker and E. J. Lavernia, *Word Congress PM2008*, Washington, D.C., USA, 06/8-12/2008, (2008).
[66] J. Kummailil, *Functionally-Adapted Biomimetic Mater. for Space Exploration,* Reports, NASA Grant NAG5-8830 (2000-2001).
[67] M. Roy, B. V. Krishna, A. Bandyopadhyay and S. Bose, *Acta Biomaterialia,* 4 (2008) 324.
[68] B. V. Krishna, S. Bose and A. Bandyopadhyay, *Acta Biomaterialia*, 3 (2007) 997.

[69] M. G. Glavicic, K. A. Sargent, P. A. Kobryn and S. L. Semiatin, *AFRL-ML-003-4103* (2003).
[70] Y. Xiong, B. Zheng, J. E. Smugeresky, L. Ajdelsztajn and J. M. Schoenung, *MS&T'05,* Pittsburgh, PA, USA, 09/25-28/2005 (2005).
[71] B. Zheng, J. E. Smugeresky, Y. Zhou, D. Baker and E. J. Lavernia, *PowerMet2007*, Denver, CO, USA, 05/13-16/2007 (2007).
[72] R. Banerjee, P. C. Collins, A. Genc and H. L. Fraser, *Mater. Sci. .Eng.* A, 358 (2003) 343.
[73] M. L. Griffith, L. D. Harwell, J. A. Romero, E. Schlienger, C. L. Atwood and J. E. Smugeresky, Multi-Material Processing by LENS®, *Proc. of the Solid Freeform Fabrication Symposium*, Austin, TX, USA, (1997) 387.
[74] L. Li, *J. Mater Sci.,* 41 (2006) 7886.
[75] R. Banerjee, C. A. Brice, S. Banerjee and H. L. Fraser, *Mater. Sci. Eng.* A, 347 (2003) 1.
[76] B. Zheng, J. E. Smugeresky, Y. Zhou and E. J. Lavernia, Advances in Powder Metal. and Particulate Mater. 2006, Proc. of the PowderMet (2006) 10/81.
[77] M. T. Ensz, M. L. Griffith and D. E. Reckaway, *Proc. of the 2002 MPIF*, Laser Metal Deposition Conference, San Antonio, TX, USA, 04/8-10/2002, (2002).
[78] W. Wei, Y. Zhou, J. He, J. E. Smugeresky and E. J. Lavernia, *Proc. of 2002 Int. Conf. on Metal Powder Deposition for Rapid Manufacturing*, San Antonio, TX, USA, (2002) 180.
[79] X. Wu, J. Liang, J. Mei, C. Mitchell, P. S. Goodwin and W. Voice, *Mater. Design*, 25 (2004) 137.
[80] B. Zheng, Y. Zhou, J. E. Smugeresky, J. M. Schoenung and E. J. Lavernia, Part II Experimental Investigation and Discussion, *Metall. Mater. Trans. A*, 39A (2008) 2237.
[81] B. P. Kashyap and K. Tangri, *Acta metall. Mater.,* 43 (1995) 3971.
[82] J. E. Smugeresky, B. Zheng, L. Ajdelsztajn, Y. Zhou, J. M. Schoenung and E. J. Lavernia, *134th TMS Annual Meeting and Exhibition 2005*, San Francisco, CA, USA, 02/13-17/2005, (2005).
[83] B. Pollard, *ASM Handbook 6*, ASM, Materials Park, OH, USA, (1993) 482.
[84] J. E. Smugeresky, B. Zheng, Y. Zhou and E. J. Lavernia, *TMS'06*, San Antonio, TX, USA, 03/12-16/2006 (2006).
[85] J. Tiley, T. Searles, E. Lee, S. Kar, R. Banerjee, J. C. Russ and H. L. Fraser, *Mater. Sci. Eng. A*, 372 (2004) 191.
[86] S. M. Kelly and S. L. Kampe, *Metall. Mater. Trans. A,* 35A (2004) 1861.
[87] S. M. Kelly and S. L. Kampe, *Metall. Mater. Trans. A* 35A, (2004) 1869.
[88] AMS Handbook, Metal powder production and characteization, 7 (1998) 426.
[89] B. Zhang, E. Grylls and H. L. Fraser, *Mater. Res. Soc. Symp. Proc.,* Materials Research Society, 552 (1999) KK5.2.1.
[90] W. Xue, B. V. Krishna, A. Bandyopadhyay and S. Bose, *Acta Biomaterialia*, 3 (2007) 1007.
[91] X. Zhao, J. Chen, X. Lin and W. Huang, *Mater. Sci. Eng.A,* In Press, Corrected Proof (2008).
[92] M. G. Glavicic, K. A. Sargent, P. A. Kobryn and S. L. Semiatin, *AFRL-ML-WP-TR-2003-4103* (2003).

[93] M. G. Glavicic, K. A. Sargent, P. A. Kobryn and S. L. Semiatin, *Interim report*, UES INC DAYTON OH, A496514, Interim rept. 5 Jan 2001-31 (2003).

[94] B. Zheng, J. E. Smugeresky, Y. Zhou, D. Baker and E. J. Lavernia, *Word Congress PM2008*, Washington, D.C. USA, 06/8-12/2008, (2008).

[95] T. W. Clyne and P. J. Withers, *An Introduction to Metal Matrix Composites*, Cambridge University Press, New York, (1993).

[96] K. U. Kainer, Metal matrix composites, *Custom-made mater. for automotive and aerospace engineering*, Wiley-VCH Verlag GmbH and Co. KgaA, Weinheim (2006).

[97] D. J. Lloyd, *Int. Mater. Rev.*, 39 (1994) 1.

[98] W. Liu and J. N. DuPont, *Metall. Mater. Trans. A*, 35A (2004) 1133.

[99] R. Banerjee, A. Genc, D. Hill, P. C. Collins and H. L. Fraser, *Scripta Materialia,* 53 (2005) 1433.

[100] A. Genc, R. Banerjee, D. Hill and H. L. Fraser, *Mater. Lett.*, 60 (2006) 859.

[101] R. Banerjee, P. C. Collins, D. Bhattacharyya, S. Banerjee and H. L. Fraser, *Acta Materialia*, 51 (2003) 3277.

[102] B. Zheng, J. E. Smugeresky, Y. Zhou, D. Baker and E. J. Lavernia, *Metall. Mater. Trans. A,* 39A (2008) 1196.

[103] J. E. Smugeresky, B. Zheng, Y. Zhou and E. J. Lavernia, *TMS'06*, San Antonio, TX, USA, 03/12-16/2006 (2006).

[104] R. Banerjee, D. Bhattacharyya, P. C. Collins, G. B. Viswanathan and H. L. Fraser, *Acta Materialia*, 52 (2004) 377.

[105] P. C. Collins, R. Banerjee, S. Banerjee and H. L. Fraser, *Mater. Sci. Eng.A*, 352 (2003) 118.

[106] B. Vamsi Krishna, W. Xue, S. Bose and A. Bandyopadhyay, *Acta Biomaterialia,* In Press, Corrected Proof (2008).

[107] W. Liu and J. N. DuPont, *Scripta Materialia,* 48 (2003) 1337.

[108] T. Glaeser, F. Klocke and T. Bergs, *Proc. of the 6th Int. Conf. on Tungsten, Refractory and Hardmetals* (Metal Powder Industries Federation, Princeton, NJ, USA, Orlando, FL, USA, 2006) 132.

[109] J. Sears and A. Costello, *Mater. Sci. Forum,* 534-536 (2007) 1537.

[110] C. P. Paul and A. Khajepour, *Optics and Laser Tech.*, in press, (2007).

[111] E. Yarrapareddy, S. Zekovic, S. Hamid and R. Kovacevic, *Proc. of the Institution of Mechanical Engineers, Part B: J. Eng. Manuf.,* 220 (2006) 1923.

[112] N. Bloembergen, *AIP Conf. Proc.*, (Eds. S. D. Ferris, H.J. Leamy and J. M. Poate) (1978).

[113] R. C. Ruhl, *Mater. Sci. Eng.,* 1 (1967) 313.

[114] W. H. Hofmeister, R. J. Bayuzick, G. Trapaga, D. M. Matson and M. C. Flemings, *Solidification,* (Eds. S. Marsh, J. Dantzig, R. Trivedi, W. Hofmeister, M. Chu, E. Lavernia, and J. Chun, TMS) (1998) 375.

[115] S. D. E. Ramati, G. J. Abbaschian and R. Mehrabian, *Meter. Trans. B*, 9B (1978).

[116] C. G. Levi and R. Mehrabian, *Meter. Trans. B*, 11B (1980) 21.

[117] O. Grong, *Metall. Modeling of Welding*, 2nd edition, London (1997) 223.

[118] G. K. Lewis and E. Schlienger, *Mater. Design,* 21 (2000) 417.

[119] S. Kou, *Welding Metall.*, John Wiley and Sons, Inc., New York, NY, USA, (1987).

[120] J. Mazumder, D. Dutta, N. Kikuchi and A. Ghosh, *Optics and Lasers in Eng.*, 34 (2000) 397.
[121] J. Mazumder, J. Choi, K. Nagarathnam, J. Koch and D. Hetzner, *JOM*, USA 49 (1997) 55.
[122] C. P. Paul, A. Jain, P. Ganesh, J. Negi and A. K. Nath, *Optics and Lasers in Eng.*, 44 (2006) 1096.
[123] W. Liu and J. N. Dupont, *Metall. Mater. Trans. A,* 35A (2004) 1133.
[124] R. A. Patterson, P. L. Martin, B. K. Damkroger and L. Christodoulou, *Welding J.*, 69 (1990) 39s.
[125] Y. Xiong, J. E. Smugeresky and J. M. Schoenung, *J. Mater. Proc. Tech.*, Submitted, (2008).
[126] J. E. Smugeresky, B. Zheng, Y. Zhou and E. J. Lavernia, *TMS'05*, San Francisco, CA, USA, 02/13-17/2005, (2005).
[127] B. Zheng, Y. Lin, J. E. Smugeresky, Y. Zhou and E. J. Lavernia, *TMS Lett.*, 4 (2005) 113.
[128] N. W. Klingbeil, S. Bontha, C. J. Brown, D. R. Gaddam, P. A. Kobryn, H. L. Fraser and J. W. Sears, *Proc. Solid Freeform Fabrication Symposium*, Austin, TX, USA, 08/2-4/2004, (2004).
[129] J. Mazumder, A. Schifferer and J. Choi, *Mater. Res. Innovations*, 3, 10, (1999) 118.
[130] E. Fearon and K. G. Watkins, Proc. of the 23rd Int. Congress on Appl. of Lasers and Electro-Optics, (2004).
[131] M. Azer, J. Deaton and H. Qi, *ICALEO'05,* Laser Institute of America, 10/31-11/3 2005, Miami, FL, USA (2005).
[132] G.-S. Cai, M. Azer, K. Harding, H. Peng, R. Tait and B.-R. Zuo, *Proc. of the ICALEO'05*, Laser Institute of America, 10/31-11/3 2005, Miami, FL, USA (2005) 856.
[133] W. Wei, Y. Zhou, R. Ye, D. Lee, J. E. Craig, J. E. Smugeresky and E. J. Lavernia, *Proc. of the Int. Conf. on Metal Powder Deposition for Rapid Manufacturing*, San Antonio, TX, USA, (2002) 128.
[134] L. Wang, S. Felicelli, Y. Gooroochurn, P. T. Wang and M. F. Horstemeyer, *Mater. Sci. Eng. A,* 474 (2008) 148.
[135] L. Wang and S. Felicelli, *Mater. Sci. Eng. A,* 435-436 (2006) 625.
[136] R. Ye, J. E. Smugeresky, B. Zheng, Y. Zhou and E. J. Lavernia, *Mater. Sci. Eng. A,* 428 (2006) 47.
[137] R. M. Bhatkal and T. Hannibal, *JOM,* (1999) 26.
[138] Y. Xiong, K. Lau, X. Zhou and J. M. Schoenung, *J. Cleaner Production*, 16 (2008) 1118.
[139] A. Zabaniotou and E. Kassidi, *J. Cleaner Production,* 11 (2003) 549.
[140] B. Rivela, M. T. Moreira, I. Munoz, J. Rieradevall and G. Feijoo, *Sci. Total Environment,* 357 (2006) 1.
[141] G. A. Keoleian, *J. Cleaner Production*, 1 (1993) 143.
[142] M. A. Curran, *Environmental Progress,* 23 (2004) 277.
[143] X. Zhou and J. M. Schoenung, *J. Environmental Management*, 83 (2006) 1.
[144] J. Schoenung, E. Lavernia and Y. Xiong, *2008 NSF Eng. Res. and Inn. Conf.*, Knoxville, TN, USA, (2008).

[145] W. L. Johnson, *Mater. Sci. Forum*, 35 (1996) 225.
[146] A. Peker and W. L. Johnson, *Appl. Phys. Lett.*, 63 (1993) 23.
[147] A. Inoue, *Mater. Trans. JIM,* 36 (1995) 866.
[148] W. L. Johnson, *MRS Bulletin,* (1999) 42.
[149] W. H. Wang, Z. X. Bao, C. X. Liu and D. Q. Zhao, *Phys. Rev. B,* 61 (2000) 3166.
[150] Z. P. Lu, C. T. Liu, J. R. Thompson and W. D. Porter, *Phys. Rev. Lett.*, 92 (2004) 245503.
[151] A. Inoue, T. Masumoto, S. Arakawa and T. Iwadachi, *Mater. Trans. JIM,* 19, (1978) 303.
[152] C. T. Liu, M. F. Chisholm and M. L. Miller, *Intermetallics*, 10 (2002) 1105.
[153] Z. P. Lu and C. T. Liu, *Phys. Rev. Lett.*, 91 (2003) 115505.
[154] J. E. Smugeresky, B. Zheng, Y. Zhou and E.J. Lavernia, *TMS'05*, San Francisco, CA, USA, 02/13-17/2005 (2005).
[155] W. Wei, Y. Zhou, R. Ye, D. Lee, J. E. Craig, J. E. Smugeresky and E. J. Lavernia, *Proc. of the Int. Conf. on Metal Powder Deposition for Rapid Manuf.*, San Antonio, TX, USA, (2002) 128.
[156] Optomec, Training Sheet and manual, (2001).
[157] P. J. Maziasz, E.A. Payzant, M.E. Schlienger and K.M. McHugh, *Scripta Materialia,* 39 (1998) 1471.
[158] H. M. Hu and E. J. Lavernia, *J. Mater. Res.*, 14 (1999) 4521.
[159] S. Ho and E. J. Lavernia, *Scripta Materialia,* 34 (1996) 527.
[160] P. Rangaswamy, T. Holden, R. Rogge and M. Griffith, *The J. Strain Analysis for Eng. Design*, 38 (2003) 519.
[161] P. Rangaswamy, M. L. Griffith, M. B. Prime, T. M. Holden, R. B. Rogge, J. M. Edwards and R. J. Sebring, *Mater. Sci. Eng. A*, 399 (2005) 72.
[162] M. Pastor, H. Zhao and T. DebRoy, *J. Laser Appl.,* 12 (2000) 91.
[163] E. Aghion and B. Bronfin, *Mater. Sci. Forum,* 350-351 (2000) 19.
[164] J. Wegrzyn, M. Mazur, A. Szymanski and B. Balcerowska, *Weld. Int.,* 2 (1987) 146.
[165] X. Cao, W. Wallace, J. P. Immarigeon and C. Poon, *Mater. Manuf. Process,* 18 (2003) 23.
[166] X. Cao, W. Wallace, C. Poon and J. P. Immarigeon, *Mater. Manuf. Process,* 18 (2003) 1.
[167] G. K. Lewis and E. Schlienger, *Mater. Design,* 21 (2000) 417.
[168] H. Haferkamp, M. Niemeyer, U. Dilthey and G. Trager, *Weld. Cutt.,* 52 (2000) 178.

In: Powder Metallurgy Research Trends
Editors: L. J. Smit and J. H. Van Dijk

ISBN: 978-1-60456-852-3

Chapter 5

FORMING MECHANISM AND BEHAVIOR OF SELECTIVE LASER SINTERING OF METAL/POLYMER COMPOSITE POWDERS

L. Jinhui, S. Yusheng, L. Zhongliang and H. Shuhuai
State Key Laboratory of Materials Processing and Die and Mould Technology; School of Material Science and Engineering
Huazhong University of Science and Technology
Wuhan, 430074, China

ABSTRACT

In order to further disclose the indirect SLS process, Forming mechanisms and behaviors of selective laser sintering of metal/polymer composite powders within one laser beam point, one laser scanning line and one laser scanning domain are mainly theoretically investigated. Some mathematical and physical models are built to explain these mechanisms and behaviors. It is indicated that all forming behaviors result from the same thermodynamic mechanism which is the decreasment of powders surface energy. The interaction between rigid powders and variable materials decides the energy states during a certain period. Mover, the variation of forming behaviors in laser beam spots, scanning lines and scanning areas are affected by energy distribution of laser beam spot, laser energy accumulation in the prolonged orientation of scanning line and powder material's effective energy density absorbed from laser respectively. The difference between the initial surface energy and final one is the indicate how stable the final mixed material system is. This work enlarges the theory of laser forming during indirect SLS and provides basis for indirect SLS process, which is a good contribution to the development of indirect SLS technology.

Keywords: selective laser sintering; metal/polymer composite powder; mechanism; behavior.

INTRODUCTION

Selective laser sintering (SLS) has been endowed with the capacity of manufacturing metal components and injection molds since it is developed in 1980s' in America(Karapatis, et al., 1998; Nelson, 1993). In indirect SLS process, polymer powders can be used as binder to be melted by laser beam during laser scanning, and the viscous polymer liquid bonds metal particles to form the metal green shapes. Although many works have been done on manufacturing metal components via indirect SLS, but most of them focus on one certain domain within SLS forming with metal/polymer composite, such as simulation of forming process, selection of forming parameters and prediction of components' physical properties. Unfortunately forming mechanism and behavior of SLS of metal/polymer composite powders, which are ultimate factor on the forming quality, are not covered completely and systematically in foretime literatures. Therefore, connected with SLS forming characters of different shapes from points, lines to planes, SLS forming mechanism is mainly discussed and the universal forming laws are posted to give a good contribution to the development of fabrication of metal parts via indirect SLS.

EFFECTS ON SUBSTANCE MADE BY LASER

Laser Energy Absorbed and Transferred by Dense Polymer and Metal

Irradiated by laser, bounded electrons in polymers are forced to vibrate by photons. Macropolarization induced by electrons vibration occurs, and the viberation become stronger owing to the superimposing of macropolarization fields and electrical fields of incident wave. Kinetic energy of polymer molecular chain is enhanced by the vibration, and the polymer will transform from solids to viscous liquid, which indicate that the laser energy is absorbed by polymer finally. Based on elastic wave theory, heats transfer from hot side to cold one recurring to the phonon vibration. The thermal conductivity coefficient (λ_t) of polymer can be:

$$\lambda_t = \frac{1}{3}\rho_m \omega l c_v \tag{1}$$

where ρ_m is the density of point unit; c_v is the specific heat at constant volume; ω is the mean sonic velocity; l is the molecule mean free path. Each parameter in equation 1 is related to temperature, hence λ_t is complicated with the variation of temperature.

Under the condition of no crystal lattice variation, photons collide with lots of outer-shell electrons in metal. The more violent the collision is, the more laser energy is absorbed by metal. The inelastic collisions result in the electron transition to upper level, which indicates that laser applies positive work to metal and provide energy to it. Many more collisions between photons and electrons will occur and photons' energy will be absorbed within the transmission depth (normally under $0.1\mu m$) of the substance corresponding to laser. Self-collisions between outer-shell electrons also exist besides the ones between photons and electrons, and the mean collision periodic is 10^{-13} second order of magnitude. Hence electrons at high energy level collides with other electrons and act on phonons to transfer laser energy in form of heats.

Compared to non-metal, out-shell electrons are also involved in thermal conduction besides the phonon vibration mechanics. In addition, the action of free electrons is beyond 30 times of that of phonons at room temperature, therefore the thermal coefficients of metals are commonly much larger than that of non-metals.

The relationship of metal absorptivity of laser (α), laser wavelength (λ) and metal electrical resistivity (ρ) can be expressed as equation 2:

$$\alpha = 0.365\sqrt{\frac{\rho}{\lambda}} \tag{2}$$

Electrical resistivity is related to temperature, the relationship of metal absorptance of laser and temperature can be depicted as equation 3 :

$$\alpha(\lambda) = \alpha(20°C) \times \left[1 + \beta(T - 20)°C\right] \tag{3}$$

where β is invariable. Obviously, the absorptance varied in the same orientation with temperature.

Table 1. The absorptivity of some powder materials

powders	CO_2 absorptivity	Nd:YAG absorptivity
Cu	0.26	0.59
Fe	0.45	0.64
Sn	0.23	0.66
Ti	0.59	0.71
Pb	—	0.79
ZnO	0.94	0.02
Al_2O_3	0.96	0.02
SiO_2	0.96	0.04
CuO	0.76	0.11
SiC	0.66	0.78
WC	0.48	0.82
NaCl	0.60	0.17
PTFE	0.73	0.05
PMMA	0.75	0.06
EP	0.94	0.09

ABSORPTION AND TRANSFER OF LASER ENERGY IN METAL/POLYMER COMPOSITE POWDERS

There is notable difference on the absorption rate between dense metals and polymers due to their different structures. While in powder state, the two have different character with their dense state. Pores in powder materials will capture laser energy besides the two absorption forms described as pretext. Shown as figure 1, the capture of laser by pores is as the same as that by black body without any omission, therefore and compared to dense materials, extra laser energy is obtained owing to the contribution of pores. Air takes the position of pores and is difficult to fluid within such narrow cavities. So it transfers heats via conduction. Vargaftik (Hu ying, et al., 1999) indicates that gas thermal coefficient can be expressed as equation 4 :

$$\lambda_t = \lambda_{t0}\left(\frac{T}{T_0}\right)^n \tag{4}$$

Equation 4 can be used on the condition of over 273K and below 1 atm. Where T is the thermodynamics temperature ; λ_{t0} is thermal conductivity coefficient at 0°C ; n is the constant. Take air for an example, $\lambda_{t0} = 2.442\times10^{-2}W/(m\bullet{}^0C)$ and $n = 0.82$, the maximum application temperature is 1000°C. According to equation 4, λ_{t0} is difficult to rise with the increment of temperature. Therefore big thermal resistance from air in the pores prevents heats from dissipation. Many factors affect the absorption of laser energy by pores, such as powder size, powder shape, powder distribution and the uniformity of the multi-phase powders mixture. Hence it is difficult to appraise the value of absorption which can only be obtained by test.

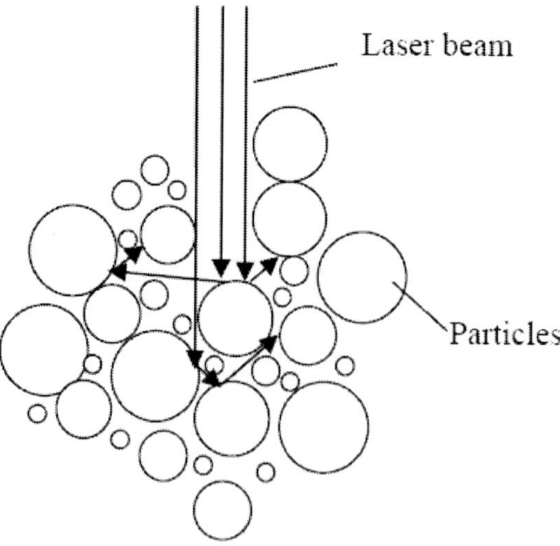

Figure 1. Capture of laser by pores in powders.

Table 1 expresses the absorptivity of powder materials for CO_2 and YAG laser[9]. It can be seen that the absorptivity of metal powders for shorter wavelength YAG laser is notable higher than that for CO_2 laser. But it is reverse towards polymer powders, and their absorptivity for CO_2 laser is one magnitude higher that for YAG laser. In addition, their absorptivity for CO_2 laser is over 0.70, especially, EP's absorptivity is high up to 0.94.

According to what the pretext depicts, the forming mechanisms of metal/polymer under the irradiation of CO_2 laser and YAG laser respectively are different. Irradiated by YAG laser, polymer particles can't be melted by the energy directly from laser beam, and most of the laser energy is absorbed by metal particles (figure 2). Energy from laser beam is transmitted to polymer particles in forms of convection and radiation when they are heated up to high temperature. Meanwhile they also transmit heat to their surrounding environment, such as air and the base binded part, which reduces the heat to polymer. Overall polymer particles melt and bond the metal particles in situ under beforementioned condition, which worsens the bonding effeciency.

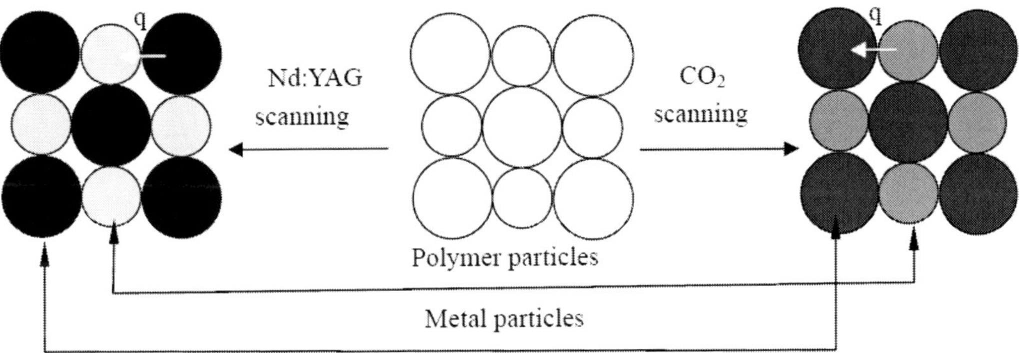

Figure 2. Thermal conditions under irradiations of two lasers.

Owing to the high absorptivity for CO_2 laser, polymer's temperature will rise relatively high. Furthermore several factors, such as not excessive low absorptivity, higher thermal conductivity coefficent and lower specific heat under constant pressure than those of polymer, not only there are notable temperature different between metal and polymer but also the instant temperature of metal particle is normally higher than polymer's. Hence polymer will receive heats from both laser beam and adjacent metal particles sequentially. Initial heats from laser beam melts the polymer into viscous liquid instantly, and liquid state is maintained for relatively long time by secondary heats from adjacent metal particles. Therefore polymer viscous liquid flows along the pores and wets the metal particles' surfaces fully under high temperature states, which makes reorgnization of metal particles.

FORMING BEHAVIOUR OF METAL/POLYMER POWDER ADMIXTURE IRRIATED BY LASER

Thermodynamics of Forming Behaviour

Laplace stress σ, expressed in equation 5, forces metal particles to move and reorganize during flowage of polymer viscous liquid in SLS process. Internal frictions among irregular shaped metal particles will block the migration of metal particles.

$$\sigma = \gamma\left(\frac{1}{r_1}+\frac{1}{r_2}\right) \tag{5}$$

In equation 5, γ is the surface tension of liquid polymer; r_1, r_2 are the radiuses of two perpendicular curvatures of curved liquid surfaces. In addition, σ is enhanced by the stress from polymer volume shrinkage(Liu Jinhui, et al., 2006) to induce the reorganization of metal particles. The forming process can be depicted as figure 3.

Supposing metal/polymer composite powders belong to a thermodynamics system, equation 6 can be obtained according to the first thermodynamics law:

$$\Delta E = Q + W \tag{6}$$

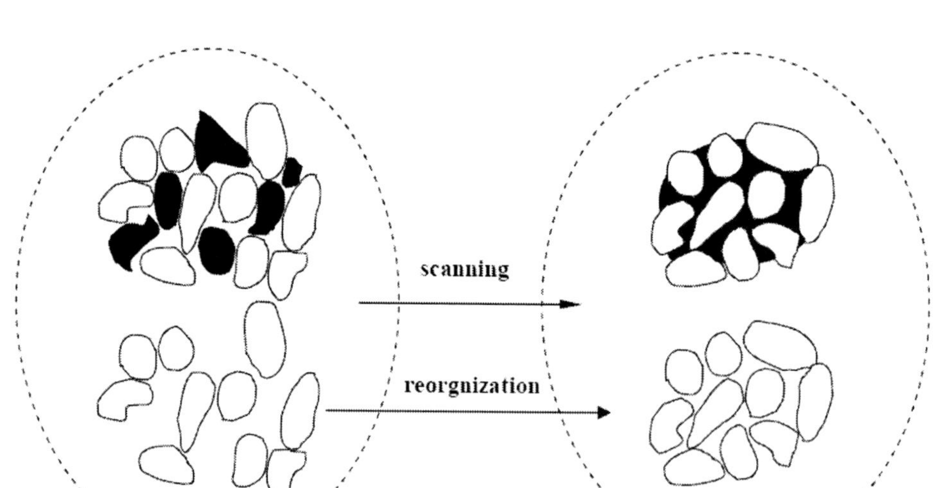

Figure 3. Variation of powder configuration before and after scanning. a) configuration before scanning; b) configuration after scanning white (black) stands for metal(polymer) particle.

The work (W) during forming process can be shown as equation 7 :

$$W = W_S + W_C \tag{7}$$

where W_S indicates the work resulted from the whole variation of surface energy induced by laser. Especially for a system composed of single metal and single polymer powders, W_S can be expressed as equation 8 :

$$W_S = \gamma_{MG} A_{MS} + \gamma_{EG} A_{ES} - \gamma_{ME} A_{ME} - \gamma_{MG} A_{MF} - \gamma_{EG} A_{EF} \quad (8)$$

where A_{MS} and A_{ES} are the total surfaces of metal particles and polymer particles before scanning respectively; A_{MF} and A_{EF} are the total surfaces of metal particles and polymer particles after scanning respectively; γ_{MG} and γ_{EG} are the surfaces energy of metal and polymer respectively; γ_{ME} is the interfacial energy between metal and polymer. Equation 8 describes the variation of surface energy during the forming processing from the individual state of metal and polymer particles before scanning to the binding state of metal particles bonded by continuous polymer after scanning.

W_C is the work from the migration of metal particles during scanning, and it is induced by laser energy. Heats (Q) can be expressed by equation 9 if the time after scanning is enough long that the system temperature returns to the initially value of environment.

$$Q = Q_{in} - Q_{out} \quad (9)$$

where Q_{in} stands for the heats from laser energy absorbed by powder system; Q_{out} stands for the heats including the section dissipated through conduction, convection and radiation and the section during the phase transformation from liquid state to solid state of polymer. On the whole, the total system energy is lowered down when the system temperature returns to the initial one before scanning due to the loss of surface energy of metal and polymer particles. Therefore the powder system becomes relative more stable after scanning. The variation of system energy ΔE can be depicted as figure 4.

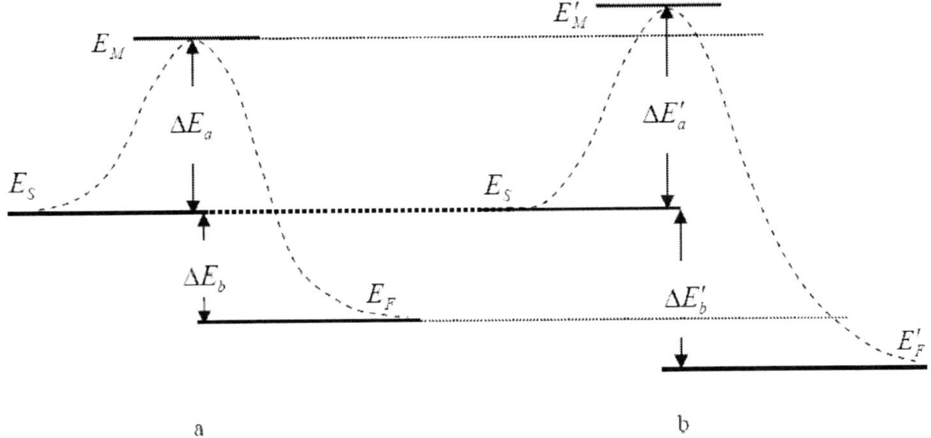

Figure 4. Scheme of energy variation process during the forming process.

In figure 4, E_S and E_F are the system energy in initial state and final state, and obviously $E_S - E_F = \Delta E_b > 0$ according the abovementioned deduction, which indicates that the forming process is non-reversible. There is a medium state whose energy is E_M between initial state and final state.

In figure 4, $E_M - E_S = \Delta E_a$, ΔE_a is the energy barrier which must be overcome by system from initial state to final state. For the system composed of metal particles and single polymer particles, ΔE_a is equal to the heats which are necessary for polymer to be heated up to its glass transformation temperature and melt. It can be expressed as equation 10:

$$\Delta E_a = \Delta H_{T_S \to T_g} + \Delta H_{S \to L}\Big|_{T=T_g} = \int_{T_S}^{T_g} C_P dT + \Delta H_{S \to L}\Big|_{T=T_g} \tag{10}$$

T_S is the initial temperature; T_g is the glass transformation temperature of polymer; C_P is the thermal capacity at constant pressure; $\Delta H_{S \to L}\Big|_{T=T_g}$ is the phase enthalpy of polymer from solid to liquid; $\Delta H_{T_S \to T_g}$ is the enthalpy of system from initial temperature to T_g without phase transformation of polymer. In practice, to prolong the time of polymer liquid and improve its fluidity, temperature target is needed to exceed T_g in order for higher energy absorbed by polymer in powder system. ΔE_b is raised with the increment of temperature target, and meanwhile many more polymer particles transform into viscous liquid. The higher temperature liquid will distribute more equably to make many more metal particles' surfaces bonded. The specific surface area is cut down sharply to make the powder system more stable due to the lower system energy. Seen as figure 4, if $\Delta E_a' > \Delta E_a$, then $\Delta E_b' > \Delta E_b$.

Forming Behavior within Beam Spot (Point State)

There are two mechanisms which are direct mechanism and indirect one respectively for polymer particles to be heated. Particles in beam spot area are heated complying with direct mechanism, while heats are transmitted to the particles outside beam area owing to the temperature gradient from the center of beam spot to outside the beam spot area (figure 5). Therefore particles outside the spot area are heated through the heats conducted inner the spot area, which is called indirect mechanism. Moreover heat currents which can be imagined as heat waves decrease along the radical direction. D in figure 5 is the diameter of beam spot. Laser energy distributes corresponding to Gauss law. The maximum energy occurs at the center of laser beam spot, and the energy decrease from center area to brim area in radial state. White and black in figure 5 stand for higher temperature and lower temperature areas respectively. Bonding state of point A internal the spot and points B□C outside the spot can be depicted as figure 5. Polymer particles at point A fully melt and spread to wet and bond the metal particles. Polymer particles at point B melt and spread partially to wet and bond the metal particles locally. Polymer particles at point C can only be softened to bond the metal particles in situ. Based on abovementioned analysis, bonding effects at point A, B and C can be arranged sequentially as A>B>C.

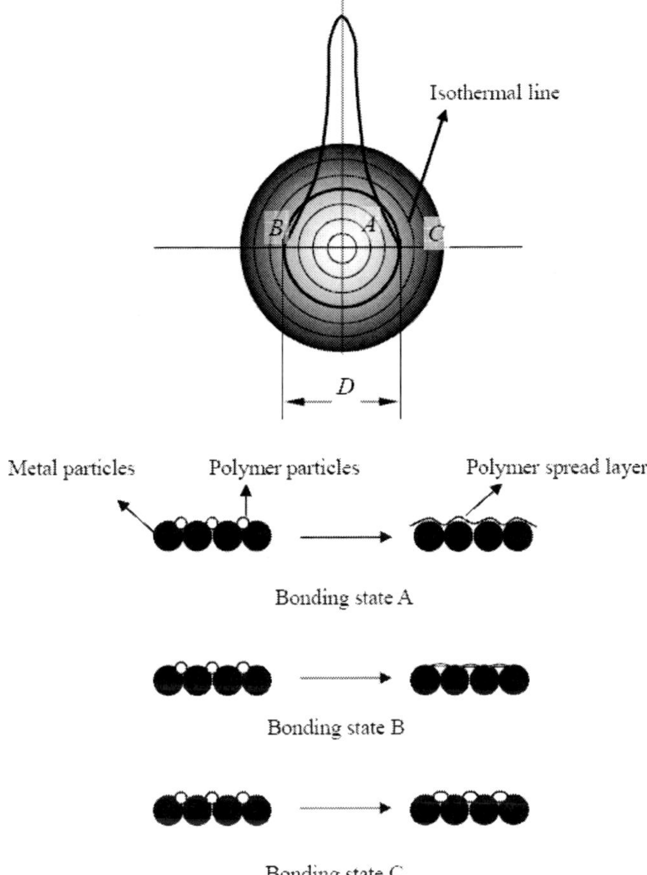

Figure 5. Bonding behaviour of polymer.

Normally bonding state better than that of point B is acceptable, therefore the sealed circle area with the diameter equal to the distance from the spot center to point B can be regarded as ideal bonding area. 3D part can be formed by composing these ideal bonding areas.

Forming Behavior within the Adjacent Scanning Lines (Plane Forming)

Seen as figure 5, polymer outside the beam spot can also be bonded by laser energy, e.g. point B, owing to the heats from spot area. Hence the ring shape domain which exists outside the spot area and in the peripheral part of ideal bonding area is defined as heat diffusion area (figure 6) (Liu Jinhui, 2003). It is obvious that the higher the laser power, the larger the heat diffusion area, the larger the ideal bonding area. D in figure 6a stands for the diameter of laser beam. Two adjacent scanning lines will overlap via ideal bonding area if the scanning space is shorter than the diameter of ideal bonding area (figure 6b). Therefore the powders with the scope of a scanning line can be bonded together, and the bonded line can also be connected into a successive area by overlapping area between adjacent lines. Under the same laser power, the shorter the scanning spacing is, the larger the overlapping area is to enhance the bonding effects.

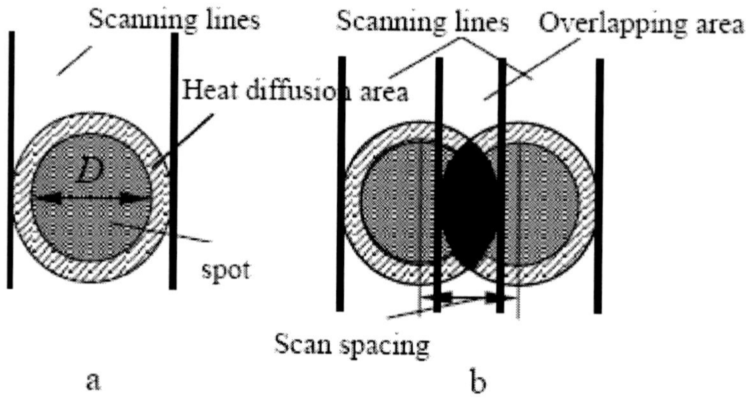

Figure 6. Scheme of the bonding state between adjacent scanning lines.

Forming Behavior Along the Prolonged Direction of Scanning Lines (Lines Forming)

Similar with the forming behavior between the adjacent scanning lines, there arc overlapping phenomenon between beam spots on scanning lines, which is not disclosed by the researchers before.

The radial distribution function of laser intensity can be expressed as $I(r)$, and r is the distance from laser beam spot center. Then the energy of laser beam spot can be calculated as:

$$P_E = \int_0^R I(r) \bullet 2\pi r dr \tag{11}$$

where R is the beam spot radius; P_E indicates the instant energy of each light impulse; under certain spot diameter. Given a laser beam spot scope, then the time (τ) during which laser beam shoots in the area when it is scanning at a certain speed can be expressed as equation 12:

$$\tau = \frac{2D}{v} \tag{12}$$

where D is the diameter of laser beam ; v is the scanning speed. Take a spot area as the investigated domain, and then the time from the entrance of laser beam on the left to the exit of laser beam on to right is just equal to τ . The moving scheme of laser beam spot is shown in figure 7. Here the whole energy inputted into the investigated domain is defined as equivalent energy.

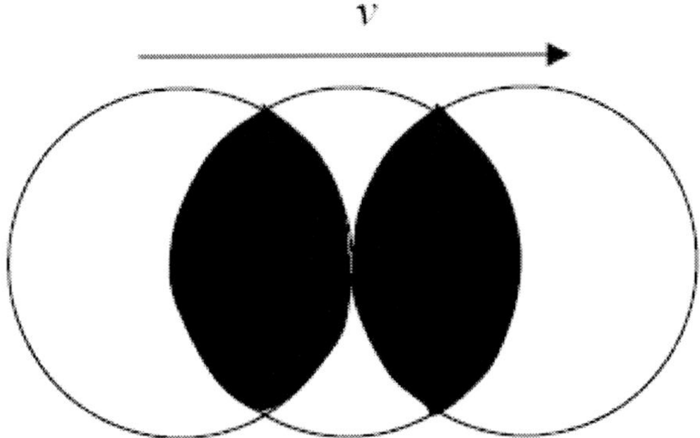

Figure 7. The moving scheme of laser beam spot.

Figure 8 describes two states of the position relationship between real spot and investigated domain. Circle 1, 2 stand investigated domain and real spot respectively. The polar coordinate equation of investigated domain peripheral curve (circle 1):

$$\rho_1 = R \tag{13}$$

Then the polar equation of beam spot peripheral curve (circle 2) is:

$$\rho_2^{\ 2} + (2R - vt)^2 - 2\rho_2(2R - vt)\cos\theta = R^2 \tag{14}$$

Circle 2 will move along the positive direction of polar to pass circle 1. The summation of spot energy at each moment can be calculated via the following equations

Combined the two circle equations, and let $\rho_1 = \rho_2 = R$, their intersect points can be obtained :

$$R^2 + (2R - vt)^2 - 2R(2R - vt)\cos\theta - R^2 \tag{15}$$

Then:

$$\theta = \arccos(1 - \frac{vt}{2R}) \tag{16}$$

It is obvious that the durations (t) in the first state of position relationship (figure 8b) and the second one (figure 8b) are $t \in \left[0, \frac{R}{v}\right]$ and $t \in \left[\frac{R}{v}, \frac{2R}{v}\right]$ respectively. On the moment of $t = \frac{2R}{v}$, the real spot will overlap the investigated domain fully and leave the domain from the other end. For the first state, the energy integral element is the partial ring area decided by thickness of $d\rho$ and the central angle of $2\theta(\pm\theta)$, and it can be expressed as: $I(\rho)\bullet\rho\bullet 2\theta\bullet d\rho$, integral by ρ:

$$P_1 = \int_{\rho} I(\rho)\rho 2\theta d\rho \tag{17}$$

where $\rho \in [R - vt, R]$, and θ is decided by circle 2 which can be shown as equation 18:

$$\theta = \arccos\frac{\rho^2 + (2R - vt) - R^2}{2\rho(2R - vt)} \tag{18}$$

Put equation 18 into equation 17 and P_1 can be expressed as equation 19 :

$$P_1 = \int_{\rho} 2\rho I(\rho)\arccos\frac{\rho^2 + (2R - vt) - R^2}{2\rho(2R - vt)} d\rho \tag{19}$$

The second state can be divided into two parts: the first energy integral area is the sealed scallop area composed of two polar radiuses corresponding to the meeting points of two circles and the arcs cut by the polar radius on circle 1. The energy integral element is similar with that in first state, but the central angle is $2\theta_0$ (figure 8c). Therefore the energy integral element of the first energy integral area can be $I(r)\bullet r\bullet 2\theta_0\bullet dr$, where r is the distance from the point on polar radius to the center of circle 1. The energy of this part can be expressed as equation 20:

$$P_2' = \int_0^R I(r) r 2\theta_0 dr \tag{20}$$

The energy integral element of second part is composed of arc $rd\theta$ and dr (figure 8c), and it can be expressed as $I(r)\bullet r\bullet d\theta\bullet dr$. Then energy of this part can be expressed as equation 21:

$$P_2'' = 2\int_{\theta}\int_{r(\theta)} I(r) r d\theta dr \tag{21}$$

where r is the distance from the point on polar radius of circle 2 to base point, and $\theta \in [\theta_0, \pi]$, $r(\theta)$ is the expression of circle 2 (r is expressed by θ).

Therefore the energy P_2 in the second state is the sum of P_2' and P_2'' :

$$P_2 = P_2' + P_2'' = \int_0^R I(r)r2\theta_0 dr + 2\int_{\theta}\int_{r(\theta)} I(r)rd\theta dr \tag{22}$$

Overall and for continuous wave laser, the energy of investigated domain can be expressed as P_{EAC} in equation 23:

$$P_{EAC} = 2\left(\int_{t_1} P_1 dt + \int_{t_2} P_2 dt\right) \tag{23}$$

while for pulse laser, pulse time and pulse interval should be considered together. One pulse periodic time (T_W) can be expressed as equation 24:

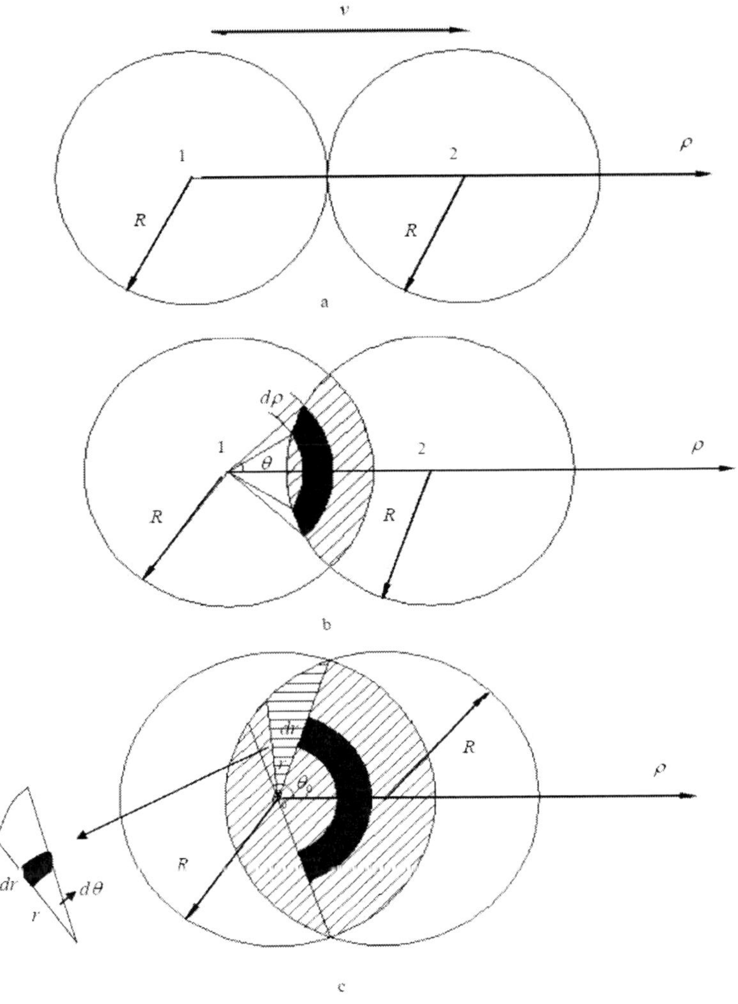

Figure 8. Two states of the position relationship between real spot and investigated domain.

$$T_W = T_{on} + T_{off} \tag{24}$$

where T_{on} stands for the time of laser irradiation; and T_{off} stands for the interval of adjacent irradiation. The initial state of laser omitted, the energy (P_{EAP}) of investigated domain can be calculated as equation 25:

$$P_{EAP} = 2\sum_{n=0}^{\text{int}(\frac{t_1}{T})} \int_{nT}^{T_{on}+nT} P_1 dt + 2\sum_{n=0}^{\text{int}(\frac{t_2}{T})} \int_{t_1+nT}^{t_1+T_{on}+nT} P_2 dt \tag{25}$$

It is indicated that there are overlapping area of beam spots in investigated domain, and overlapping extent can seriously affect the bonding effects of powders in scanning lines. Obviously within a certain scope, the higher the energy in investigated domain, the higher the bonding strength of the powders is.

Forming Behavior Affected by Laser Energy Transmission Depth

The variation of laser energy can induce the variation of its transmission depth which will further affect the bonding depth between the layers. The relation between bonding depth (h_s) and layer thickness (d_T) is shown in figure 9 (Liu jinhui, et al., 2006). The premise that adjacent layers can be bonded together is $d_T \le h_s$. The polymer in the part close to the upper suface of the layer can be remelted by laser energy penetrating the next layer, and this interweaved part is shown as dark black in figure 9b. The next layer is bonded with the former one though this interweaved part. In addition, all the layers can be bonded together according to this mechanism. It is obvious that the higher the laser power is, the higher the bonding strength between the layers is on condition of the same layer thickness.

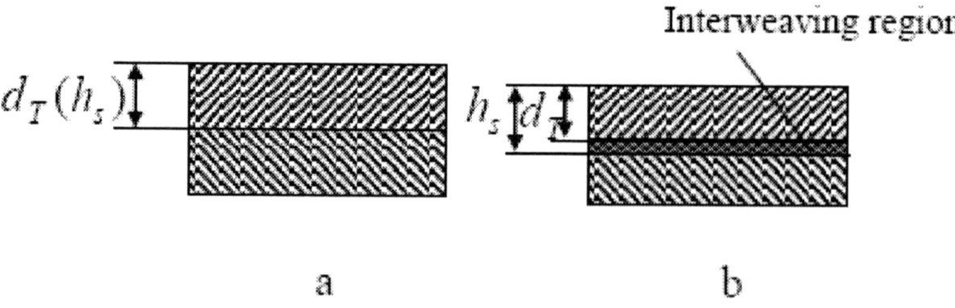

Figure 9. Bonding behavior between the layers.

In practice, the variation of laser energy transmission depth is controlled by energy provided by laser within a short period. On normal condition, one laser scanning line heats the powders to a certain temperature, and the temperature increment resulted from this scanning line will be damaged by the subsequent loss of heats via convection and emission ways. But the temperature variation induced by the next scanning line must be affected by the thermal effect of anterior scanning line. Furthermore the temperature within the anterior line will also be varied by the next scanning line. Deduced by analogy, the distribution of temperature field can't be uniform and it must be the function of time.

Effects of time and scanning lines interaction neglected, laser energy density is brought forward by Nelson(Nelson, 1993):

$$\rho = \frac{P}{Hv} \tag{27}$$

Equation 27 shows the laser energy density of a laser scanning area with the scanning spacing of H, the mean laser power of P and scanning speed of v. Considering the absorption of laser by powder material, we introduce powder absorptivity (α) into equation 27, and defining the effective absorption energy density (ρ_E) shown in equation 28 to explain the phenomenon of interlayer forming behavior. Obviously h_S in figure 9 will increase to enhance the interlayer strength with the rise of ρ_E.

$$\rho_E = \frac{\alpha P}{Hv} \tag{28}$$

Decomposition of Polymer

Polymer decomposes if laser energy is high enough. Seen as figure 10, the strength of green part increases slower during section b than that during the hinder part of section a and then decreases during section c, which is induced by polymer decomposition when laser power is high enough and other parameters are constant(Liu Jinhui, 2006). And the strength of the green part decreases due to the decreasement of polymer weted metal surfaces.

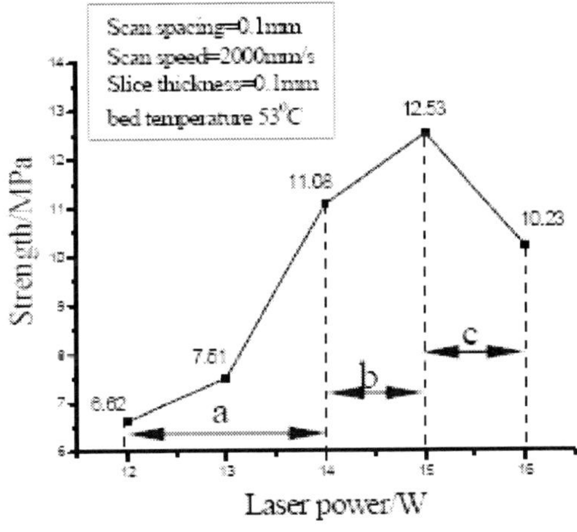

Figure 10. The relationship of the strength of the specimens and laser power.

SUMMARY AND CONCLUSION

Irradiated by laser beam during SLS process, the forming behavior of metal/polymer composite powders can be divided into three states, including the point state forming, line state forming and area state forming. These forming behaviors are all drived by the same thermodynamic mechanism which is the decreasment process of powders surface energy. Mover, the interaction between rigid powder and variable materials decides the energy states during a certain period. The variation of forming behaviors in beam spot, scanning lines and scanning areas are affected by laser energy, laser energy accumulation and laser energy distribution within a certain scope respectively. The final strength of green part is decided by forming behavior, which can decided by the difference between the initial surface energy and final one of the mixed materials system and can also be considered as stable extent from the angle of thermodynamic. In addition, polymer depcompsition will infuence the forming behaviors and affected the final strength of green part.

On the whole, the forming behavior of metal/polymer composite powders can be varied via adjusting the laser energy state which is controlled by SLS process parameters essentially. Therefore the investigation of forming mechanism and behavior during SLS of metal/polymer composite powders will be benefit to the choosing of process parameters and this work can further do good contribution to the application of SLS technology in the domain of manufacturing metal components and mould.

ACKNOWLEDGEMENTS

This work is financially supported by thc Chinese Postdoctor Science Foundation (20070410277), Chinese Medium and Mini Science-type Enterprise Innovatory Foundation (05C26214201059) and Postdoctor Foundation of Huazhong University of Sci. and Tech. (20070024).

REFERRENCES

N.P.; V. G.; G.. R. *RAPID PROTOTYPING J.* 1998, Vol.4, No.2, 77-89.

N.J. *Ph.D. dissertation "Selective laser sintering: A definition of the process and an empirical sintering model"*, The University of Texas at Austin, 1993.

H.Y.; L. R.; L. G. *Physical Chemistry.* advanced education press: Beijing, China, 1999, Second edition, pp 132-134.

L. J.; S.Y.; W. G. *J. Huazhong Univ. of Sci. and Tech.* 2006, Vol. 34, No.3, 54-57.

J. L.; Y. S.; Z.L. *ADV ENG MATER.* 2006, Vol.8, No.10, 988-994.

[169]

In: Powder Metallurgy Research Trends
Editors: L. J. Smit and J. H. Van Dijk
ISBN: 978-1-60456-852-3

Chapter 6

MECHANICAL PROPERTIES AND FRACTURE TOUGHNESS OF CERAMIC-METAL SINTERED NON-GRADED COMPOSITES AND FUNCTIONALLY GRADED MATERIALS

Keiichiro Tohgo*
Department of Mechanical Engineering, Shizuoka University,
3-5-1, Johoku, Nakaku, Hamamatsu, 432-8561, Japan

ABSTRACT

In order to evaluate distributions of mechanical properties and fracture toughness in ceramic-metal FGMs, mechanical properties and fracture toughness have been investigated on ceramic-metal non-graded composites (non-FGMs) and FGMs. The materials were fabricated by powder metallurgy using partially stabilized zirconia (PSZ) and austenitic stainless steel (SUS 304). Vickers hardness, Young's modulus and bending fracture strength are examined on smooth specimens of non-FGMs. The Vickers hardness continuously decreases with an increase in a volume fraction of SUS 304 metal phase, while the Young's modulus and fracture strength exhibit the low value in the non-FGMs with balanced composition of each phase. This suggests that the interfacial strength between the ceramic and metal phases is very low. The fracture toughness is determined by conventional tests for several non-FGMs with each material composition and by a method utilizing stable crack growth in FGMs. In contrast with the Young's modulus and fracture strength, the fracture toughness increases with an increase in a content of SUS 304 on both FGMs and non-FGMs and it is higher in the FGMs than in the non-FGMs. Under the assumption that the high fracture toughness in FGMs is attributed to the residual compressive stress in the ceramic-rich region created in the fabrication process, the residual stress distribution in the FGMs is estimated from the difference in fracture toughness between the FGMs and non-FGMs.

* Keiichiro Tohgo: Tel./Fax: +53-478-1027, e-mail: *tmktoug@ipc.shizuoka.ac.jp*

Keywords: Functionally graded materials (FGMs), Powder metallurgy, Mechanical properties, Fracture toughness, Stable crack growth, R-curve behavior, Residual stress.

1. INTRODUCTION

In functionally graded materials (FGMs), the material components such as metals, ceramics, plastics, pores and so on are graded continuously in some direction to derive the unique mechanical, thermal and electrical performances different from those of homogeneous or joined dissimilar materials [1, 2]. Figure 1(a) shows a schematic illustration of a ceramic-metal FGM plate in which a ceramic content varies gradually from 100% at the ceramic surface to 0% inside the plate. Due to the gradation of material composition, the FGM plate can be designed to reduce thermal stress with taking advantage of the heat and corrosion resistance of ceramics and the mechanical strength of metals. This concept of ceramic-metal FGMs are promising in thermal and structural applications such as thermal barrier coatings, wear and corrosion resistant coatings, ceramic/metal joinings and cutting tools [1, 2]. Mechanical performance of FGMs should be made clear to ensure their reliability and to extend their applications.

As shown in figures 1(b) and (c), the mechanical properties describing deformation (Young's modulus, Poisson's ration, yield stress, etc) and strength (tensile strength, ductility, fracture toughness, etc.) are also graded with the gradation of material composition. Therefore, the mechanical performance and fracture behavior of FGMs are very complicated. As shown in figure 2, surface cracks emanating at the ceramic surface of ceramic-metal FGM plates behave in several ways depending on the material gradation and loading condition. Multiple cracks, crack arrest, crack bending are well observed under thermal shock or thermal fatigue [3-6]. The crack branching and flaking occur as a result of crack growth along the interface in multi-layered FGMs [3-7]. In the analysis of the complicated fracture behavior in FGMs and in the evaluation of the mechanical performance of FGMs, the distributions of mechanical properties in FGMs as in figure 1 are indispensable.

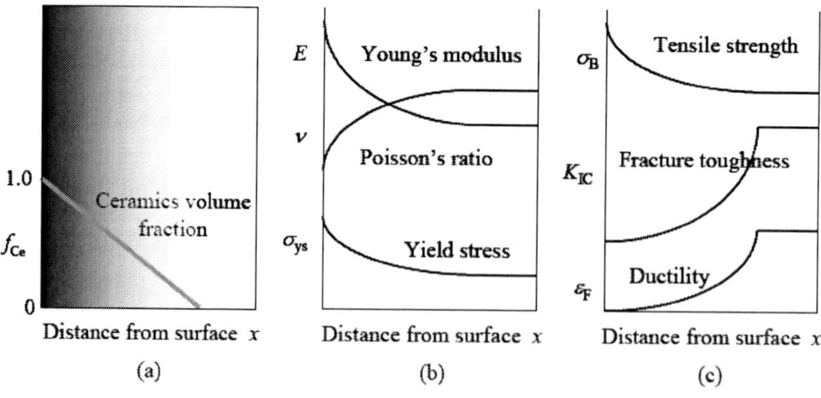

Figure 1. Distributions of mechanical properties in FGMs.

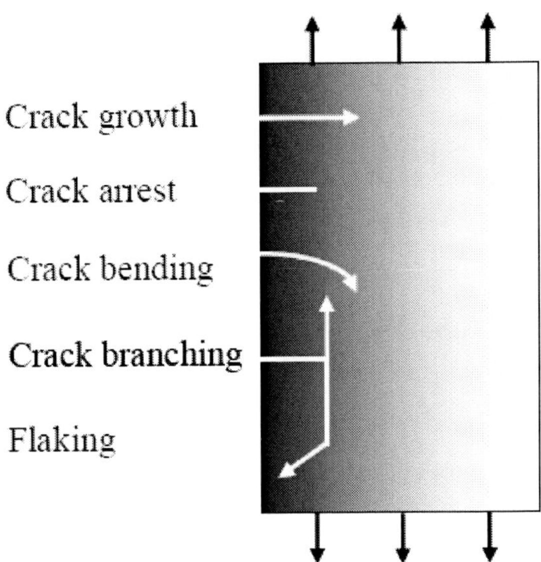

Figure 2. Cracking behavior in FGMs.

As FGMs are regarded as a kind of composite from their microstructure, models developed for composites can be applied to predict the distribution of deformation properties in FGMs by changing the material composition [8-12]. Owing to luck of appropriate methods to predict the strength of composites, a rule of mixture between the properties of constituent materials was applied as a simple way to evaluate the distributions of strength properties in FGMs. For the distribution of fracture toughness or *R*-curve behavior, prediction models based on the crack bridging model or cohesive model were proposed, where the metal phase in ceramic-metal FGMs plays an important role [13-16].

Ceramic-metal FGMs can be fabricated by physical or chemical vapor deposition, powder metallurgy, plasma spray, self-propagating high temperature synthesis, and so on, combined with some techniques to create the gradation of material composition [1, 2]. The produced FGMs may contain pores and micro cracks, week interfaces between the constituent materials, chemical reaction phases and residual stresses due to the mismatch of thermal expansion coefficients between the constituent materials, depending on the fabrication method and condition. Therefore, the actual distributions of mechanical properties are not necessarily consistent with the prediction based on the ideal models. The distributions of the mechanical properties in a given FGM are basically determined by experiments of non-graded composites with each material composition in the FGM or by the FGM itself. The experimental investigations on the mechanical properties and fracture behavior of FGMs were relatively limited compared with the theoretical and numerical investigations. For the fracture toughness, several investigations have been reported using ultraviolet-irradiation hardened polymer [17] or glass-particle reinforced epoxy [18], ceramic-metal FGMs [19-24], and ceramic-ceramic FGMs [25].

In the present investigation, non-graded composites (non-FGMs) with each material composition and FGMs are fabricated by powder metallurgy using partially stabilized zirconia (PSZ) and austenitic stainless steel (SUS 304), and the distributions of mechanical properties and fracture toughness are evaluated. Vickers hardness, Young's modulus and bending fracture strength were examined on smooth specimens of the non-FGMs. The fracture toughness is determined by conventional tests for several non-FGMs and by a method using stable crack growth for FGMs. Based on the experimental results, the influence of material composition on the mechanical properties and fracture toughness and the distribution of fracture toughness in the FGMs are discussed. Finally, the residual stress in FGMs created in the fabrication process is estimated from the difference in fracture toughness between the FGMs and non-FGMs.

2. Materials and Experimental Procedure

2.1. Materials

Non-graded composites (non-FGMs) and FGMs were fabricated by powder metallurgy by using commercially available partially stabilized zirconia (ZrO_2-3molY_2O_3, PSZ, manufactured by Atmix Corp.) and two kinds of austenitic stainless steel (SUS 304, manufactured by KCM Corp.) powders. The mean particle sizes of powders were 0.32μm for the PSZ and 45μm and 10μm for the SUS 304, respectively. The PSZ and SUS 304 powders were mixed in volume ratios of 10 to 0, 8 to 2, 6 to 4, 4 to 6, 2 to 8, 1 to 9 and 0 to 10, respectively, and each mixture was suspended in isopropyl alcohol, milled for three hours by a vibrational ball mill (Nissin Giken Corp.) and dried. The powder blends were put to form a non-graded composition or were layered to form a graded composition in a graphite die with 30mm in diameter. The powder compacts were pressed up to 14MPa at a room temperature, and then sintered under the condition of a temperature of 1200°C and compression of 30MPa for one hour in a vacuum by a hot-press machine (Shimazu Corp.). As a result, seven kinds of non-FGMs and one FGM for each combination of powders were obtained in the form of disk with 30mm in diameter and 6mm in height. These materials are referred to as non-FGM(45μm), FGM(45μm), non-FGM(10μm) and FGM(10μm) by using the mean particle size of SUS 304 powders, respectively. Figure 3 shows microstructures of the non-FGM(45μm) and non-FGM(10μm). The white and gray regions exhibit the SUS 304 and PSZ phases, respectively. The non-FGM(10μm) and non-FGM(45μm) have fine and coarse microstructures, respectively. As the particle size of the PSZ powder was considerably smaller than that of the SUS 304 powders, PSZ particles would have entered gaps between SUS 304 particles on the fabrication. As a result, the SUS 304 particles are dispersed in the PSZ matrix in the material with high content of PSZ (80%PSZ and 60%PSZ), while the small PSZ particles and platelets are dispersed in the SUS 304 matrix in the material with low content of PSZ (20%PSZ and 10%PSZ). The materials with intermediate content of PSZ (40%PSZ) exhibit the microstructures where the SUS 304 particles are dispersed in the PSZ matrix in some regions and both PSZ and SUS 304 phases interpenetrate each other in other regions. Area fractions of PSZ and SUS 304 phases in the microstructures were measured by image processing, and are listed in table 1 for the non-FGM(45μm) and non-FGM(10μm). The area fractions will be used as effective volume fractions instead of the nominal values in the fabrication. The FGMs have a functionally graded layer (FGM layer) of nominally 2mm in thickness on a SUS 304 substrate. Figure 4 shows a microstructure of a FGM layer in the FGM(45μm). The microstructure in the FGM layer varies stepwise from the monolithic PSZ on the surface to the monolithic SUS 304 on the substrate. A total thickness of the FGM layer and thickness of each layer in the FGM layer are slightly different from specimen to specimen.

Table 1. Area fractions of PSZ measured on non-FGMs

Materials (Nominal volume fraction of PSZ)		80%PSZ	60%PSZ	40%PSZ	20%PSZ	10%PSZ
Area fraction of PSZ (%)	Non-FGM (45μm)	79.2	60.6	38.1	17.0	10.4
	Non-FGM (10μm)	83.9	68.1	52.9	40.6	26.6

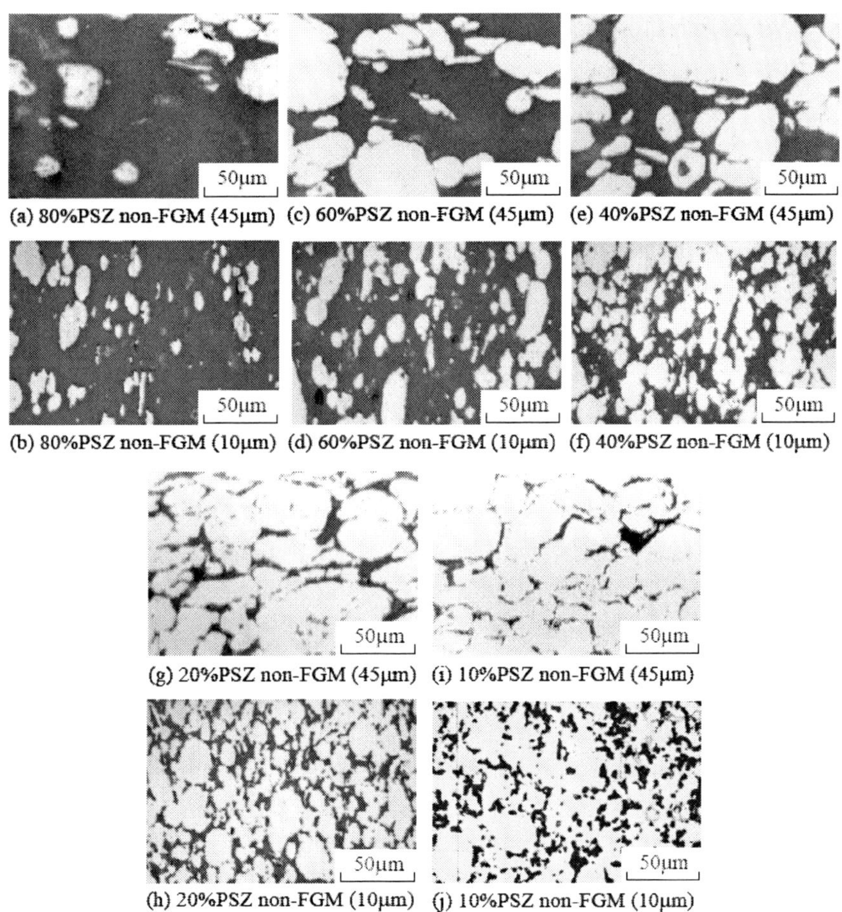

Figure 3. Microstructures of non-FGMs.

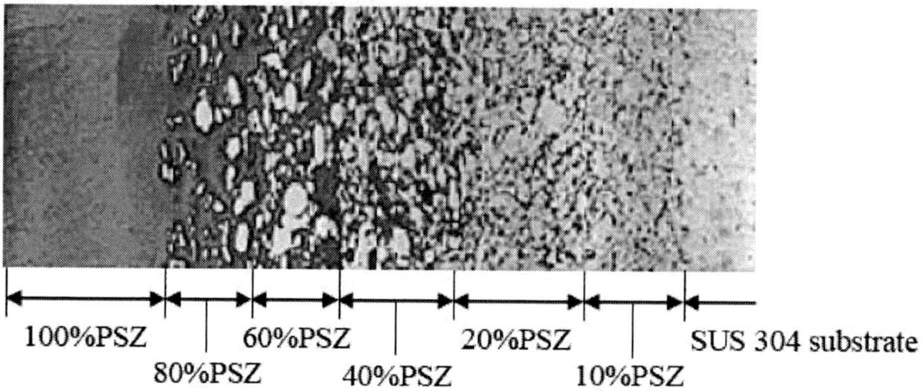

Figure 4. Microstruture of FGM(45μm).

2.2. Experimental Procedure

Rectangular specimens were cut out from the sintered disks and polished with diamond powder paste (particle size: 5μm). On these specimens, Vickers hardness, Young's modulus, bending fracture strength and fracture toughness were examined at a room temperature and air. The Vickers hardness was measured under the condition of applied load of 19.6N for the non-FGMs and 9.8N for the FGMs and holding time of 15 seconds on the side surface of the specimens. Four-point-bending tests and three-point-bending tests were carried out at a cross head speed of 0.05mm/min for the smooth non-FGM specimens as shown in figures 5 (a) and (b) to obtain the Young's modulus and bending fracture strength, respectively. The Young's modulus was determined from a linear relation between the bending stress and strains measured by strain gauges pasted on both surfaces of four-point-bending specimens. Fracture toughness tests of the non-FGMs were carried out on three-point-bending specimens with a sharp edge notch of 2mm in notch length and 15μm in notch root radius as in figure 5(c). To evaluate the fracture toughness distribution in the FGMs, stable crack growth tests were conducted on three-point-bending specimen with a starter-crack due to a series of median cracks introduced by Vickers indentation on the ceramic surface as in figure 5(d) [24]. The depth of the starter-crack was about 300μm. These tests were conducted at a cross head speed of 0.05mm/min. During the fracture toughness tests, the relationship between load (P) and load-point displacement (δ) was recorded, and the image of crack initiation and growth on the side-surface of the specimens was monitored and recorded in video recorder through a CCD camera. The test and video recording were started at the same time and the time was measured through the test so that the fracture process could be investigated in detail after the test.

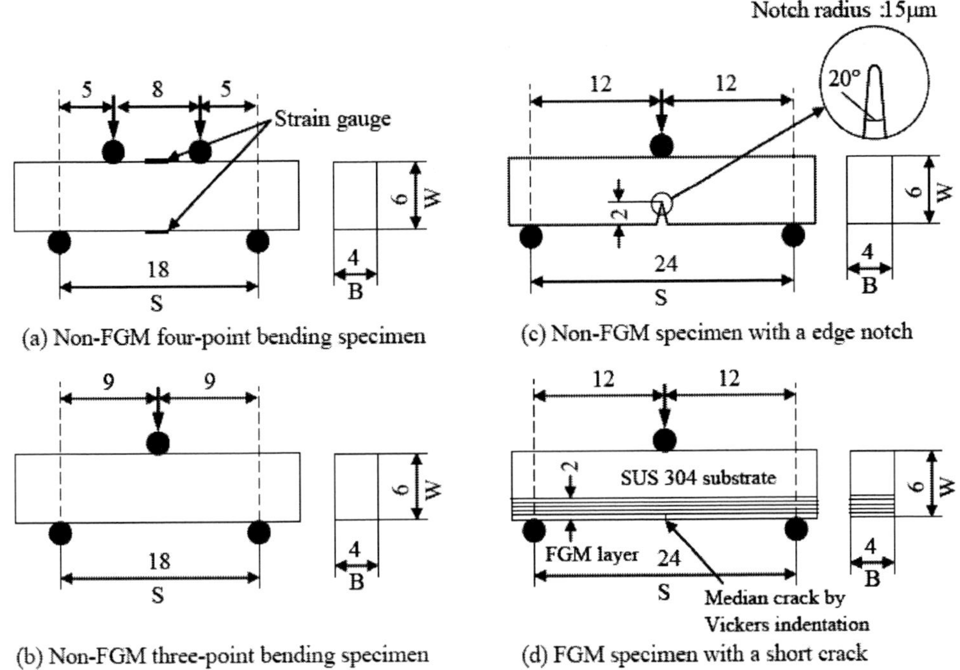

Figure 5. Specimen configurations. (dimensions in mm); (a) Non-FGM four-point bending specimen; (b) Non-FGM three-point bending specimen; (c) Non-FGM specimen with an edge notch; (d) FGM specimen with a short crack.

3. Vickers Hardness, Young's Modulus and Fracture Strength

Figures 6, 7 and 8 show the Vickers hardness, Young's modulus and bending fracture strength as functions of material composition on the non-FGM(45μm), respectively. In figure 6, each plot represents the mean value of five measurements for each non-FGM(45μm) and each region of the FGM(45μm). Although the Vickers hardness of the non-FGM and FGM continuously deceases from 13GPa for the PSZ to 2GPa for the SUS304 with an increase in a volume fraction of SUS 304, it is slightly higher in the FGM than in the non-FGM.

As the Young's moduli of the monolithic PSZ and SUS 304 are almost the same; namely, 190GPa for PSZ and 200GPa for SUS 304, almost constant values are expected in a full range of the material composition. However, as shown in figure 7, the Young's modulus exhibits low value in the composites with intermediate composition in the non-FGM(45μm). This trend was also observed in the non-FGM(10μm). This suggests that the initial defects are contained in the composites or the interfacial strength between both phases is very low. In composites with ceramic-rich composition (or metal-rich composition), if the interface between both phases is completely debonded, the composites behave as porous ceramic (or porous metal). In figure 7, the estimation of the Young's moduli for porous ceramic and porous metal is also plotted. The estimation was obtained by constitutive relation of particle-reinforced composites based on the Eshelby's equivalent inclusion method and Mori-Tanaka mean field concepts [26]. From the comparison between the experimental results and estimation for porous materials, it is suggested that the low Young's modulus in composites is attributed to the interfacial debonding between both phases, but not all interfaces are debonded.

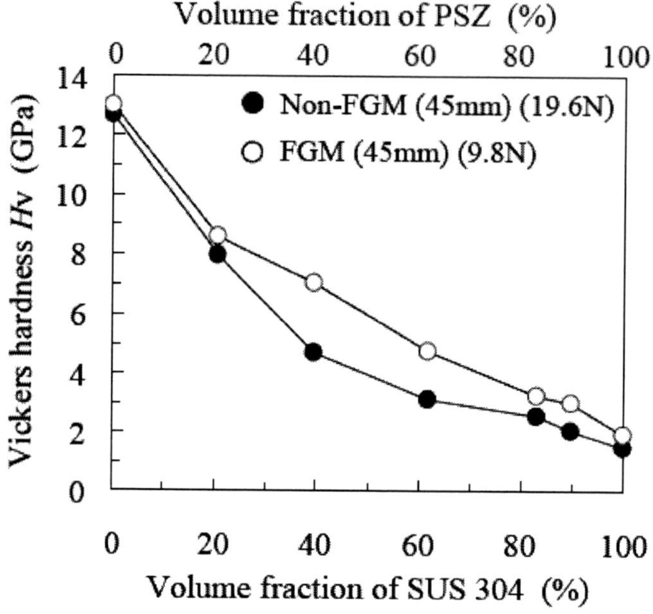

Figure 6. Vickers hardness as a function of material composition in non-FGM(45μm) and FGM(45μm).

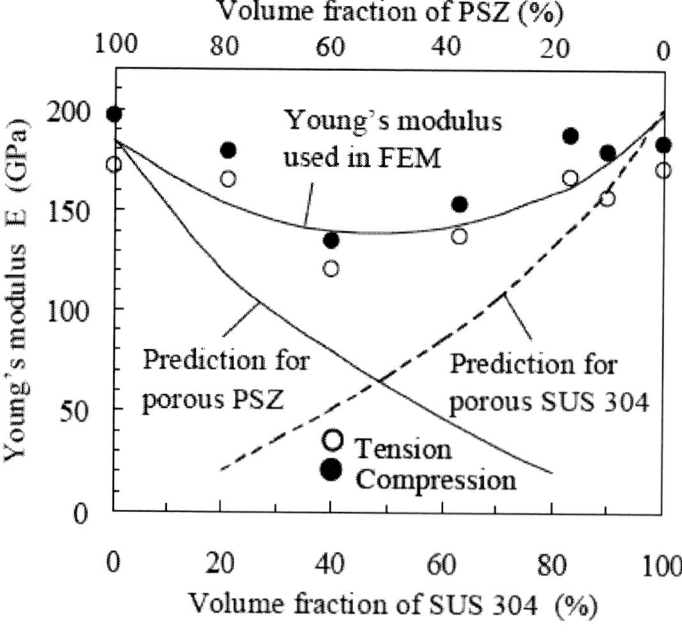

Figure 7. Young's modulus as a function of material composition in non-FGM(45μm).

The bending fracture strength for the materials with metal rich composition could not be obtained because the tests were terminated before final fracture owing to large plastic deformation. As shown in figure 8, the fracture strength for the monolithic PSZ is very high, and it considerably drops to around 300MPa on the composites with 20%, 40%, 60% and 80% SUS 304. This is also attributed to the weak interface between both phases in the composite system.

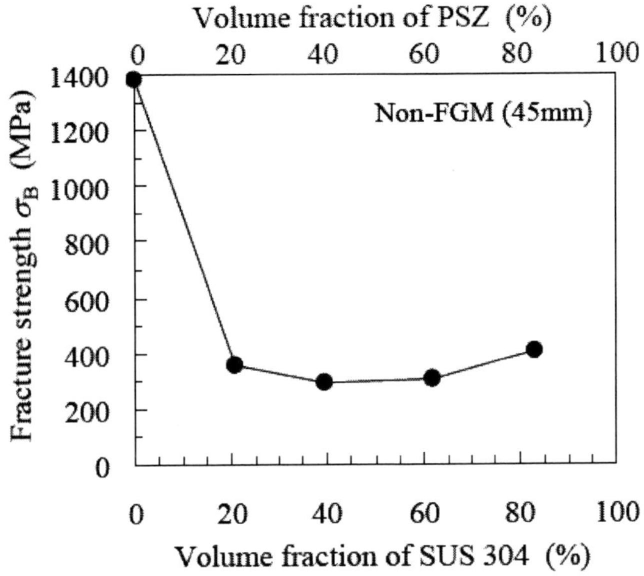

Figure 8. Bending fracture strength as a function of material composition in non-FGM(45μm).

4. Fracture Toughness of non-FGM

The load vs. load-point displacement relations obtained by the fracture toughness tests of the non-FGM(45μm) are shown in figure 9. In this figure, fracture process around a notch tip in the 10%PSZ non-FGM is schematically illustrated based on the observation by a CCD camera. Fracture process accompanied with the load-displacement relation on the 10%PSZ non-FGM is described as follows: The applied load increases linearly against the displacement up to a point A in figure 9, and then the load-displacement relation becomes nonlinear and a dent of the side surface corresponding to the plastic deformation around a notch-tip is observed. In the non-FGMs with low PSZ content, the matrix material is SUS 304. The plastic deformation seems to occur in the matrix around a notch-tip at this stage. At a point B before the maximum load (point C) a crack is initiated from a notch tip, and the load decreases with crack growth after the point C. Although the knee point corresponding to plastic deformation was not observed on the load-displacement relations of the non-FGMs with higher PSZ content than 40%, the crack initiation and extension before the maximum load were observed in all non-FGMs except for the monolithic PSZ. The monolithic PSZ exhibited unstable fracture at the maximum load. In the non-FGM(10μm), the load-displacement relations and fracture process were almost the same as in the non-FGM(45μm) except that the knee point in the load-displacement relation was not observed even in the 10%PSZ and 20%PSZ non-FGMs.

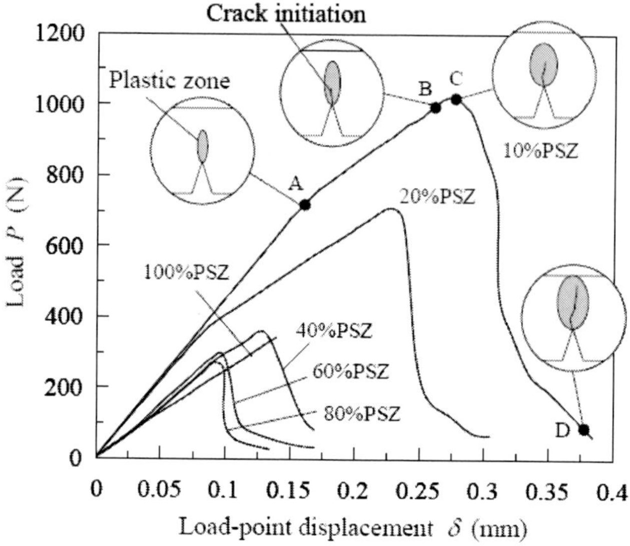

Figure 9. Load vs. load-point displacement relations obtained by fracture toughness tests of non-FGM(45μm).

The crack initiation before the maximum load suggests that the non-FGMs exhibit the *R*-curve behavior on their crack resistance, namely the fracture toughness for crack extension is higher than that for crack initiation. The *R*-curve behavior in metal-matrix composites is generally caused by crack bridging behind a crack tip due to frictional contact or unbroken ligaments and by plastic deformation of the metal phase [13-16]. Figure 10 shows the appearance of the crack growth observed on the side surface of 20%PSZ non-FGM(45μm) specimen. The crack bridging due to unbroken ligament is observed on the crack wake. The critical stress intensity factor K_R as the fracture toughness for crack extension was obtained from the maximum load and corresponding crack length.

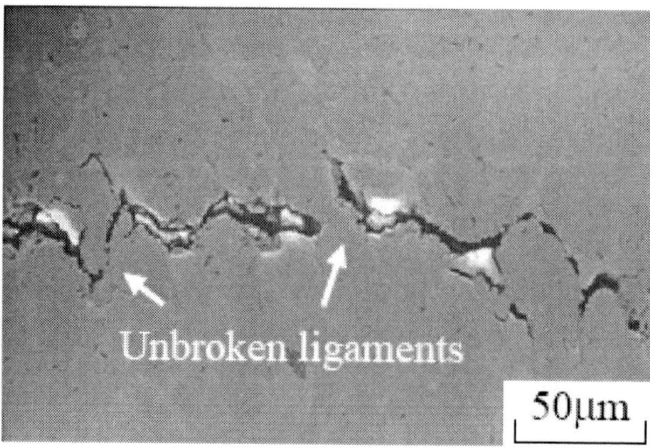

Figure 10. Crack bridging behind a crack tip unbroken ligaments in the 20%PSZ non-FGM(45μm).

The K_R values are plotted against the material composition of the non-FGM(45μm) and non-FGM(10μm) in figure 11. On the 100%PSZ specimen, the stable crack initiation and growth were not observed and unstable fracture from a notch occurred at the maximum load. This means that a sharp notch with notch root radius of 15μm still can not be regarded as a crack for 100%PSZ. Therefore, the fracture toughness obtained by the indentation fracture method [27] is plotted for the monolithic PSZ. Although the fracture toughness due to the indentation fracture method is also plotted for non-FGM(10μm) with 80%PSZ, it is slightly lower than that due to three-point-bending test. The fracture toughness increases with an increase in a metal phase content in both non-FGM(45μm) and non-FGM(10μm), and a little difference between both non-FGMs suggests a little influence of microstructure on the fracture toughness. A solid line in figure 11 represents the data points of both non-FGM(45μm) and non-FGM(10μm). The fracture toughness of both non-FGMs is at most $30\,\mathrm{MPa}\sqrt{\mathrm{m}}$ at 90% SUS 304, and is very low compared with the fracture toughness of 100% SUS 304 (more than $100\,\mathrm{MPa}\sqrt{\mathrm{m}}$). This implies that the fracture toughness of the PSZ-SUS 304 composites abruptly decreases between 100% SUS 304 and 90% SUS 304.

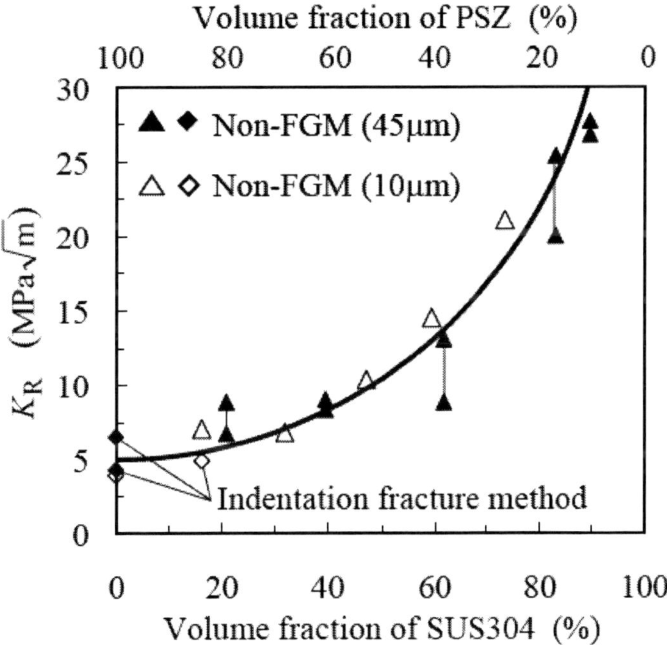

Figure 11. Fracture toughness as a function of material composition in non-FGMs. A solid line represents the data points of both non-FGMs.

Figure 12 shows optical micrographs of the side surfaces of the fractured specimens. On the PSZ matrix composites with high PSZ content, the crack path consists of flat fracture of PSZ matrix and interfacial debonding between PSZ and SUS 304. On the other hand, on the SUS 304 matrix composites with low PSZ content, the crack path is created by ductile fracture of SUS 304 along a sequence of PSZ particles. SEM micrographs of the fracture surfaces are shown in figure 13. On the non-FGMs with high PSZ content as shown in figures

13(b) and (c), the debonded SUS 304 particles and their traces are observed in the brittle fracture surface of the PSZ matrix. On the other hand, the metal-rich non-FGMs exhibit the brittle fracture surface of PSZ phase in the ductile fracture surface such as dimple-pattern of the SUS 304 matrix as shown in figures 13(d), (e) and (f). The area of ductile fracture surface increases with an increase in SUS 304 content. From these observations in figures 12 and 13, it is confirmed that the interfacial strength between the SUS 304 and PSZ is relatively low. However, the fracture toughness is still improved by an increase in a volume fraction of SUS 304 phase as in figure 11.

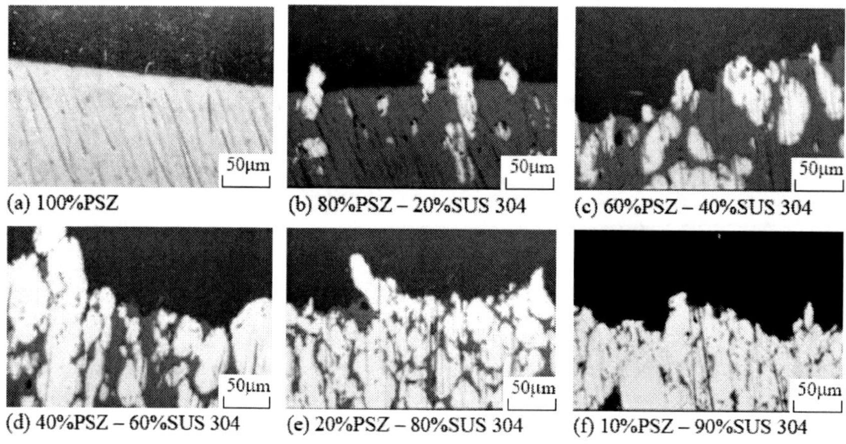

Figure 12. Micrographs of side surfaces of fractured non-FGM(45μm) specimens.

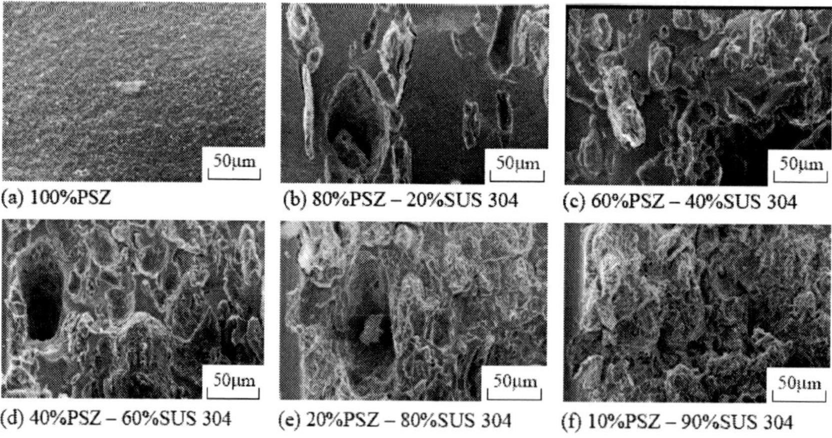

Figure 13. Micrographs of fracture surfaces of fractured non-FGM(45μm) specimens.

5. Fracture Toughness of FGM

5.1. Stress Intensity Factor of a Crack in FGM

Since the elastic moduli are graded corresponding to gradation of material composition in FGMs, the stress intensity factor of a crack in FGMs should be analyzed for each FGM even if the specimen configuration and loading condition are fixed. In the present material system, the gradation of Young's modulus was imagined not to be significant because of the slight difference in the Young's modulus between PSZ and SUS 304. However, the Young's modulus obtained by the experiment varied with the material composition as mentioned in Section 3.

Therefore, the stress intensity factor of a crack in the three-point-bending FGM specimens was analyzed with a finite element method taking into account the variation of Young's modulus corresponding to material composition in FGMs. Figure 14 shows the finite element mesh of three-point-bending specimen, where the Young's modulus is assigned to the Gauss points of each mesh based on the material composition and figure 7. The stepwise gradation of material composition in the FGM layer is set as 0.4 mm for 100% PSZ layer and 0.32mm for layers from 80% PSZ to 10% PSZ, and numerical analyses were carried out under the plane strain condition for the cases that a crack tip reaches to the center of each layer and to the SUS 304 substrate. The stress intensity factor K is determined by extrapolation of $K' \left(= \sigma_y \sqrt{2\pi x}\right)$ to a crack tip, where σ_y is the normal stress in y direction at the position of x ahead of a crack tip. The correction factor is defined by

$$F = \frac{K}{\sigma_0 \sqrt{\pi a}} \tag{1}$$

$$\sigma_0 = \frac{3PS}{2W^2 B} \tag{2}$$

where P, a, S, W and B are the applied load, crack length, span length, width and thickness of the edge-cracked specimen under three-point-bending, respectively.

Figure 15 shows the correction factors as functions of normalized crack length α (=a/W) for a crack in the three-point-bending FGM specimens. In the figure the correction factors for the homogeneous materials are also shown. As sown in figure 15, the correction factors for the FGM specimens decrease in the FGM layer, and they are consistent with those for homogeneous materials when the crack length is beyond the FGM layer.

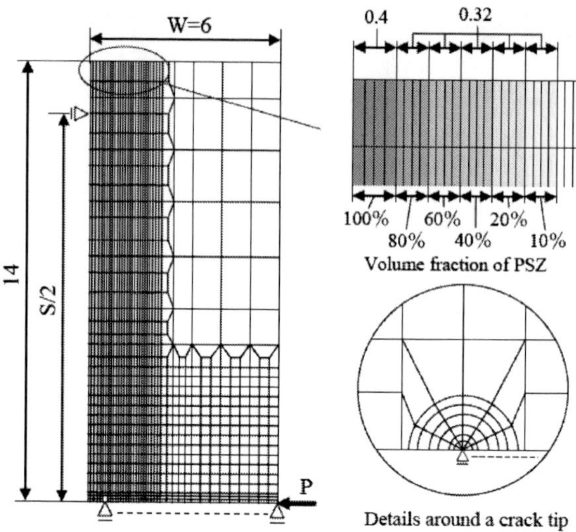

Figure 14. Finite element mesh of three-point bending specimen.

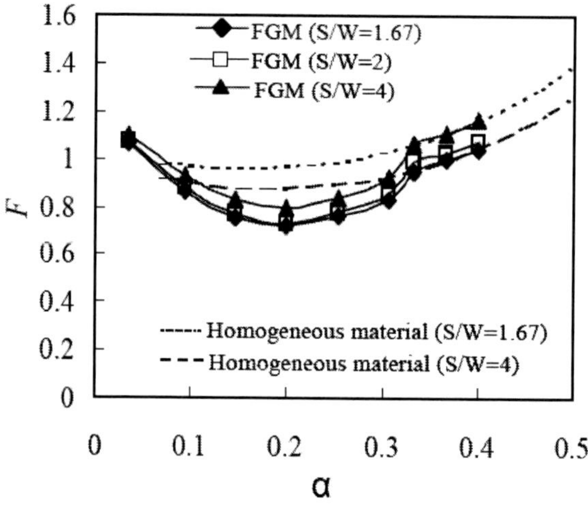

Figure 15. Correction factors of stress intensity in three-point bending FGM specimens.

5.2. Fracture Toughness of FGM

On the three-point-bending test of a FGM specimen with a short crack in the ceramic side surface as in figure 5(d), stable crack growth in a FGM layer can be realized depending on the distribution of material composition, specimen size, etc. Since the stress intensity factor during stable crack growth exhibits the fracture resistance at the crack-tip region, the distribution of fracture toughness in the FGM layer is obtained by one specimen [24].

On all FGM(45μm) specimens the stable crack growth in a FGM layer was realized. Figure 16 shows an example of load-displacement relations obtained by three-point-bending tests of FGM(45μm). The crack length measured on the video image after the test is also plotted in figure 16. When the applied load reaches to a point A in figure 16, the unstable crack growth occurs from a starter-crack to the 80% PSZ layer. However, any signal such as load drop is not observed on the load-displacement relation. Then, the crack propagates stably through the FGM layer with an increase in the applied load. The load-displacement relation is slightly nonlinear before the crack tip reaches to the 20% PSZ layer (Point C). At this stage, a plastic zone spreads out from the crack tip into the 10% PSZ layer and SUS304 substrate, and the applied load once drops then increases with an increase in the displacement. The crack propagates gradually into the plastically deformed 10% PSZ layer and finally is arrested at the SUS 304 substrate (Point D).

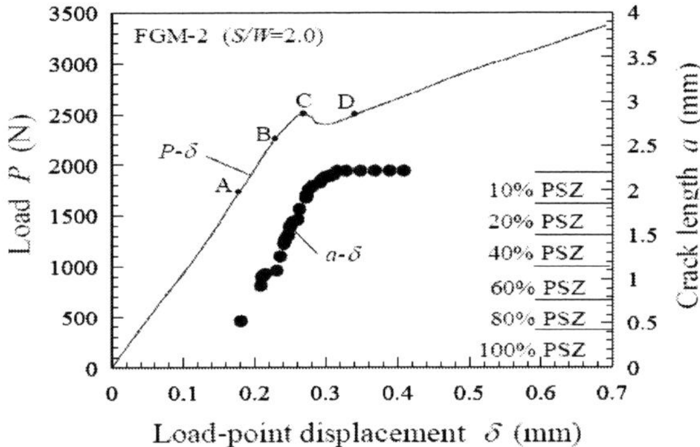

Figure 16. Load and crack length vs. load-point displacement in FGM(45μm).

On the other hand, on the FGM(10μm) specimen the stable crack growth was hard to occur and was realized in only one specimen out of several specimens. Figure 17 shows the load and crack length vs. load-point displacement relation of the specimen showing stable crack growth. When the applied load reaches to a point A in figure 17, the stable crack growth starts from a starter-crack. After the crack stably extends to 1.4mm in crack length at a point B, unstable crack growth occurs to 1.9mm in crack length and the load drops down to a point C. Then, the crack stably extends again to the interface between the FGM layer and the SUS 304 substrate, and the large plastic deformation spreads into the substrate.

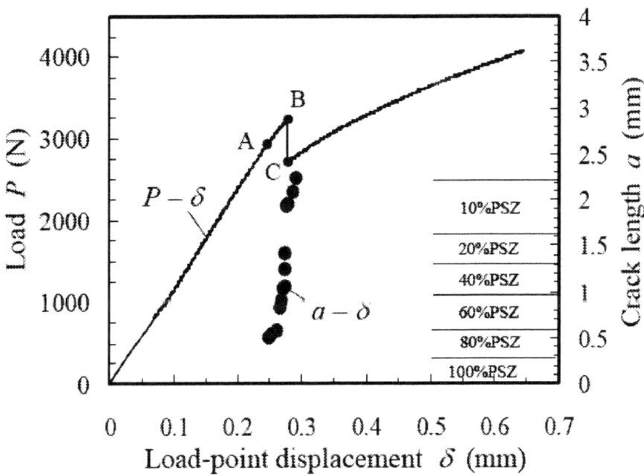

Figure 17. Load and crack length vs. load-point displacement in FGM(10μm).

From the relationship between applied load and crack length during the stable crack growth, the critical stress intensity factor (K_R) was calculated by using the correction factor in figure 15. Figures 18 and 19 show the fracture toughness K_R as a function of the crack length for the FGM(45μm) specimens and FGM(10μm) specimen, respectively. These figures describe the distributions of fracture toughness through the FGM layers, as the crack length means the distance from the ceramics surface. The fracture toughness increases with an increase in the distance from ceramics surface through the FGM layer. In figure 18 for the FGM(45μm), a slight difference in the distribution is observed among the specimens even in the same kind of FGMs. This may be attributed to the difference in the thickness of FGM layers, gradation of material composition, residual stress created in the fabrication, and so on. In figure 19 for the FGM(10μm), the fracture toughness between 1.4mm and 1.9mm in crack length is not obtained because of unstable crack growth. From the comparison between figures 18 and 19, it is found that the gradient of the fracture toughness in the FGM layer for the FGM(10μm) is more gentle than that for the FGM(45μm). This implies that the stable crack growth was harder to occur in the FGM(10μm) than in the FGM(45μm).

From these data, the relationship between the fracture toughness and material composition is obtained as in figure 20 for both FGM(10μm) and FGM(45μm). Although the fracture toughness increases with an increase in the SUS 304 volume fraction in both FGMs, it is higher in the FGM(10μm) than in the FGM(45μm) in contrast to the case of the non-FGMs.

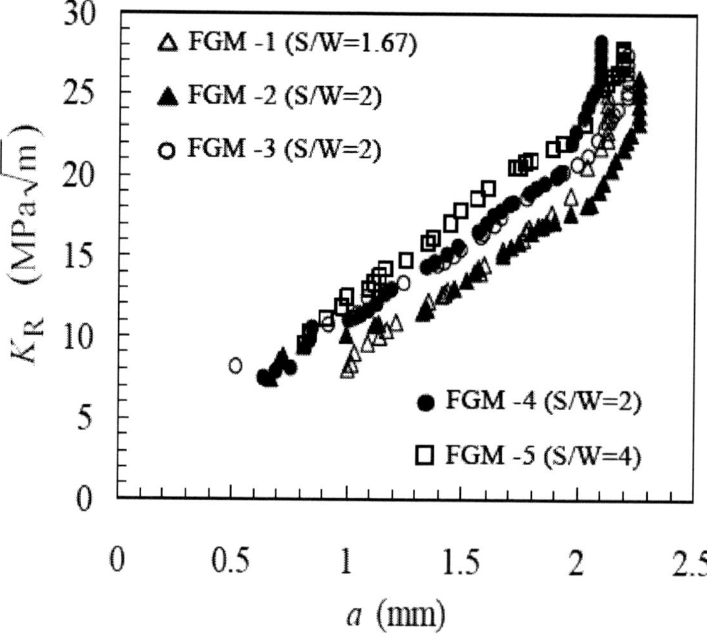

Figure 18. Distributions of fracture toughness in FGM(45μm).

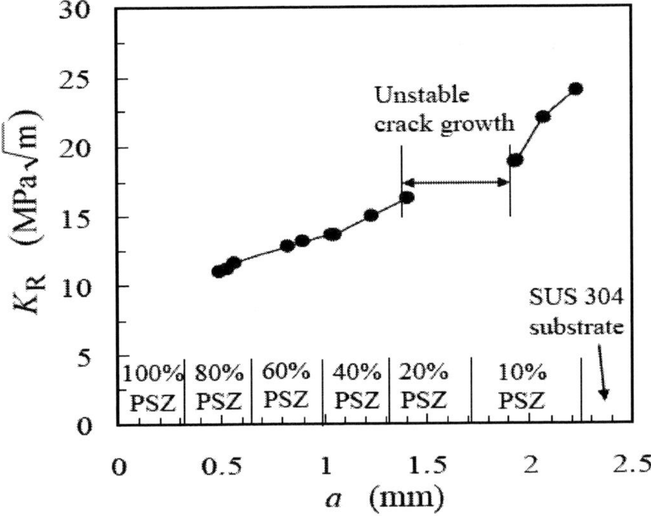

Figure 19. Distribution of fracture toughness in FGM(10μm).

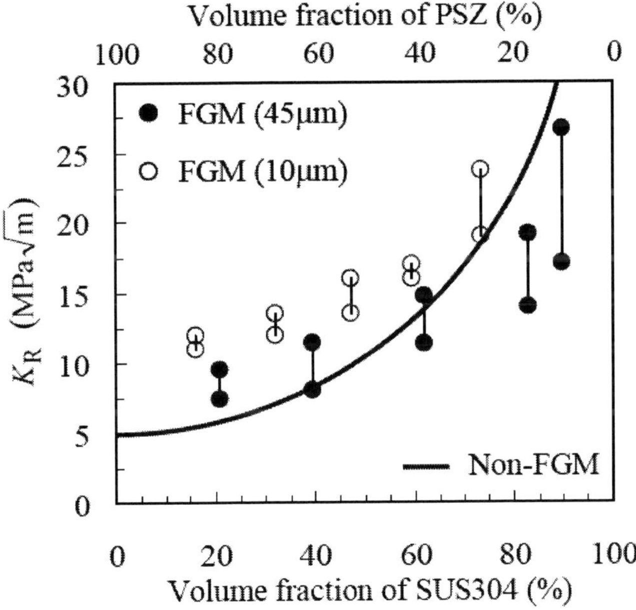

Figure 20. Fracture toughness as a function of material composition and comparison of fracture toughness among FGM(45μm), FGM(10μm) and non-FGMs.

5.3. Comparison of Fracture Toughness between the Non-FGMs and FGMs

In figure 20, the fracture toughness of both non-FGM(45μm) and FGM(10μm) is also shown by a solid line. The comparison of the fracture toughness between the non-FGM and FGM for the cases of 45μm and 10μm in particle size of SUS 304 powder is described as follows: In the case of 45μm, the fracture toughness of the FGM is slightly higher in the ceramic-rich region and slightly lower in the metal-rich region than that of the non-FGM. On the other hand, in the case of 10μm, the fracture toughness of the FGM is much higher than that of the non-FGM.

6. Estimation of Residual Stress in FGM

In inhomogeneous materials such as the non-FGM and FGM, microscopic and macroscopic residual stresses are created in a fabrication process due to mismatch of the thermal expansion coefficients between ceramic and metal phases, and their magnitude seems to be at most the level of yield stress because the plastic yielding can occur in the metal phase. The microscopic residual stress in the PSZ-SUS 304 system is compression in the ceramic phase and tension in the metal phase. Since the macroscopic residual stress is obtained by averaging the microscopic residual stress in a representative volume of microstructure, it vanishes in the non-FGM, and is compression in the ceramic-rich region and tension in the metal-rich region in the FGM. It is supposed that the high fracture toughness of FGMs is caused by the compressive macroscopic residual stress in the ceramic-rich region and this is one of advantages in FGMs. In this section, the macroscopic residual stress in the FGMs is estimated from the difference in fracture toughness between the FGMs and non-FGMs.

Figures 21 and 22 show fracture toughness distributions in the FGM-5(45μm) specimen in figure 18 and FGM(10μm) specimen in figure 19, respectively. $K_R^{FGM}(x)$ is an actual distribution and $K_R^{Non\text{-}FGM}(x)$ is obtained based on the fracture toughness of non-FGM in figure 11 and graded material composition. In the estimation of the residual stress a smoothing curve of stepwise $K_R^{Non\text{-}FGM}(x)$ is used. The difference in fracture toughness distributions is given as

$$K_{Res}(x) = K_R^{Non-FGM}(x) - K_R^{FGM}(x) \quad (3)$$

It is assumed that $K_{Res}(x)$ is due to the residual stress $\sigma_{Res}(x)$. If the specimen is large enough as compared with a crack length, the stress intensity factor of a crack under residual stress distribution is obtained by the one of a crack subjected to the same stress distribution on crack surfaces from the principle of superposition as shown in figure 23. In order to estimate the stress intensity factor of a crack in figure 23 (c), the result for an edge crack in a semi-infinite plate subjected to concentrated forces on the crack surfaces as in figure 24(a) is used [28]. The stress intensity factor is given as

$$K = \frac{2P}{\sqrt{\pi a}} \frac{F(b/a)}{\sqrt{1-(b/a)^2}} \quad (4)$$

$$F(\beta) = 1 + \left(1-\beta^2\right)\left(0.2945 - 0.3912\beta^2 + 0.7685\beta^4 - 0.9942\beta^6 + 0.5094\beta^8\right) \quad (5)$$

$$\beta = b / a \tag{6}$$

The residual stress distribution, namely the stress distribution on the crack surfaces in figure 23 is discretized as shown in figure 24(b). When the crack length is x_k, the stress intensity factor of a crack subjected to the stress distribution on the crack surfaces is obtained as

$$K_{\mathrm{Res}}(x_k) = \sum_{i=1}^{k} \frac{2\,\sigma_i\,\Delta x\,F(b_i / x_k)}{\sqrt{\pi\,x_k}\sqrt{1-(b_i / x_k)^2}} \tag{7}$$

From the above equation, the residual stress σ_k at the position b_k is described by the following relation:

$$\sigma_k = \frac{\sqrt{\pi\,x_k}\sqrt{1-(b_k / x_k)^2}}{2\,\Delta x\,F(b_k / x_k)} \left\{ K_{\mathrm{Res}}(x_k) - \sum_{i=1}^{k-1} \frac{2\,\sigma_i\,\Delta x\,F(b_i / x_k)}{\sqrt{\pi\,x_k}\sqrt{1-(b_i / x_k)^2}} \right\} \tag{8}$$

As $K_{\mathrm{Res}}(x_k)$ is known, the residual stress σ_1 at b_1 is given as

$$\sigma_1 = \frac{\sqrt{\pi\,x_1}\sqrt{1-(b_1 / x_1)^2}}{2\,\Delta x\,F(b_1 / x_1)} K_{\mathrm{Res}}(x_1) \tag{9}$$

Then, as σ_1 is obtained by Eq. (9), the σ_2 at b_2 is given by

$$\sigma_2 = \frac{\sqrt{\pi\,x_2}\sqrt{1-(b_2 / x_2)^2}}{2\,\Delta x\,F(b_2 / x_2)} \left\{ K_{\mathrm{Res}}(x_2) - \frac{2\,\sigma_1\,\Delta x\,F(b_1 / x_2)}{\sqrt{\pi\,x_2}\sqrt{1-(b_1 / x_2)^2}} \right\} \tag{10}$$

By continuing this procedure up to i=n, the residual stress distribution is obtained.

When we applied the above method to a semi-finite plate with an edge crack under uniform tensile stress, the uniform stress distribution was predicted except for the edge region of the plate corresponding to first several descretized segments.

In the numerical calculation, the distributions of $K_{\mathrm{Res}}(x)$ in figures 21 and 22 are used for the FGM(45μm) and FGM(10μm), and n=220 is set for the FGM layer of 2.2mm. The obtained results are shown in figure 25. The reasonable results are obtained: the residual stress in the FGMs is compression at the ceramic-rich surface region, and decreases and turns to tension with going through the inside. The compressive residual stress at the surface region is much higher in the FGM(10μm) than in the FGM(45μm). This difference in residual stress might be attributed to the size effect of microstructure on the flow stress and to the gradation of material composition in both FGMs.

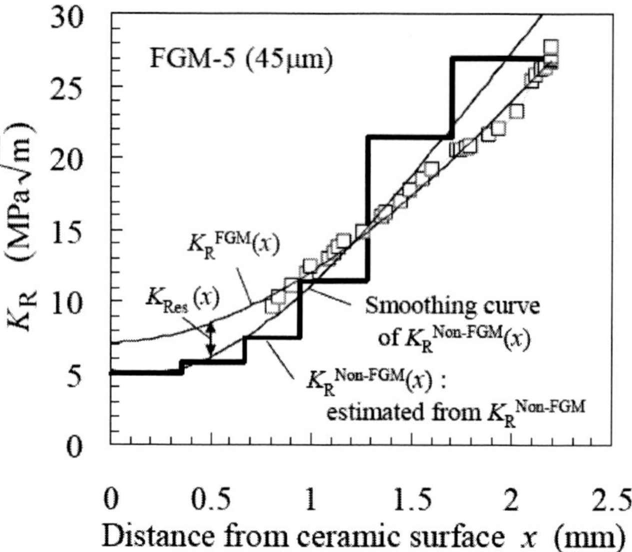

Figure 21. Fracture toughness distribution in a FGM-5(45□m). KRFGM(x) is an actual distribution and KRNon-FGM(x) is obtained based on the fracture toughness of non-FGM and graded material composition.

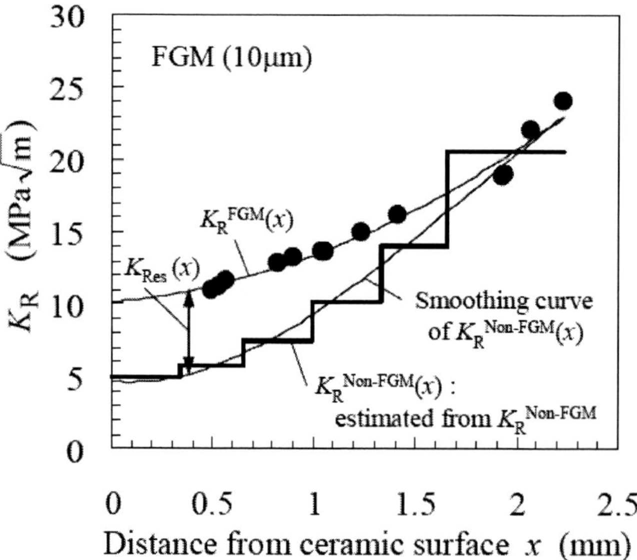

Figure 22. Fracture toughness distribution in a FGM(10μm). $K_R^{FGM}(x)$ is an actual distribution and $K_R^{Non\text{-}FGM}(x)$ is obtained based on the fracture toughness of non-FGM and graded material composition.

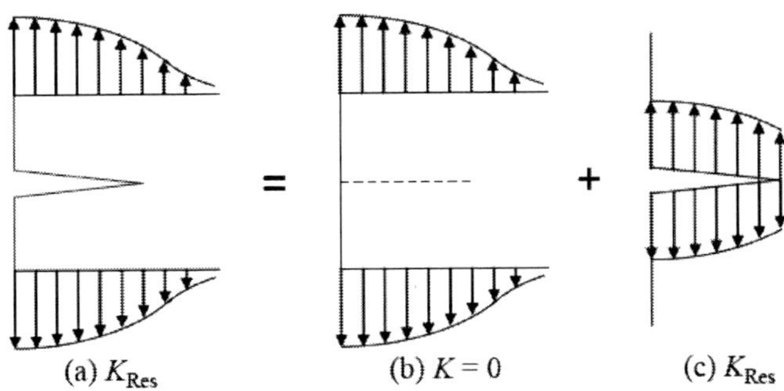

Figure 23. Principle of superposition for a crack under residual stress distribution.

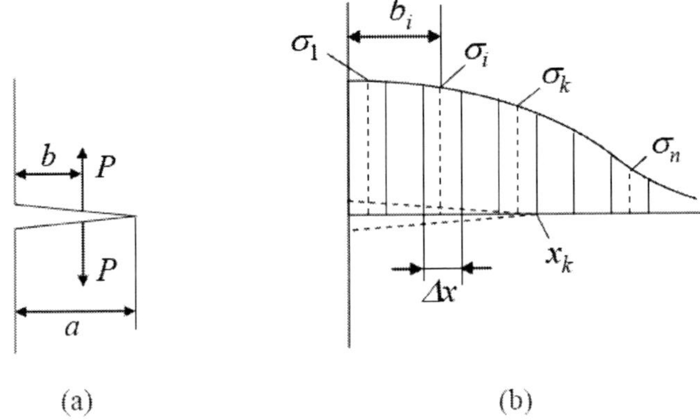

Figure 24. An edge crack in semi-infinite plate subjected to concentrated forces on the crack surfaces (a) and an edge crack under residual stress distribution (b).

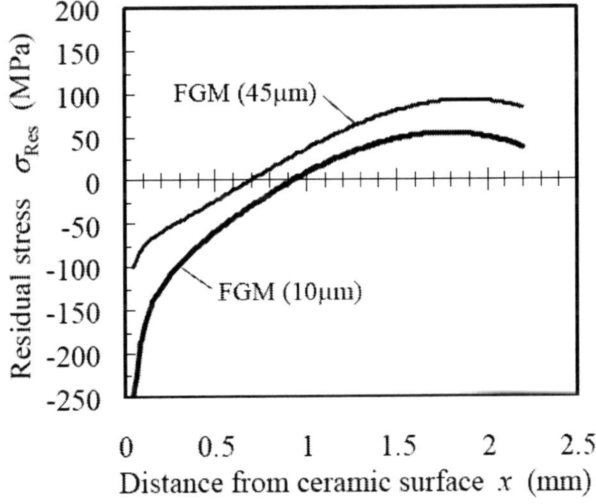

Figure 25. Distributions of residual stress in FGMs.

7. CONCLUSION

In order to evaluate the distribution of mechanical properties and fracture toughness in the PSZ-SUS 304 FGMs, mechanical properties and fracture toughness have been investigated on the sintered non-graded composites (non-FGMs) and FGMs. The obtained results are summarized as follows:

(1) The Vickers hardness continuously decreases with an increase in a volume fraction of SUS 304 metal phase, while the Young's modulus and fracture strength exhibit the low value in the non-FGMs with balanced composition of each phase. This suggests that the interfacial strength between the ceramic and metal phases is very low.

(2) In contrast with the Young's modulus and fracture strength, the fracture toughness increases with an increase in a content of SUS 304 on both FGMs and non-FGMs

(3) The influence of microstructure on the fracture toughness of the non-FGMs is negligible, while the fracture toughness is higher in the FGM with fine microstructure than in the FGM with coarse microstructure.

(4) The fracture toughness of the FGMs is higher than that of the non-FGM especially in the case of fine microstructure.

(5) The residual stress in the FGMs is estimated from the difference in fracture toughness between the FGMs and non-FGMs. As a result, it is found that high fracture toughness in FGMs is attributed to the compressive residual stress in the ceramic-rich surface region in FGMs, and that the compressive residual stress at the surface region is much higher in the FGM with fine microstructure than in the FGM with coarse microstructure.

REFERENCES

[1] Edited by Functionally Graded Material Forum of Japan. *Functionally Graded Materials*; Kogyo Chosakai Publishing Inc.: Tokyo, Japan, 1999.

[2] Miyamoto, Y.; Kaysser, W. A.; Rabin, B. H.; Kawasaki, A.; Ford, R. G. *Functionally Graded Materials: Design, Processing and Applications*; Kluwer Academic Publishers, 1999.

[3] Kawasaki, A.; Watanabe, R.; Yuuki, M.; Nakanishi, Y.; Onabe, H. Effects of microstructure on thermal shock cracking of functionally graded thermal barrier coatings studied by burner heating test. *Materials. Trans. JIM* 1996, 37, 788-795.

[4] Kawasaki, A.; Watanabe, R. Thermal fracture behavior of metal/ceramic functionally graded materials. *Eng. Frac. Mech.* 2002, 69, 1713-1728.

[5] Rangaraj, S.; Kokini, K. Estimating the fracture resistance of functionally graded thermal barrier coatings from thermal shock tests. *Surf. Coat. Technol.* 2003, 173, 201-212.

[6] Bahr, H. A.; Balke, H.; Fett, T.; Hofinger, I.; Kirchhoff, G.; Munz, D.; Neubrand, A.; Semenov, A. S.; Weiss, H. J.; Yang, Y. Y. Cracks in functionally graded materials. *Mater. Sci. Eng.* 2003, A362, 2-16.

[7] Tohgo, K.; Hadano, A. Characterization of fracture process in ceramic-metal functionally graded material under three-point-bending. *JSME Int. J. Ser. A* 2006, 49, 321-330.
[8] Hashin, Z. Analysis of composite materials - a survey. *Trans. ASME J. Appl. Mech.* 1983, 50, 481-505.
[9] Wakashima, K.; Tsukamopto, H. Mean-field micromechanics model and its application to the analysis of thermomechanical behaviour of composite materials. *Mater. Sci. Eng.* 1991, A146, 291-316.
[10] Mura, T. *Micromechanics of Defects in Solids*; Martinus Nijhoff Publishers: The Hague, 1982.
[11] Leβle, P.; Dong, M.; Schmauder, S. Self-consistent matricity model to simulate the mechanical behavior of interpenetrating microstructures. *Comput. Mater. Sci.* 1999, 15, 455-465.
[12] Tohgo, K.; Masunari, A.; Yoshida, M. Two-phase composite model taking into account the matricity of microstructure and its application to functionally graded materials. *Composites: Part A* 2006, 37, 1688-1695.
[13] Jin, Z. H.; Batra, R. C. Some basic fracture mechanics concepts in functionally graded materials. *J. Mech. Phys. Solids* 1996, 44, 1221-1235.
[14] Jin, Z. H.; Paulino, G. H.; Dodds, Jr. R. H. Cohesive fracture modeling of elastic-plastic crack growth in functionally graded materials. *Eng. Frac. Mech.* 2003, 70, 1885-1912.
[15] Jin, Z. H.; Batra, R. C. R-curve and strength behavior of a functionally graded material. *Mater. Sci. Eng.* 1998, A242, 70-76.
[16] Cai, H.; Bao, G. Crack bridging in functionally graded coatings. *Int. J. Solids Struct.* 1998, 35, 701-717.
[17] Li, H.; Lambros, J.; Cheeseman, B. A.; Santare, M. H. Experimental investigation of the quasi-static fracture of functionally graded materials. *Int. J. Solids Struct.* 2000, 37, 3715.
[18] Rousseau, C. E.; Tippur, H. V. Compositionally graded materials with cracks normal to the elastic gradient. *Acta Mater.* 2000, 48, 4021-4033.
[19] Tomsia, A. P.; Saiz, E.; Ishibashi, H.; Diaz, M.; Requena, J.; Moya, J. S. Powder Processing of Mullite/Mo functionally graded materials. *J. Eur. Ceram. Soc.* 1998, 18, 1365-1371.
[20] Rodriguez-Castro, R.; Wetherhold, R. C.; Kelestemur, M. H. Microstructure and mechanical behavior of functionally graded Al A359/SiC_p composite. *Mater. Sci. Eng.* 2002, A323, 445-456.
[21] Lin, J. S.; Miyamoto, Y. Notch effect of surface compression and the toughening of graded Al_2O_3/TiC/Ni materials. *Acta Mater.* 2000, 48, 767-775.
[22] Lin, C. Y.; McShane, H. B.; Rawlings, R. D. Structure and properties of functionally gradient aluminium alloy 2124/SiC composites. *Mater. Sci. Technol.* 1994, 10, 659-664.
[23] Fukui, Y.; Bowen, P. Fatigue crack propagation in an in-situ Al-Al_3Ni functionally gradient material. *Trans. Jpn. Soc. Mech. Eng. Ser. A* 1994, 60, 2048-2053.
[24] Tohgo, K.; Suzuki, T.; Araki, H. Evaluation of R-curve behavior of ceramic-metal functionally graded materials by stable crack growth. *Eng. Frac. Mech.* 2005, 72, 2359-2372.

[25] Moon, R. J.; Hoffman, M.; Hilden, J.; Bowman, K.J.; Trumble, K. P.; Rodel, J. R-curve behavior in alumina-zirconia composites with repeating graded layers. *Eng. Frac. Mech.* 2002, 69, 1647-1665.

[26] Japanese Industrial Standards. *Testing methods for fracture toughness of fine ceramics*; Japanese Standard Association, JIS R1607, 1995.

[27] Tohgo, K.; Chou, T. W. Incremental theory of particulate-reinforced composites including debonding damage. *JSME Int. J. Ser. A* 1996, 39, 389-397.

[28] Hartranft, R. J.; Sih, G. C. *Method of Analysis and Solutions of Crack Problems*, Sih, G. C.; Ed.; Noordhoff, Holland, 1973.

In: Powder Metallurgy Research Trends
Editors: L. J. Smit and J. H. Van Dijk

ISBN: 978-1-60456-852-3

Chapter 7

EFFECT OF MICROSTRUCTURAL PARAMETERS ON THE WEAR PROPERTIES OF HIGH VOLUME CERAMIC REINFORCED METAL MATRIX COMPOSITES

Farid Akhtar

Institute of Inorganic Chemistry, Stockholm University, Stockholm 10691, Sweden.
Department of Metallurgical and Materials Engineering,
University of Engineering and Technology Lahore,
Lahore 54890, Pakistan

ABSTRACT

Conventionally, hardness is a primary parameter to determine the wear resistance of high volume ceramic reinforced metal matrix composites. However, this study reports that the friction coefficient and wear characteristics of these metal matrix composites are strongly controlled by their microstructural parameters and are independent of their hardness. Ratio of average grain size to the mean free path of binder phase was found to be a controlling microstructural parameter for wear performance. Metal matrix ceramic reinforced composites with higher ratio of grain size to mean free path of the binder phase can exhibit higher wear resistance and lower friction coefficient.

Keywords: Metal Matrix Composites; Ceramic reinforcements; Microstructure; Wear; Hardness; Friction.

INTRODUCTION

Hardness is a primary parameter for estimating the wear resistance of a material. According to the previous research the wear loss varies inversely with the hardness [1]. In two phase metal matrix composites, wear resistance seems to be influenced by the hardness of the composite [2-5]. However, several researchers have found that materials with lower hardness exhibit better wear resistance [6-8]. Besides hardness, grain size and mean free path of the ceramic reinforcements and binder phase affect the wear performance of metal matrix composites [3,4,9]. The mean free path of the binder phase alters with the size of reinforcements having equal/same volume fraction of ceramic reinforcements in a composite. The effect of these microstructural variables on the wear behavior of the ceramic reinforced composite is not well understood and mostly the wear performance of these composites is linked with their hardness. In the present study, a physical relation between the two parameters, average Grain size of ceramic reinforcements and mean free path of binder phase, is described to predict the wear behavior as well as the operative wear mechanisms.

EXPERIMENTAL

TiC was chosen as reinforcement and 465 martensitic stainless steel/Cu as the binder phase. The TiC reinforced steel/Cu matrix composites were prepared by powder metallurgy [10]. The sintered composites were polished upto 0.5μm diamond finish for metallographically. Kevex Leo- 1450 SEM was used to observe microstructure. Quantitative metallography was used to determine the microstructural parameters (Average TiC grain size and mean free path of the binder). At least five micrographs taken from different areas were used to get the accurate measurements of microstructural parameters. Hardness test was performed on Shimadzu microhardness tester. Reciprocating sliding wear tests were performed on SRV tribometer. The wear tests of the composites against hardened high speed steel balls (10mm diameter) of hardness 62.5HRC were performed. The amplitude and frequency was 4.0mm and 40HZ respectively. The tests were performed up to 125000 oscillations. In all the tests, weight loss was measured. The applied loads were 100N and 300N. The values for the tangential force and the displacement pass through an analogue/digital card into a computer where, at specified intervals, friction force displacement loops were stored. The coefficient of friction was calculated as the dissipated energy per cycle divided by the normal load and twice the stroke.

RESULTS AND DISCUSSION

Three kinds of composites containing 78.7, 70.4 and 61.3vol% TiC reinforcements in martensitic stainless steel matrix are prepared. The average grain size of TiC reinforcements measured is 2.6, 3.1 and 3.7μm respectively. The grain size is changing with the vol% of binder phase because the microstructure of steel matrix TiC reinforced composites is evolved by continual process of dissolution of TiC in steel binder phase and then reprecipitation on TiC grains [10,11]. The mean free path for 78.7, 70.4 and 61.3vol% TiC reinforced metal matrix composites is 0.93, 1.2 and 1.75μm respectively. In two- phase metal matrix composites, soft and hard phases exist simultaneously. Thus, two types of wear conditions may prevail depending on the nature of contact and hardness of the counter body. The two types of wear are (i) equal wear of each of the two phases and (ii) variable wear (less/negligible for harder one) under constant applied load on each of the phases [12]. The first type of wear is more dominant in the case of hard body compared to the ceramic reinforcements while the variable wear applies more effectively in the case of hard ceramic reinforcements compared to the counter body. In our case, variable wear condition becomes more effective since the counter body has lower hardness than TiC reinforcements and thus wear loss happens due to the removal of the softer metallic matrix [13-15]. The counter body which is softer than the reinforcing phase but harder than the metallic matrix plough and groove across the matrix and push matrix phase into ridges along the sides of grooves and cause the matrix loss from the composite surface [11-13]. The contribution of ploughing and grooving mechanisms is restricted by the mean free path of the binder phase. Table-1 summarizes the effect of ratio of average TiC grain size (GS_{TiC}) to mean free path of the binder phase (MFP_{binder}) on the wear loss at 100N and 300N loads. It shows that the composite with higher ratio of av. TiC grain size to mean free path of the binder phase exhibits smaller wear loss. It indicates that the binder phase which has much lower hardness than the counter body is protected by the hard TiC particles. Thus, the applied load is primarily carried by the hard and rigid TiC reinforcements and the softer metallic matrix is protected. The counter body is unable to plough across the binder phase and at this stage the wear mechanism is polishing wear [2,13]. When GS_{TiC}/MFP_{binder} ratio is decreased, sudden increase in wear loss is observed. Now, the applied load is distributed over hard TiC reinforcements and metallic matrix. The metallic matrix is no longer protected by the reinforcements. Here, the operative wear mechanisms are ploughing and grooving and cause the observed wear loss of the matrix phase. Figure 1(a), (b) and (c) shows the worn surfaces of TiC reinforced steel matrix composites at 100N applied load with decreasing GS_{TiC}/MFP_{binder} ratio respectively.

Table 1. Effect of ratio (GS_{TiC}/MFP_{binder}) on the wear performance of the TiC reinforced steel matrix composite

TiC Vol%	Av. TiC Grain Size	MFP	Ratio of Av. TiC Grain Size to MFP	Composite Hardness	Binder Hardness	Wt loss at 100N	Wt loss at 300N
78.7	2.7ΜΜ	0.93ΜΜ	2.90	1411HV	392HV	0.4mg	1.2mg
70.4	3.1ΜΜ	1.23ΜΜ	2.52	1064HV	401HV	1.1mg	2.6mg
61.3	3.8ΜΜ	1.75ΜΜ	2.17	887HV	397HV	2.4mg	3.7mg

From these micrographs the relation between the GS_{TiC}/MFP_{binder} ratio and operative wear mechanisms is quite clear. Composite with the highest GS_{TiC}/MFP_{binder} ratio (1a) shows that the operative mechanism is polishing wear as all TiC particles are visible. This condition can be related to the polishing of samples for metallography. When the GS_{TiC}/MFP_{binder} ratio is decreased comparatively (1b) the polishing wear is accompanied with the plastic poughing of the matrix phase. At this stage, some of the TiC particles are covered with the binder phase. Composite with least GS_{TiC}/MFP_{binder} ratio (1c) shows that the operative wear mechanisms are plastic ploughing and grooving of the matrix phase and the surface of the composite is covered with the binder phase and no reinforcing particles are visible. The variation in friction coefficient with the sliding distance and GS_{TiC}/MFP_{binder} ratio is given in figure 2 and 3 respectively. It is following the same trend of wear resistance and showing dependency on the microstructure of the composite. The composite with GS_{TiC}/MFP_{binder} ratio shows lower coefficient of friction and composite with lower GS_{TiC}/MFP_{binder} ratio shows higher coefficient of friction. In the composites with lower GS_{TiC}/MFP_{binder} ratio, the friction coefficients are approaching to one another at higher load and are independent of the hardness and volume fraction of the reinforcing particles (figure 2b). A low friction coefficient and high wear resistance can be achieved in composites having high GS_{TiC}/MFP_{binder} ratio.

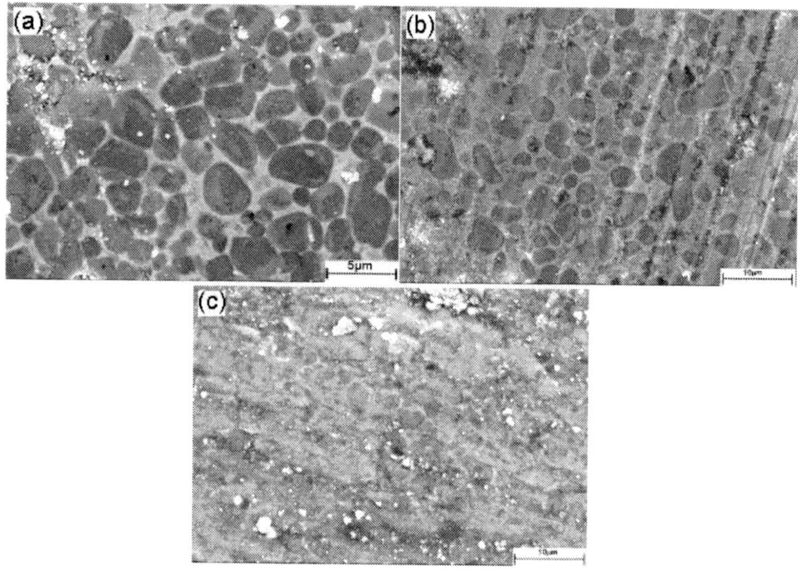

Figure 1. Worn surfaces of TiC reinforced 465 martensitic stainless steel matrix composites at 100N applied load with decreasing GS_{TiC}/MFP_{binder} ratio (a) 78.7 vol% TiC (b) 70.4 vol% TiC (c) 61.3 vol% TiC.

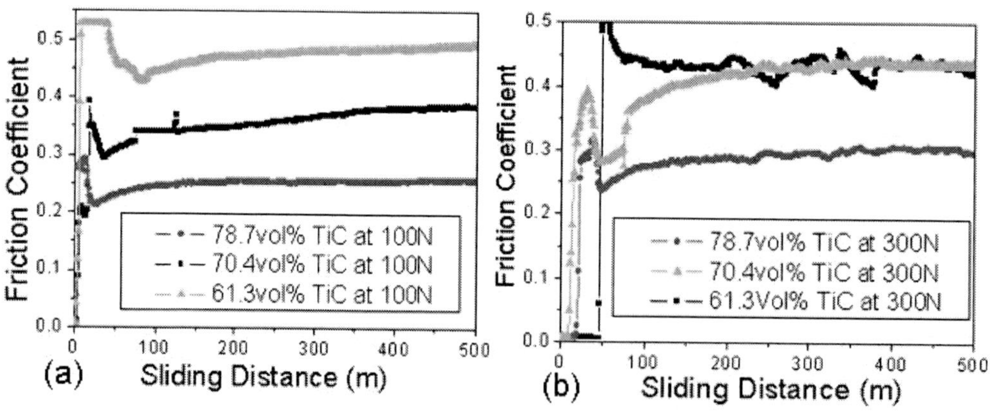

Figure 2. Variation in friction coefficient of TiC reinforced 465 martensitic stainless steel matrix composites (a) at 100N applied load (b) at 300N applied load.

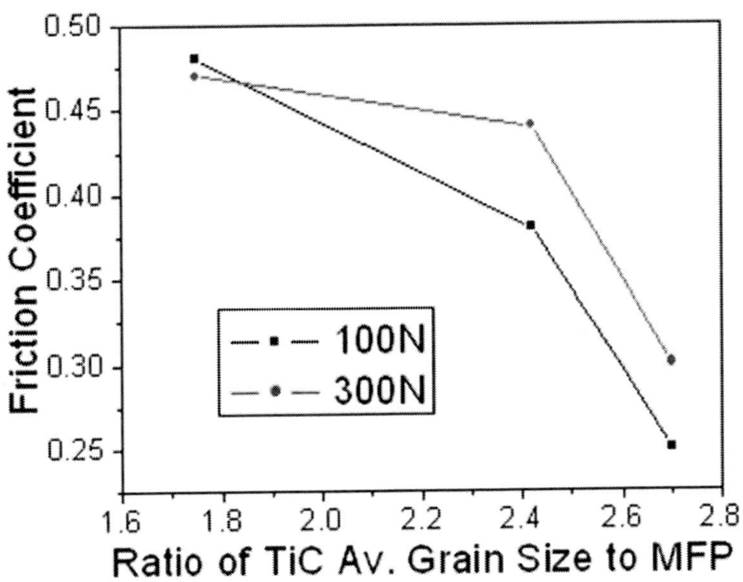

Figure 3. Variation in friction coefficient of TiC reinforced 465 martensitic stainless steel matrix composites with GS_{TiC}/MFP_{binder} ratio.

Table 2. Effect of ratio (GS_{TiC}/MFP_{binder}) on the wear performance of the TiC reinforced Cu matrix composite

TiC Vol%	Av. TiC Grain Size	MFP	Ratio of Av. TiC Grain Size to MFP	Composite Hardness	Binder Hardness	Wt loss at100N	Wt loss At 300N
78	2.2мm	0.9мm	2.44	719HV	101HV	0.5mg	1.4mg
69	2.3мm	1.25мm	1.84	518HV	86HV	6.1mg	14.1mg

Table 2 summaries the effect of GS_{TiC}/MFP_{binder} ratio on the wear performance of the TiC reinforced copper matrix composite. It confirms that the microstructural parameters strongly control the wear performance/mechanisms of metal matrix composites. The hardness of TiC reinforced copper matrix composite is much lower than the hardness of the TiC reinforced steel matrix composite but at higher GS_{TiC}/MFP_{binder} ratio the wear loss is almost the same against the same counter body and sliding wear conditions because the operative mechanism is only the polishing wear. When GS_{TiC}/MFP_{binder} ratio is decreased the wear loss is increased rapidly because the operative wear mechanisms are plastic ploughing and grooving of the binder phase [16,17]. The wear loss of TiC reinforced copper matrix composite compared to steel matrix composite is very high at lower GS_{TiC}/MFP_{binder} ratio. This is due to the inherent property of the binder phase. Copper is a soft metal and can be easily ploughed and grooved by the counter body (62.5HRC) compared to the martensitic stainless steel matrix. According to the Archard's equation, the volume loss per sliding distance is linearly proportional to the applied normal load and inversely proportional to hardness [1]. It is not valid for wear of two phase metal matrix composites. Therefore, the metal matrix composites can show a higher wear resistance even with lower hardness.

CONCLUSIONS

Finally, in metal matrix composites having reinforcements harder than the counter wear materials, the wear resistance is independent of the hardness of the composite and can be determined by the microstructure of the composite. The ratio GS_{TiC}/MFP_{binder} controls the wear performance of the composite and determines the operative wear mechanism under specific wear conditions. Control of microstructural parameters is the effective method to develop and assess the wear performance of high volume reinforced metal matrix composites.

REFERENCES

[1] Hutchings I M; Tribology: Friction and Wear of Engineering Materials, USA, CRC Press Inc., 1992.

[2] Skolianos S; Kattamis T Z; Chen M; B. Chambers V. Cast microstructure and tribological properties of particulate TiC-reinforced Ni-base or stainless steel matrix composites. *Mat. Sci. Engg. A,* 1994, 183 195-204.

[3] Reshetnyak H; Kubarsepp J. Structure sensitivity of wear resistance of hardmetals. *Int. J. Ref. Met. and Hardmater.* 1997, 15: 89-95.

[4] Tjong S C; Lau K C. Abrasive wear behavior of TiB_2 particle-reinforced copper matrix composites. *Mat. Sci. Engg. A.* 2000, 282,183-186.

[5] Kubarsepp J; Klaasen H; Pirso J. Behavior of TiC-base cermets in different wear conditions. *Wear.* 2001, 249, 229-234.

[6] Leyland A; Matthews A. On the significance of the H/E ratio in wear control: a nanocomposite coating approach to optimized tribological behavior. *Wear.* 2000, 246, 1-11.

[7] Leyland A; Matthews A. Design criteria for wear-resistant nanostructured and glassy-metal coatings. *Surf. Coat Technol.* 2004, 177-178, 317-324.

[8] Ni W; Cheng Y; Lukitsch M J; Weiner A M; Lev L C; Grummon D S. Effects of the ratio of hardness to Young's modulus on the friction and wear behavior of bilayer coatings. *App. Phys. Letts.* 2004, 85, 4028-4030.

[9] Klaasen H; Kubarsepp J. Abrasive wear performance of carbide composites. *Wear.* 2006, 261, 520-526.

[10] Akhtar F; Guo S J. On the processing, microstructure, mechanical and wear properties of cermet/stainless steel layer composites. *Acta Mater.* 2007, 55, 1467-1477.

[11] Hussainova I. Effect of microstructure on the erosive wear of titanium carbide-based cermets. *Wear.* 2003, 255, 121-128.

[12] Modi O P; Prasad B K; Jha A K; Dasgupta R; Yegneswaran A H. Low-Stress abrasive wear behaviour of a 0.2%C steel: Influence of microstructure and test parameters. *Tribo Letts.* 2003, 15, 249-255.

[13] Pirso J; Viljus M; Letunovits S. Friction and dry sliding wear behavior of cermets. *Wear.* 2006, 260, 815-824.

[14] Tjong S C; Lau K C. Properties and abrasive wear of TiB_2/Al-4%Cu composites produced by hot isostatic pressing. *Comp. Sci. Tech.* 1999, 59, 2005-2013.

[15] Rodriguez J; Martin A; Llorca J. Modeling the effect of temperature on the wear resistance of metals reinforced with ceramic particles *Acta Mater.* 2000, 48, 993-1003.

[16] Xu J; Liu W J. Wear characteristic of in situ synthetic TiB_2 particulate reinforced Al matrix composite formed by laser cladding. *Wear.* 2006, 260, 486-492.

[17] Tjong S C, Lau K C. Abrasion resistance of stainless-steel composites reinforced with hard TiB_2 particles. *Comp. Sci. Tech.* 2000, 60, 1141-1146.

Reviewed by
Professor Guo Shiju
University of Science and Technology Beijing, China.
[170]

In: Powder Metallurgy Research Trends
Editors: L. J. Smit and J. H. Van Dijk

ISBN: 978-1-60456-852-3

Chapter 8

REACTIVE HOT PRESSING OF IN-SITU METAL MATRIX COMPOSITES

S. C. Tjong[*]
Department of Physics and Materials Science,
City University of Hong Kong,
Tat Chee Avenue, Kowloon, Hong Kong

ABSTRACT

Reactive hot pressing (RHP) is widely recognized as an effective technique to prepare dense metal matrix composites (MMCs) reinforced with in-situ ceramic particulates/whiskers in one processing step. Ultrafine ceramic reinforcements are formed in-situ by exothermic reactions between elemental and/or compound powders during hot pressing. They are distributed uniformly within the metal matrix of MMCs. Moreover, the reinforcement-matrix interfaces are clean, resulting in a strong interfacial bonding. These enable effective load transfer from the metal matrix to ceramic reinforcements during mechanical loadings. Inherently, the in-situ MMCs exhibit improved mechanical strength, rendering them potential candidate materials for structural engineering applications. In the present review article, recent developments in the structure and mechanical properties of the RHP prepared aluminum, copper and titanium-based composites reinforced with in-situ ceramic particulates/whiskers are addressed and discussed. Particular attention is paid to the beneficial strengthening effects of in-situ TiB_2 particulates or TiB whiskers in the composites.

INTRODUCTION

In recent years, the demand for advanced structural materials for engineering applications is increasing owing to the fact that conventional metallic alloys cannot meet the needs of industrial sectors. In this regard, the composite material design by using ceramic reinforcement appears to be an effective strategy in improving the strength and reliability of

[*] E-mail: aptjong@cityu.edu.hk

metallic alloys. The improvement in mechanical strength derives from the load bearing capacity of the ceramic reinforcements. Metal matrix composites (MMCs) can be reinforced with continuous ceramic fibers, whiskers or particulates. Particulate-reinforced MMCs possess distinct advantages over fiber-reinforced composites in terms of low cost, ease of fabrication and isotropic mechanical property. The ceramic particulate- and whisker-reinforced MMCs have attracted increasing interest in the aerospace, automotive and chemical industries because of their high specific strength and stiffness, excellent wear and creep resistances compared to their monolithic alloy counterparts [1-5].

MMCs have been produced by a number of processing techniques including melting/casting and powder metallurgy (PM). Ingot casting offers substantial cost reduction advantages over the PM route. However, many structural effects such as gas porosities, clustering of ceramic reinforcements and formation of interfacial products are their distinct disadvantages. Such defects can affect the mechanical performances of MMCs [6]. In recent studies, much attention has been paid to the PM route with the advantages in controlling microstructure, improving ceramic particulate distribution, and obtaining the near-net-shape products. Generally, the ceramic particulates are prepared independently prior to the composite fabrication. The particulates are introduced into the metal matrix via ingot casting or PM process. The products can be viewed as ex-situ MMCs. The properties of ex-situ MMCs are influenced by the morphology of the particulates, their spatial distribution and interface interaction. The high cost of ceramic particulates is the key issue in making ex-situ MMCs. Furthermore, ex-situ ceramic particulates generally have a size ranging from a few to several hundred micrometers. Large ceramic particulates often act as stress concentrators and fracture readily during mechanical loadings. To overcome the inherent problems that are associated with conventionally processed composites, novel processing techniques based on the in-situ production of ceramic reinforcements have been developed. These techniques include exothermic dispersion (XD) [7], reactive hot pressing (RHP) [8-10], combustion synthesis (CS) or self-propagating high temperature synthesis (SHS) [11,12] and direct reaction synthesis [13]. In these techniques, the reinforcing phase is formed in the metallic matrix via in-situ chemical reaction between the matrix and the precursor elements or compounds during the composite fabrication. These composites are termed as in-situ MMCs. Among these, RHP process is attractive due to its simplicity, flexibility and ease of fabrication of consolidated composites. Ultrafine ceramic particulates are formed by the exothermic reaction between the element constituents under hot pressing conditions. The in-situ reinforcing particles are finer in size, and dispersed more uniformly in the host matrix. The reduction in the particle size is generally known to exhibit a beneficial effect in improving the mechanical strength of MMCs. In-situ formed ceramic particulates exhibit thermodynamic compatibility at the metal-ceramic interface, resulting in stronger metal-ceramic bonding. Such unique properties make in-situ MMCs possess more excellent mechanical properties than their ex-situ counterparts. It is worth-noting that advanced composite materials with matrices based on ceramics and intermetallics can also be prepared by RHP [14-17].

ALUMINUM-BASED COMPOSITES

Aluminum alloys are the most popular matrices for the MMCs due to their low density and adequate corrosion resistance. The density of pure aluminum is 2.70 g/cm^3. Aluminum is inexpensive when compared with other light metals such as magnesium and titanium. However, pure aluminum has shortcomings such as low melting temperature (933 K) and low mechanical strength. The strength of aluminum can be improved considerably by alloying in order to form fine precipitates. The structure and mechanical properties of ex-situ aluminum matrix composites have been extensively studied for the past two decades, particularly for those reinforced with SiC particles. However, SiC reinforcement could react with aluminum during composite fabrication, thereby forming chemical product at the matrix-ceramic interfacial region. Thus, efforts have been spent to search for other ceramic materials showing immunity to aluminum during processing. Titanium carbide and titanium boride are attractive ceramics for use as a reinforcing phase in aluminum matrix due to the fact that they exhibit outstanding features such as high modulus, high hardness, high melting temperature and inertness to aluminum. For example, titanium diboride (TiB_2) possesses a hardness of 33 GPa, modulus of 530 GPa and melting temperature of 2790 °C [18]. These reinforcing materials can be formed in-situ in aluminum matrix using RHP and melt reactive synthesis techniques. In the RHP route, the sizes of in-situ particles range from submicrometer to a few micrometers depending on the processing conditions and original size of raw powder materials. The submicron in-situ particles are effective in strengthening aluminum matrix via Orowan mechanism. It is noted that a non-ceramic phase can also be formed in the host matrix via manipulation of exothermic reactions in powder mixtures. More recently, Roy et al. reported that intermetallic compounds, i.e titanium aluminide, and iron aluminide, can be formed in-situ in the Al-based MMCs by means of the RHP process [19-22].

In-Situ Tic/Al-Based Composites

In-situ TiC reinforced Al-based composites have been fabricated from Al or Al alloy, Ti and C precursors by several researchers using SHS [11,12], RHP [23]and melt reactive synthesis [24,25]. The disadvantage of SHS process to prepare in-situ composite is the porous nature of synthesized products. Thus RHP and melt reactive synthesis are more effective techniques to prepare consolidated composites. For the Al-Ti-C system, exothermic reactions take place between elemental powder precursors, leading to the formation of TiC particles (TiC_p). However, brittle Al_3Ti phase and Al_4C_3 compound could be produced, depending on the processing techniques employed. For TiC/Al composites prepared by melt reactive synthesis of Ti, Al and C precursors, Al_3Ti was the first phase formed:

$$Ti + 3Al \rightarrow Al_3Ti \quad (1)$$

Subsequently Al_3Ti reacted with C to form the more thermodynamically stable TiC according to the following possible reactions [22]:

$$2\, Al_3Ti + C \rightarrow Ti_2AlC + 5\, Al \quad (2)$$

$$Ti_2AlC + C \rightarrow 2TiC + Al \quad (3)$$

From the X-ray diffraction (XRD) and microstructural examinations, Lu et al. reported that the Al_3Ti formation is favorable at 850 °C while Al_3Ti becomes unstable at reaction temperatures above 1000 °C. Thus, in-situ reaction of TiC must be carried out above 1000 °C.

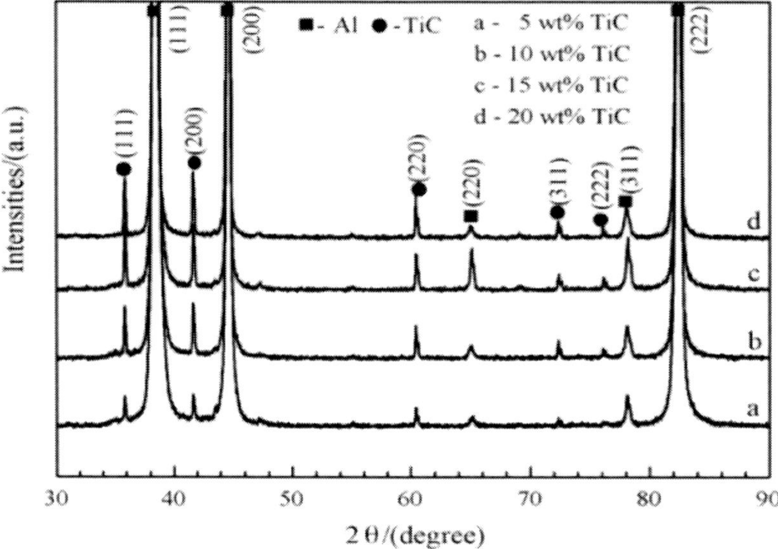

Figure 1. XRD patterns of in-situ TiC/Al-Cu-Mg composites Reprinted from [23] with permission of Elsevier.

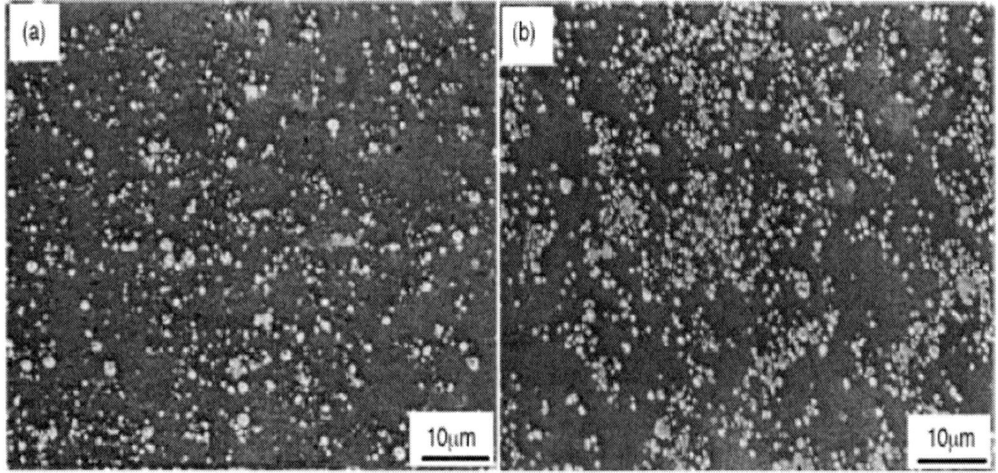

Figure 2. SEM micrographs of in-situ TiC/Al-Cu-Mg composites: (a) 5 wt% TiC and (b) 10 wt% TiC. Reprinted from [23] with permission of Elsevier.

From the literature, brittle Al_3Ti phase can be eliminated by optimizing the ratio of precursor materials in the reaction system, thereby leading to the change in thermodynamics and kinetics of the system. More recently, Zeng et al. prepared in-situ TiC/Al-Cu-Mg (LY12) composites by using RHP route [23]. In the process, TiC, aluminum alloy and graphite powders of different ratios (80, 85, 90 and 95 wt% Al, respectively, with C/Ti =1) were used as precursors. Al_3Ti was not detectable in the XRD patterns (figure 1) and SEM micrographs (figure 2) of the in-situ composite produced. The TiC particles with sizes of about 0.5-1.5 μm are finely distributed in the Al-based alloy matrix. The size of in-situ TiC particles is much smaller than that of ex-situ particles in discontinuously reinforced composites [1]. After T6 treatment, the yield stress, tensile stress and Young's modulus of the composites increased with the increase of TiC content. However, the tensile elongation decreased sharply with increasing filler content.

In-Situ Tib_2/Al-Based Composites

The in-situ TiB_2/Al-based composites have been successfully prepared from the Al-Ti-B system using melt reactive synthesis [26,27] and RHP [9] processes. In the former route, TiB_2 particulates can be synthesized in-situ from the reaction between Al-alloy with Ti salt (K_2TiF_6) and boron bearing salt (KBF_4). Formation of Al_3Ti phase can be suppressed by careful control of reaction temperature, reaction time and the Ti:B ratio. Yi et al. [26] reported that the TiB_2 particles synthesized in-situ in the Al-ally matrix (Al-12%Si-1.2%Cu-1%Ni-1%Mg; designated as ZL109) exhibit cubic and hexagonal morphologies, with sizes ranging from 0.2 to 0.8 μm (figure 3). The tensile strength of the base alloy and the in-situ composites at room and high temperatures are listed in table 1. Apparently, the in-situ TiB_2 particulates are very effective to improve the tensile strength of ZL109 alloy.

Ma and Tjong prepared the in-situ 15% TiB_2/Al composites by reactive hot pressing of Al (40 μm), Ti (60 μm) and B (2 μm) powders [9]. The ratio of these powders was properly adjusted so that 15 vol% TiB_2 was generated, assuming that the exothermic reactions occurred completely. In the process, the cold compacted powder mixture was heated above 800 °C in vacuum and maintained for 2 minutes, then cooled down to 600 °C and hot pressed [9]. For the Al-Ti-B system, Reaction (1) and the following reaction may take place during RHP process:

$$Ti + 2B \rightarrow TiB_2 \quad (4)$$

Table 1. Ultimate tensile strength (MPa) of the ZL109 alloy and its composites prepared by melt reactive synthesis. Reprinted from [26] with permission of Elsevier

Specimen	25 °C	150 °C	205 °C	260 °C	315 °C	400 °C
ZL109 (T6)	331	311	265	166	119	57
ZL109 (Cast)	241	---	180	---	66	---
1.1vol%/TiB_2/ZL109 (T6)	363	346	311	240	167	73
1.1vol%/TiB_2/ZL109 (Cast)	262	---	231	---	85	---
5vol%/TiB_2/ZL109 (T6)	376	361	343	271	189	84
5vol%/TiB_2/ZL109 (Cast)	289	---	259	---	109	---

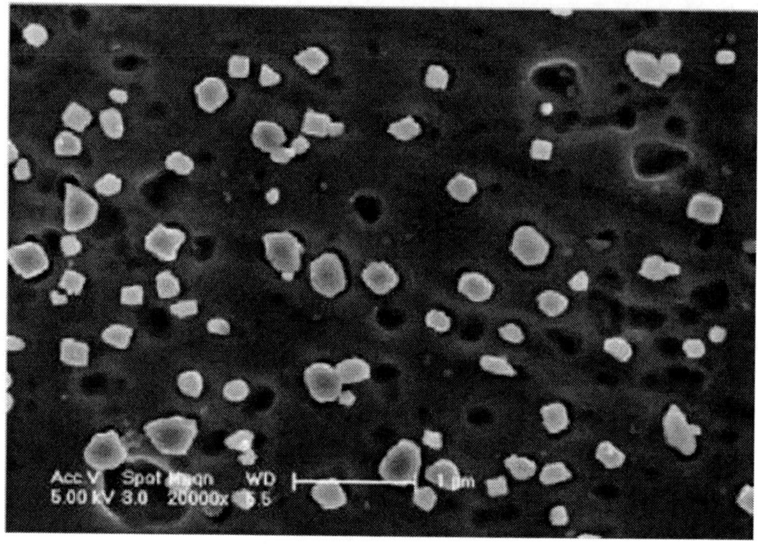

Figure 3. SEM micrograph of as-prepared 5vol% TiB_2/ZL109 (T6) composite. Reprinted from [26] with permission of Elsevier.

According to the thermodynamic data [29], Reaction (4) occurs more easily than Reaction (1). Liquid aluminum may induce in the Al-Ti-B reactive system containing low melting point constituent like aluminum during RHP. After the nucleation of TiB_2 in molten aluminum, the growth of the TiB_2 is controlled by the diffusion of Ti. Figure 4 shows the XRD pattern of the synthesized in-situ composite. The pattern reveals the presence of Al, TiB_2 and Al_3Ti peaks, indicating that TiB_2 and Al_3Ti are simultaneously formed in the composite. The microstructure consists of large Al_3Ti blocks with a size of several tens of micrometers and finer TiB_2 particulates. TEM micrographs (not shown) reveal that the in-situ TiB_2 particles exhibit cubic and rectangular morphologies [9].

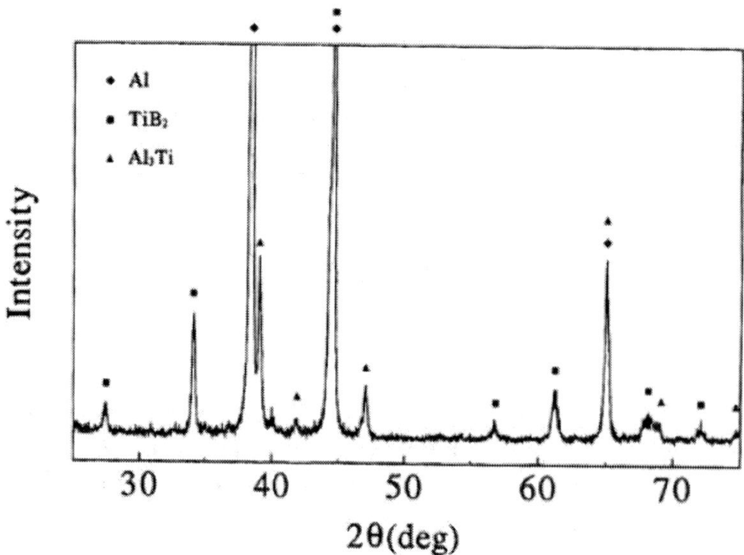

Figure 4. XRD pattern of in-situ 15 vol% TiB_2/Al composite [9].

The in-situ composite prepared from the Al-Ti-B system via RHP route contains large Al_3Ti blocks. It is necessary to eliminate this brittle phase in order to enhance the mechanical performances of the composite. This can be achieved by controlling the Ti:B ratio in the Al-TiO_2-B system [9,28]. Three kinds of in-situ reinforcements can form in the Al-TiO_2-B system during RHP process, i.e. TiB_2, Al_3Ti and Al_2O_3. When the B/TiO_2 ratio in the Al-TiO_2-B system is less than 2, the reaction between Al, TiO_2 and B takes place as follows:

$$3\ TiO_2 + 7Al + 4B \rightarrow 2\ Al_2O_3 + 2\ TiB_2 + Al_3Ti \qquad (5)$$

As the B/TiO_2 molecular ratio reaches 2, the reaction between Al, TiO_2 and B proceeds as follows:

$$3\ TiO_2 + 4Al + 6B \rightarrow 2\ Al_2O_3 + 3TiB_2 \qquad (6)$$

Table 2. Nominal chemical composition (vol.%) of *in-situ* phases and tensile properties of composites prepared from the Al-TiO_2-B system [28]

Specimen	TiB_2	Al_2O_3	Al_3Ti	Elongation, %	Yield strength, MPa	Tensile strength, MPa
(Al_2O_3 +TiB_2+Al_3Ti)/ Al	7.9	10.5	4.0	1.8	400	480
(Al_2O_3 +TiB_2)/Al	9.5	10.5	-----	4.0	545	638
Pure Al	-----	-----	-----	59.8	49	82

Figure 5 shows the XRD pattern of 22 vol% (Al_2O_3 +TiB_2+Al_3Ti)/Al and 20 vol% (Al_2O_3 +TiB_2)/Al composites fabricated from the Al-TiO_2-B system. For the former composite prepared with B/TiO_2 molecular weight ratio of 5/3, the XRD trace clearly reveals the presence of TiB_2, Al_2O_3, Al_3Ti and Al phases. Increasing the B/TiO_2 molecular weight ratio to 6/3, the diffracting peak intensity of TiB_2 phase tends to increase whilst that of Al_3Ti appears to decrease considerably. Figure 6(a) shows the microstructure of 22 vol% (Al_2O_3 +TiB_2+Al_3Ti)/Al. Extremely fine in-situ TiB_2 and Al_2O_3 particles (~ 0.3 μm) can be seen to disperse uniformly in aluminum matrix. Large Al_3Ti blocks with sizes ranging from ~ 10 μm are observed to distribute non-uniformly in the matrix. When the B/TiO_2 molecular ratio reaches 2, Al_3Ti blocks are nearly eliminated in 20 vol% (Al_2O_3 +TiB_2)/Al composite (figure 6(b)). Because of the formation of extremely fine in-situ TiB_2 and Al_2O_3 particles and elimination of Al_3Ti blocks, 20 vol% (Al_2O_3 +TiB_2)/Al composite exhibited higher tensile strength and tensile ductility than the 22 vol% (Al_2O_3 +TiB_2+Al_3Ti)/Al (table 2). Thus, Orowan mechanism is responsible for the strengthening of this composite. It should be noted the presence of *in-situ* reinforcements leads to a dramatic decrease in tensile ductility of aluminum. This is the typical characteristics of MMCs.

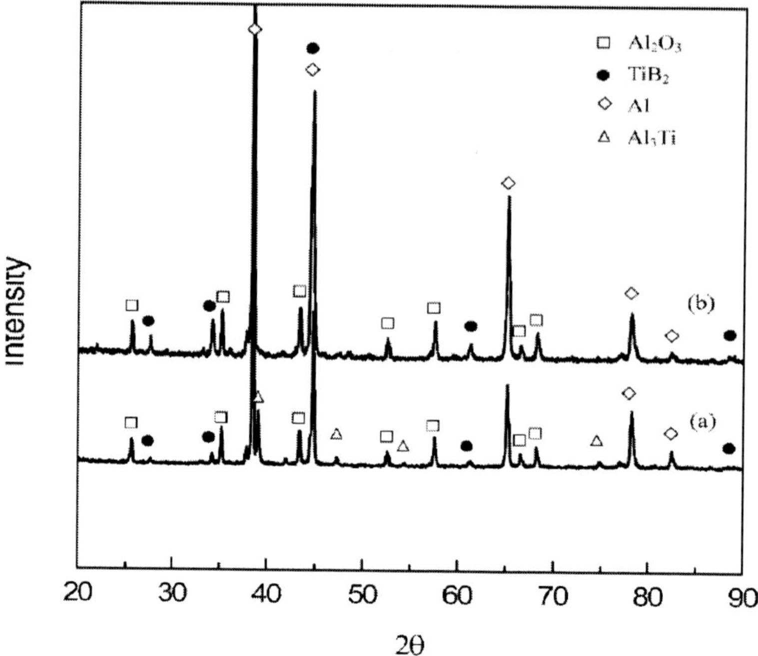

Figure 5. XRD patterns of (a) 22 vol% (Al_2O_3+TiB_2+Al_3Ti)/Al and (b) 20 vol% (Al_2O_3 + TiB_2)/Al composites [28].

Formation of the brittle Al_3Ti phase can also be suppressed by adding C in the Al-Ti-B and Al-TiO_2-B systems [30, 31]. By incorporating carbon into Ti and B powders and Al wire during melt reactive synthesis, the Al_3Ti phase would react with C to form TiC preferentially [30]:

$$Al_3Ti + C \rightarrow 3Al + TiC \tag{7}$$

Figure 7 shows the XRD traces of Al-15 vol% TiB_2 composite by incorporating 2 and 3 wt% C into the Ti and B mixture during melt reactive synthesis. Prominent TiC peak can be readily seen in addition to Al , TiB_2 and Al_3Ti peaks for the composite containing 2 wt% C. The intensity of Al_3Ti peak diminishes by increasing the carbon content to 3 wt%. Figure 8 shows the microstructure of the composite with the addition of 3 wt% carbon. The TiB_2 and TiC particles are observed to be homogeneously dispersed in the Al matrix. Tee et al also demonstrated that the addition of 3 wt% C to the Al-4.5Cu alloy also facilitates formation of in-situ TiC and TiB_2 particles. The tensile stress (258 MPa) and elongation (10.3%) of resulting 15vol%(TiC+TiB_2)/AlCu composite are higher than the 15vol%TiB_2/AlCu composite having tensile strength of 237 MPa and elongation of 6.6 %. Similarly, Tjong and Wang reported that the elimination of Al_3Ti phase in the composite prepared from the TiO_2-Al-B-C system by RHP technique is beneficial in improving the tensile ductility [31].

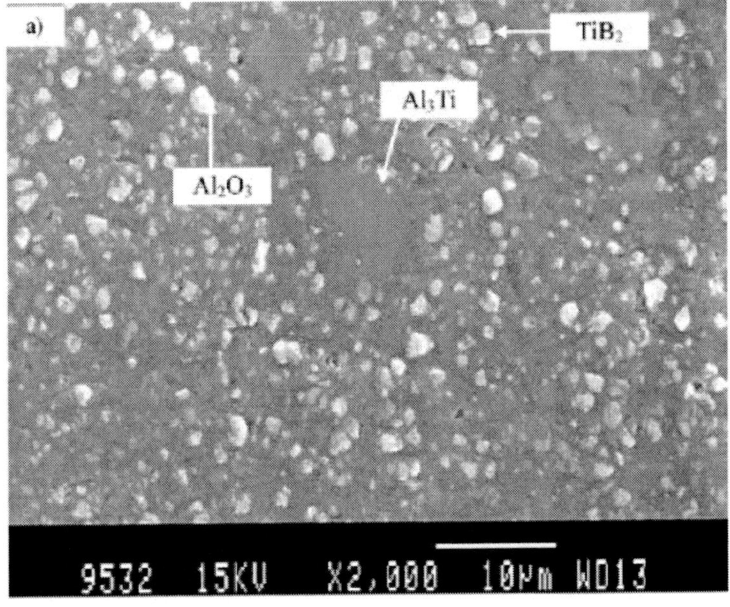

Figure 6. SEM micrographs showing the microstructure of (a) 22 vol% (Al_2O_3 +TiB_2+Al_3Ti)/Al and (b) 20 vol% (Al_2O_3 + TiB_2)/Al composites [28].

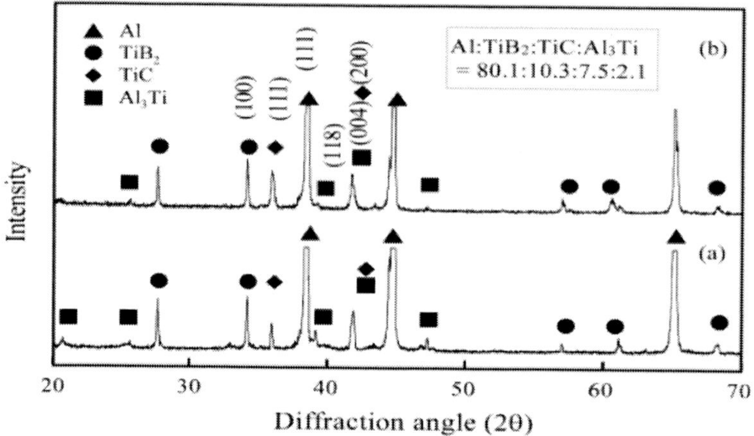

Figure 7. XRD patterns of Al-15 vol% TiB_2 with (a) 2 wt% and (b) 3 wt% carbon [30].

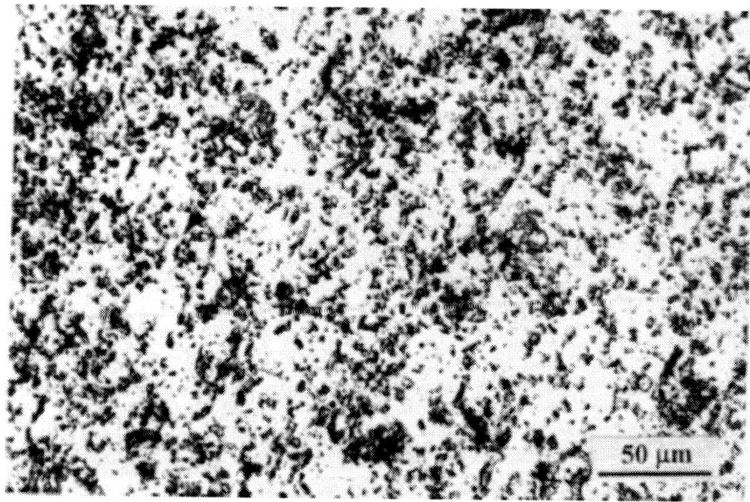

Figure 8. SEM micrograph of Al-15 vol% TiB_2 with the addition of 3 wt% carbon [30].

Structural materials are often subjected to cyclic loading during their service lives. As in-situ MMCs show potential applications in automotive and aerospace sectors, their fatigue properties must be considered. For ex-situ MMCs, the fatigue failure mechanisms are more complicated, and are strongly affected by several factors. These factors include: (1) the intrinsic properties of the matrix, e.g. composition, microstructure, aging treatment; (2) intrinsic properties of the particulate reinforcement phase, e.g. chemistry, size, volume fraction, distribution, and bonding conditions with the matrix; (3) the influence of test parameters such as strain- or stress-controlled, type and frequency of loading, temperature and (4) processing routes [32-36]. Al-based MMCs reinforced with ex-situ particulates generally exhibit superior room-temperature fatigue and endurance limit in high cycle fatigue (HCF), but not in low cycle fatigue (LCF). The improvement in the fatigue life of ex-situ Al-based MMCs over aluminum under stress-controlled condition is attributed to an increase in fatigue strength. Moreover, stress-controlled fatigue lives of Al-based MMCs has been reported to increase with decreased particle size, increased volume fraction and increased matrix tensile strength for a given stress level. Poor LCF life is resulted from the low ductility of the composites associated with the incorporation of ceramic reinforcements. In-situ Al-based composites reinforced with fined TiB_2 and Al_2O_3 particles generally exhibit excellent fatigue endurance limit and life under stress-controlled condition.

Figures 9(a)-(b) show the stress response curves of the 22 vol% (Al_2O_3 +TiB_2+Al_3Ti)/Al and 20 vol% (Al_2O_3 +TiB_2)/Al composites fabricated from the Al-TiO_2-B system under various total strain amplitudes. The stress amplitude was taken as the average of the peak values of the stress in tension and in compression during cyclic loading. At low total strain amplitudes ($\Delta\varepsilon_t/2$; $\Delta\varepsilon_t$: total strain range) of 0.1 and 0.2%, both composites exhibit essentially stable behavior, i.e. it shows no cyclic hardening or softening. At higher total strain amplitudes, a gradual progressive softening from the onset of cyclic deformation is observed in the 20vol% (Al_2O_3 +TiB_2)/Al composite. A stable cyclic response is commonly observed in aluminum alloys strengthened by non-shearable precipitates cycled at low strain amplitudes. The 22 vol% (Al_2O_3 +TiB_2+Al_3Ti)/Al composite fails immediately after cycling for one cycle at higher total strain amplitude of 0.6% (Figure 9(a)). A rapid drop in stress carrying capability at 0.6% strain amplitude is resulted from the formation of microcracks in brittle Al_3Ti blocks. Figure 10(a) shows the SEM fractograph of the 22 vol% (Al_2O_3 +TiB_2+Al_3Ti)/Al composite cycled to failure under total strain amplitude of 0.1 %. Microcracks can be observed in large Al_3Ti block due its brittle nature. Fine *in-situ* TiB_2 and Al_2O_3 particles remain intact with the matrix due to a strong interfacial particle-matrix bonding (Figure 10(b)). Figure 11 is the SEM backscattered electron micrograph showing the surface deformation behavior of the 22 vol%(Al_2O_3 +TiB_2+Al_3Ti)/Al composite fatigued for 2000 cycles at a total strain amplitude of 0.2%. Apparently, microcracks are originated from brittle Al_3Ti blocks. These cracks then propagate along the metal matrix until final failure.

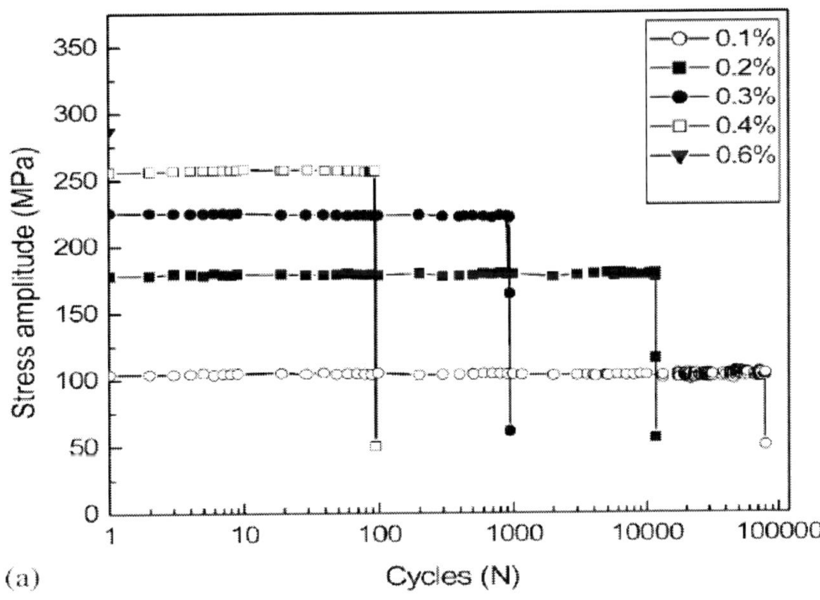

Figure 9 (Continued)

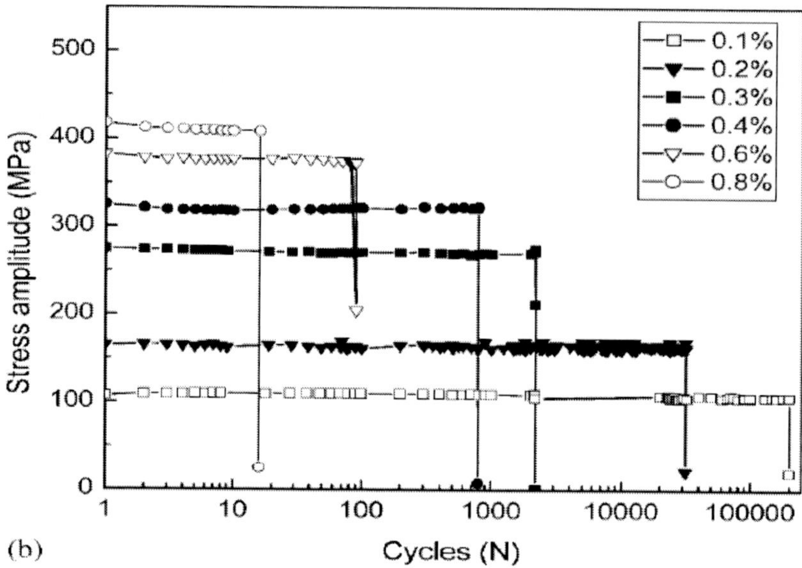

Figure 9. LCF stress response curves for (a) 22 vol% (Al_2O_3 +TiB_2+Al_3Ti)/Al and (b) 20 vol% (Al_2O_3 + TiB_2)/Al composites [28].

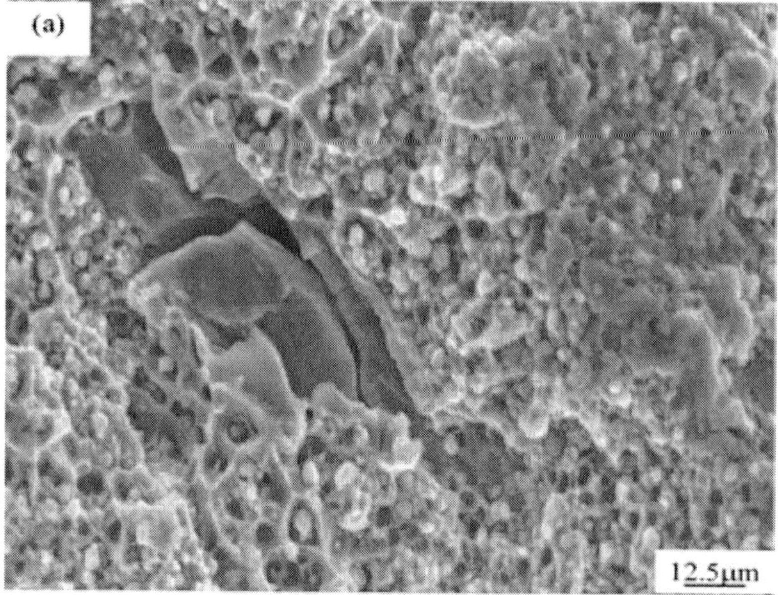

Figure 10 (Continued)

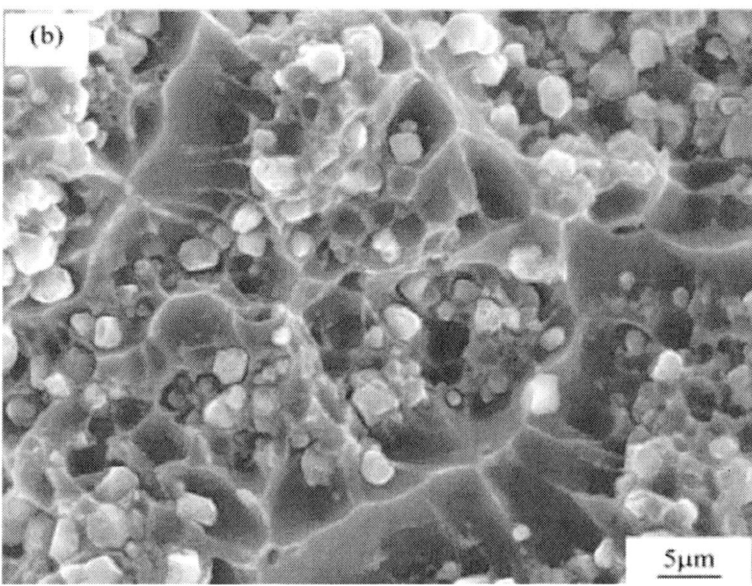

Figure 10. SEM fractographs of 22 vol% (Al_2O_3 +TiB_2+Al_3Ti)/Al composite cycled to failure at total strain amplitude of 0.1 %. (a) Cracking of Al_3Ti blocks and (b) higher magnification showing brittle feature of Al_3Ti blocks [28].

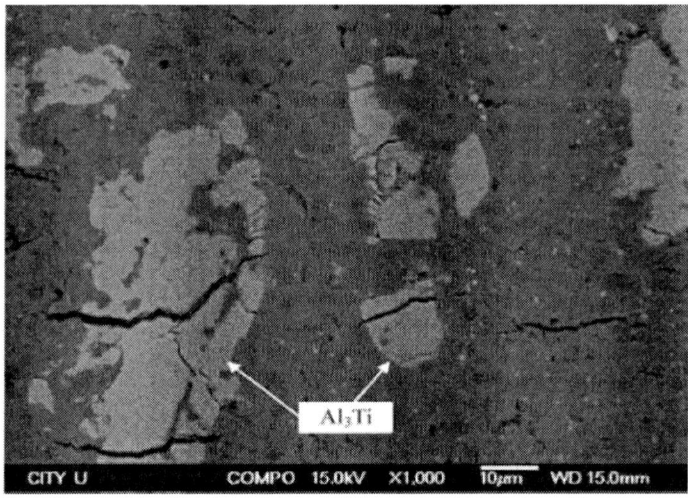

Figure 11. Backscattered electron micrograph of 22 vol% (Al_2O_3 +TiB_2+Al_3Ti)/Al composite surface after 2000 cycles at total strain amplitude of 0.2% showing crack initiation in Al_3Ti blocks [28].

It is noted from figures 9(a)-(b) that the low cycle fatigue life is strongly dependent on cyclic strain amplitudes. For a given fixed total strain amplitude, it is apparent that the 20 vol% (Al_2O_3 +TiB_2)/Al composite exhibits a substantial longer LCF life than the 22 vol% (Al_2O_3 +TiB_2+Al_3Ti)/Al composite containing Al_3Ti blocks. The fatigue lives of the 20 vol% (Al_2O_3 +TiB_2)/Al composite are even longer than the *ex-situ* 20 vol.% SiC_p (10 μm)/Al under given total strain amplitudes [36] as listed in table 3. The excellent fatigue resistance of the in situ 20 vol% (Al_2O_3 +TiB_2)/Al composite is attributed to the finer size of particulates and to a good interfacial matrix/particle bonding.

Figure 12 shows the fatigue S-N curves of pure Al, 22 vol% (Al_2O_3 +TiB_2+Al_3Ti)/Al and 20 vol% (Al_2O_3 +TiB_2)/Al composites fabricated from the Al-TiO_2-B system [37]. The incorporation of *in-situ* ceramic reinforcements in Al leads to a substantial improvement of its endurance limit from 40 to 170 MPa. The endurance stresses of 22 vol% (Al_2O_3 +TiB_2+Al_3Ti)/Al and 20 vol% (Al_2O_3 +TiB_2)/Al composites are 120 and 170 MPa, respectively. In MMCs, most of the applied load is carried by the high strength and stiffness ceramic reinforcement, thus the MMCs experience a lower average strain than the unreinforced metal under a given stress. At a constant stress, the formation of *in-situ* ceramic particles improves the fatigue endurance limit and lifetime over those of unreinforced Al.

Table 3. LCF life of in-situ 20 vol% (Al_2O_3 +TiB_2)/Al composite and ex-situ 20 vol.% SiC_p/Al [28]

Specimen	Total strain amplitude (%)	LCF life (cycle)
	0.1	200000
20 vol% (Al_2O_3 +TiB_2)/Al	0.2	31498
	0.3	2177
	0.4	792
	0.10	7000
20 vol% SiCp (10μm)/Al	0.2	1400
	0.39	110

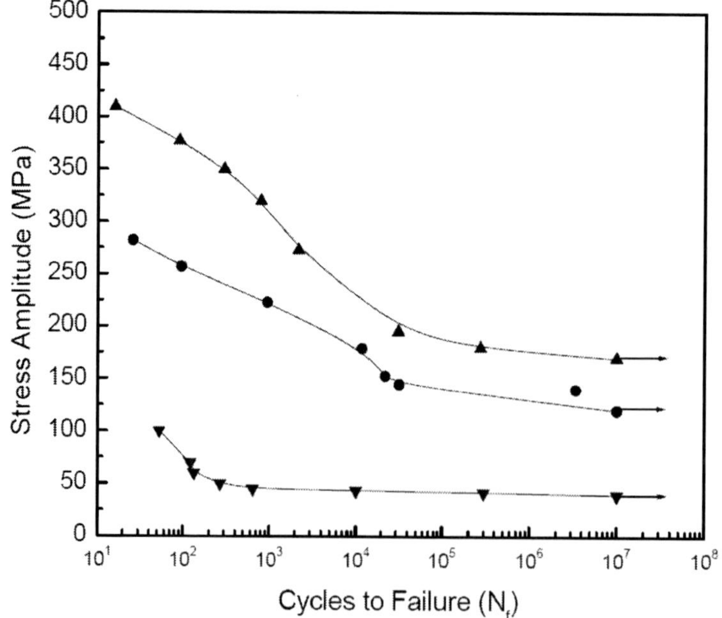

Figure 12. Fatigue S-N curves of pure Al (▼), 22 vol% (Al_2O_3 +TiB_2+Al_3Ti)/Al (●) and 20 vol% (Al_2O_3 + TiB_2)/Al (▲) composites [37].

In-Situ Aluminide/Al-Based Composites

Intermetallics such as iron aluminide and titanium aluminide have been considered as candidates of novel structural materials because they offer low density, high specific strength, high melting point and excellent oxidation resistance. However, the inherent brittleness of aluminides limits their practical applications in industrial sectors. Nevertheless, aluminides of titanium and iron having attractive physical properties show potential applications as reinforcing materials for Al-based composites. Such reinforced aluminides are still in very early stages of development. Very recently, Roy et al. prepared Fe-aluminide and Ti-aluminide reinforced Al-based MMCs by reactive hot pressing of aluminum powder and nano-powder of Fe_2O_3 or TiO_2 [19, 20]. The size of reinforcing phase can be reduced considerably by using raw powders of nanometer scale. Possible in-situ reactions take place for Fe-aluminide reinforced composites can be described as:

$$Fe_2O_3 + 2Al \rightarrow Al_2O_3 + 2Fe \quad (8)$$

$$x\,Fe + y\,Al \rightarrow Fe_xAl_y \text{ (x,y can be 1,2 or 3)} \quad (9)$$

Figure 13 shows the XRD patterns of Al-based composite reinforced with 30 vol.% (Al_2O_3 + Fe-aluminide) by hot pressing at different temperatures. XRD pattern analysis reveals the presence of iron aluminides (e.g. $FeAl_3$, $FeAl_3$), Al_2O_3, unreacted Fe_2O_3 and Al in the composite samples. The peak intensity of unreacted species decreases with the increase in pressing temperature. The SEM images of Al matrix composites reinforced with 10 vol% and 30 vol% (Al_2O_3 + Fe-aluminide) are shown in figures 14(a)-(b). The reinforcing particles with submicrometer size are well distributed in the host aluminum matrix. The formation of in-situ aluminide reinforcement leads to an increase of hardness of composites (figure 15). The hardness increase is more for the composites processed at higher hot pressing temperature. Analogously, Ti-aluminide reinforced composites can be prepared by reactive hot pressing of Al and TiO_2. Possible reactions are given as follows:

$$3TiO_2 + 4Al \rightarrow 2Al_2O_3 + 3Ti \quad (10)$$

$$x\,Ti + y\,Al \rightarrow Ti_xAl_y \quad (11)$$

Figure 16 show the variation of Vickers hardness with the reinforcement volume content for the composites prepared by RHP at 700 and 900 °C. The increase in hardness can be correlated with higher amounts of in-situ aluminide particles formed. The increment in hardness of the composites reduces their wear volume as expected. The wear resistance of the composites is improved by increasing the hot pressing temperature (figure 17(b)). It is noted the composite with 40% filler content exhibits higher wear volume compared to the composite with 20% reinforcement (figure 17(a)). This severe wear is due to the spallation of brittle in-situ formed particles from the composite. The spalled particles adhered to the steel counterbody of wear test facility can abrade the composite material, producing high wear volume accordingly.

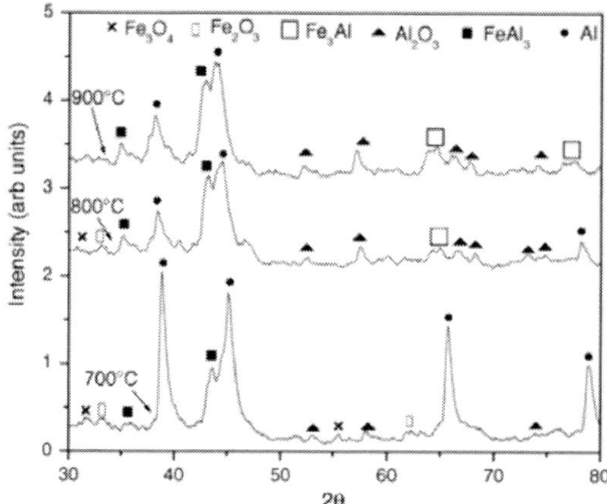

Figure 13. XRD patterns of Al matrix composite containing 30vol% (Al_2O_3 + Fe-aluminide) prepared by RHP at different temperatures. Reprinted from [19] with permission of Elsevier.

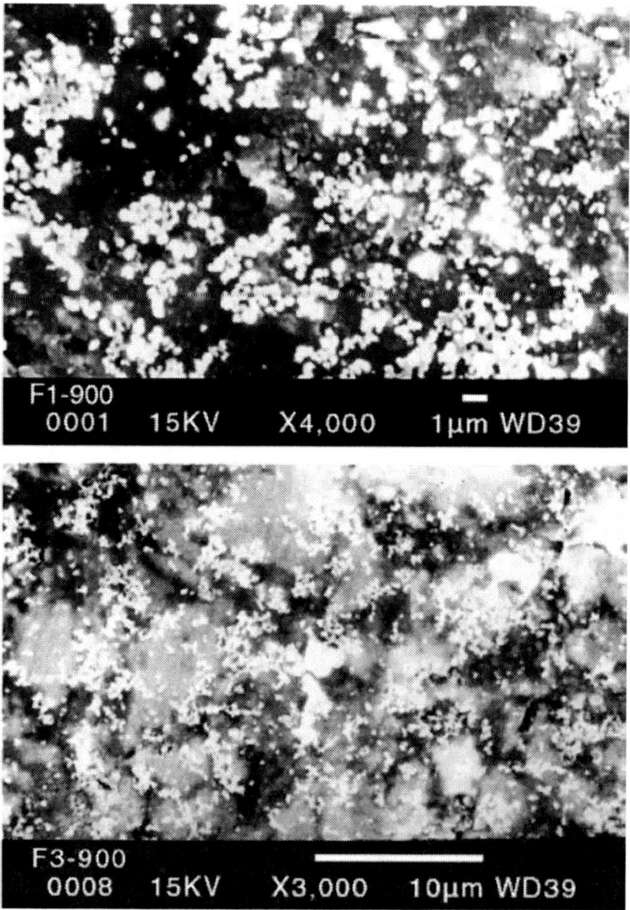

Figure 14. SEM micrographs of Al matrix composite containing (a) 10 vol% (Al_2O_3 + Fe-aluminide) and (b) 30 vol% (Al_2O_3 + Fe-aluminide) prepared by RHP at 900 °C. Reprinted from [19] with permission of Elsevier.

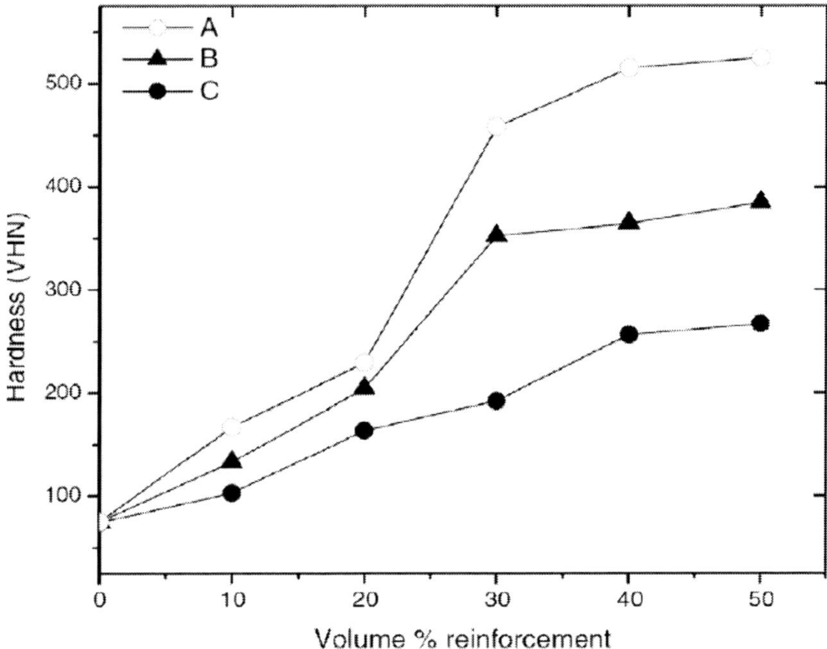

Figure 15. Vickers hardness variation with respect to the volume content of reinforcement for the composites prepared by RHP at (A) 900 °C, (B) 800 °C, (C) 700 °C. Reprinted from [21] with permission of Elsevier.

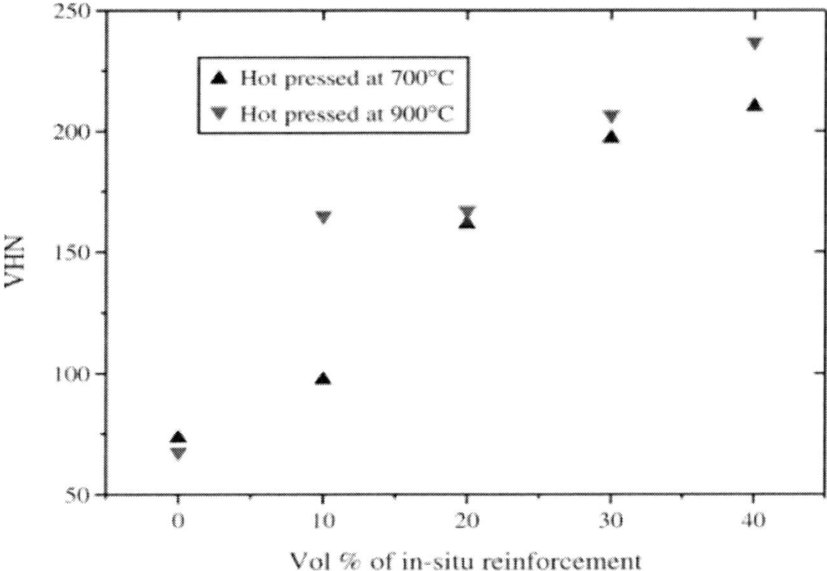

Figure 16. Variation of Vickers hardness of (Al_2O_3 + Ti-aluminide)/Al composites prepared by RHP at 700 °C and 900 °C with volume content. Reprinted from [22] with permission of Elsevier.

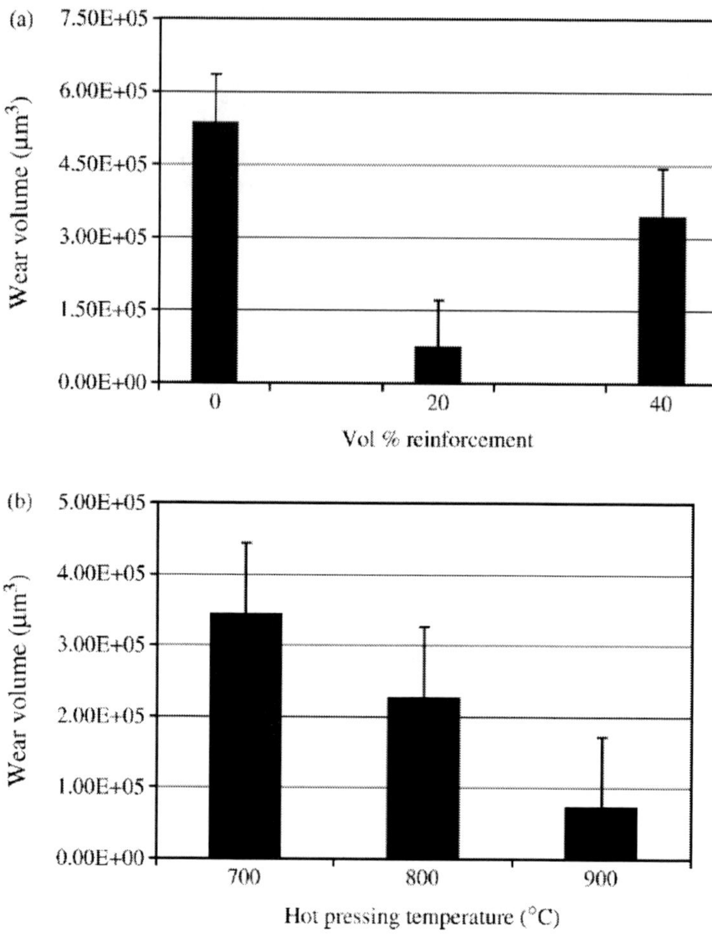

Figure 17. (a) Variation of wear volume of the composites (hot pressed at 900 °C) with reinforcement volume content; (b) Wear volume of 20 vol% reinforcement vs. hot pressing temperature. Reprinted from [22] with permission of Elsevier.

COPPER-BASED COMPOSITES

Copper is a heavy metal (density of 8.96 g/cm^3) with very good thermal and electrical conductivities, but exhibits poor wear and elevated temperature mechanical properties. Such shortcomings can lead to premature failure of components in practice. The addition of ceramic reinforcement to copper reduces density, increases the tensile stiffness and strength as well as the wear resistance [38-40]. Cu-based composites find applications as materials for electrical discharge machining electrodes, heat sinks for fusion, and contacts for relays and switches. Compared to the in-situ Al-based composites, less information is available on the microstructure and mechanical behavior of in-situ copper composites. Recently, Zarrinfar et al. prepared in-situ Cu-based composites via melt reactive synthesis of Ti, C and Cu powder mixtures to produce the Cu-TiC_x master alloy. The master alloy was then introduced into molten copper to form the TiC_x/Cu composite [41].

In general, TiB_2 is more attractive as a reinforcing material for Cu-based composites because it is electrical and thermal conductive [18]. The TiB_2 particulates can be formed *in situ* in the copper matrix via mechanical alloying (MA) and RHP routes. Biselli et al. have fabricated *in situ* TiB_2/Cu composite from Cu, Ti and B powder precursors using MA process followed by a suitable treatment [42]. They reported that the *in situ* formed TiB_2 particulates are resistant to coarsening during annealing, thereby retaining the mechanical strength of the composite. Furthermore, the yield strength of *in situ* TiB_2/Cu composite is higher than that of the Al_2O_3/Cu dispersion alloy. The enhancement in yield strength results from good microstructural stability and chemical stability of the TiB_2 phase. More recently, Ma and Tjong prepared in-situ Cu-based MMC reinforced with 15 vol% TiB_2 particulates via reactive hot pressing of Cu, Ti and B powders [40, 43]. The LCF and high-temperature creep characteristics high temperature of 15 vol% TiB_2/Cu composites were investigated. They reported that the creep resistance of in-situ TiB_2/Cu composite is several orders of magnitude higher than that of pure copper. Figures 18 and 19 show the XRD pattern and SEM micrograph of in- situ 15 vol%TiB_2/Cu composite formed by the RHP process, respectively. The in-situ TiB_2 particulates exhibit irregular shape, and their sizes vary from 0. 167 to 4.24 μm. The size distribution of TiB_2 particulates determined from an image analysis is shown in figure 20. Very few particles of 2-4 μm (less than 2%) are formed in-situ during reactive hot pressing. The sizes of most TiB_2 particles are well below one micron as expected. The volume content of TiB_2 particles with sizes below 0.54 μm is 70.85 %. Such submicron particles can enhance the yield strength of the composite via Orowan bowing mechanism. The tensile properties of the *in situ* TiB_2/Cu composite and unreinforced Cu at room temperature are listed in table 4.

Table 4. Tensile properties of pure Cu and 15 vol%TiB_2/Cu composite at room temperature [43]

Specimen	Elastic Modulus (GPa)	UTS (MPa)	YS (MPa)	El. (%)
Pure Cu	90	304.8	75.0	43.8
15 vol%TiB_2/Cu	120	734.7	655.6	9.6

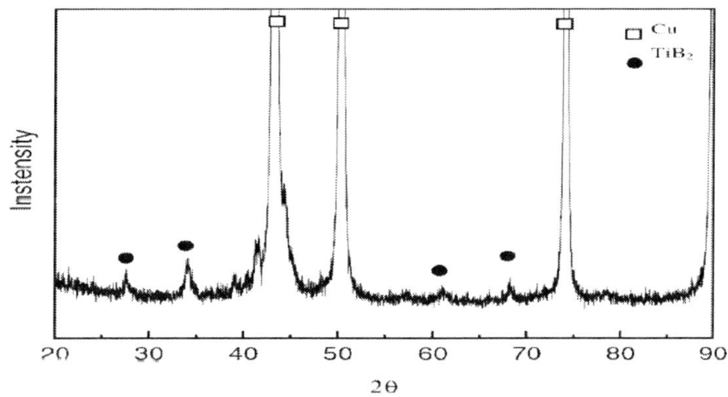

Figure 18. XRD pattern of in-situ 15 vol%TiB_2/Cu composite [43].

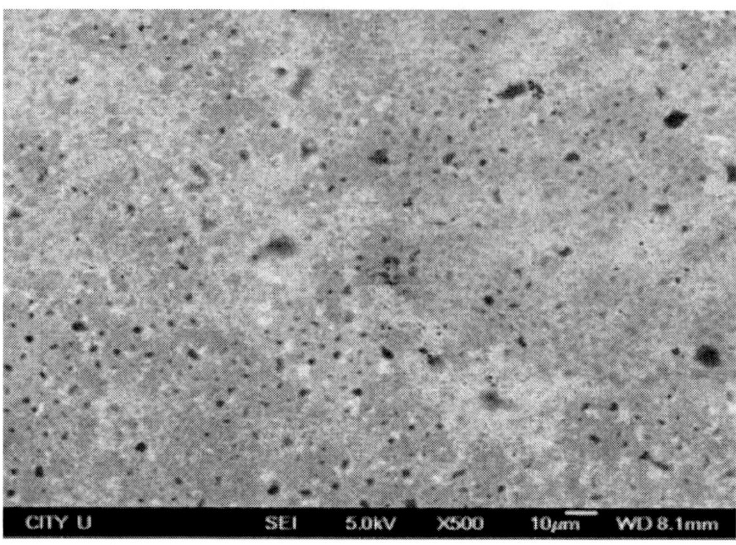

Figure 19. SEM micrograph of in-situ 15 vol%TiB_2/Cu composite [43].

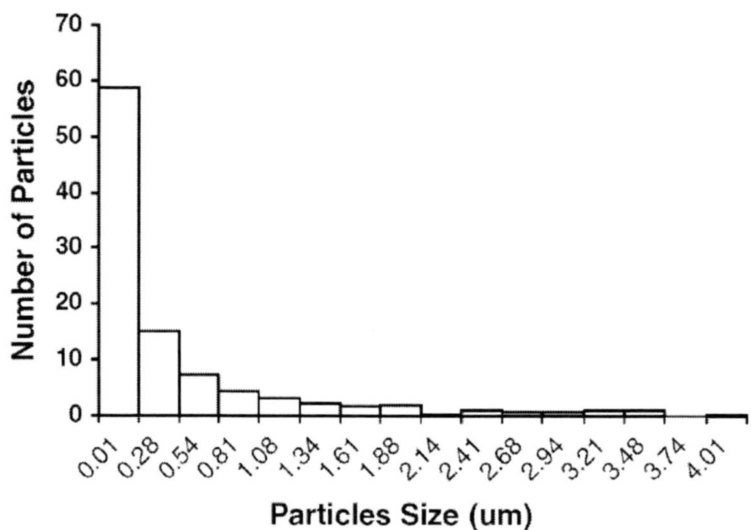

Figure 20. Size distribution of in-situ TiB_2 particles in Cu matrix of 15 vol% TiB_2/Cu composite determined from the image analysis [43].

Figure 21 shows the cyclic stress response curves of the in-situ 15 vol%TiB_2/Cu composite under various total strain amplitudes. For comparison, the cyclic stress response curves of pure Cu under fixed total strain amplitudes ($\Delta\varepsilon_t/2$) of 0.3 and 0.4% are shown in figure 22. The cyclic stress response was determined by monitoring the stress amplitude during strain-controlled fatigue measurements. In this regard, the cyclic stress response curves represent the variation of stress amplitude with number of cycles (N) at fixed total strain amplitudes. From figure 22, soft Cu exhibits cyclic hardening up to final fracture as expected. However, in-situ 15vol% TiB_2/Cu composite exhibits essentially stable cyclic behavior at low total strain amplitudes of 0.1 to 0.3%. The composite exhibits very slight hardening under a strain amplitude of 0.4 %. Such cyclic hardening is more pronounced by increasing the strain amplitude to 0.6 %. It is generally known that the cyclic strain hardening of pure Cu is derived from the generation, multiplication and interaction of dislocations during cyclic deformation. Such dislocations tend to arrange themselves into low energy dislocation structures such as loop patches and persistent slip bands. Thus the dislocation structures generated during cycling can provide useful mechanistic information on the cyclic deformation of metals and composites. In this aspect, TEM is a powerful tool to observe the dislocation structures developed in metals during cycling.

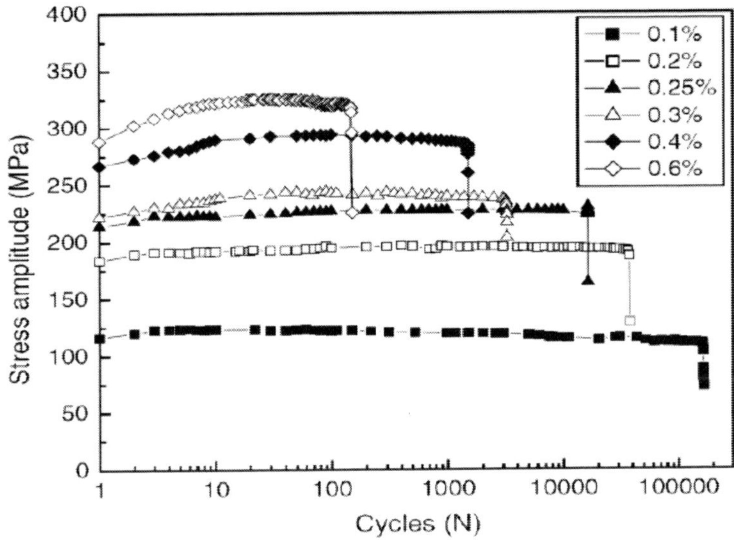

Figure 21. Cyclic stress response of *in situ* TiB_2/Cu composite cycled at various total strain amplitudes [43].

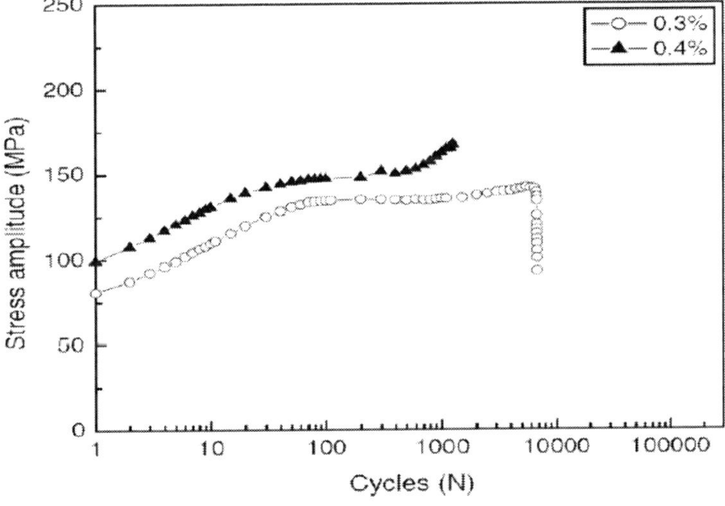

Figure 22. Cyclic stress response of pure Cu composite cycled at total strain amplitudes of 0.3 and 0.4% [43].

Figure 23 shows a typical TEM micrograph of in-situ 15vol% TiB_2/Cu composite prior to cycling. High density of dislocations is observed in the Cu matrix after extruding the TiB_2/Cu composite at 1173 K. The dislocations are generated upon cooling in-situ 15vol% TiB_2/Cu composite from fabrication temperature due to a large difference in thermal expansion coefficients (CTE) between the TiB_2 ceramic reinforcement and copper matrix. During cycling, new dislocations are developed in 15vol% TiB_2/Cu composite at 0.4%. The dislocation motion is impeded by fine TiB_2 submicron particles such that the dislocations appear as tangles and loop patches (figure 24). At larger strain amplitude of 0.6%, bundles of loop patches link together to form the wall of dislocation cells (figure 25(a)) and dislocation networks (figures 25(b)-(c)). It is considered that cyclic hardening of the 15 vol% TiB_2/Cu composite deformed at strain amplitudes of 0.4 and 0.6% results from the dislocation multiplication and accumulation.

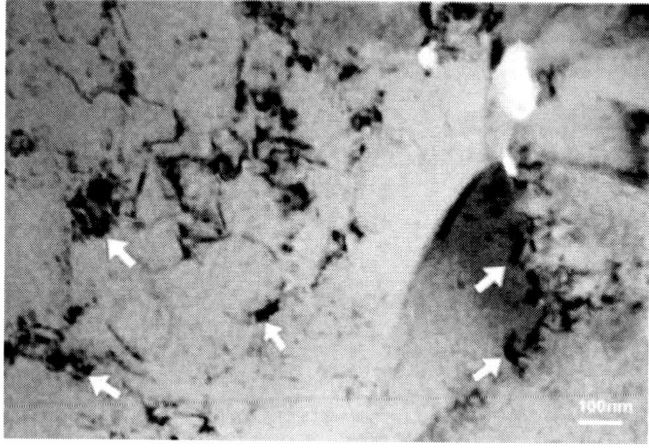

Figure 23. TEM micrograph of RHP prepared 15 vol%TiB_2/Cu composite prior to cyclic deformation. White arrows indicate in- situ TiB_2 particles [43].

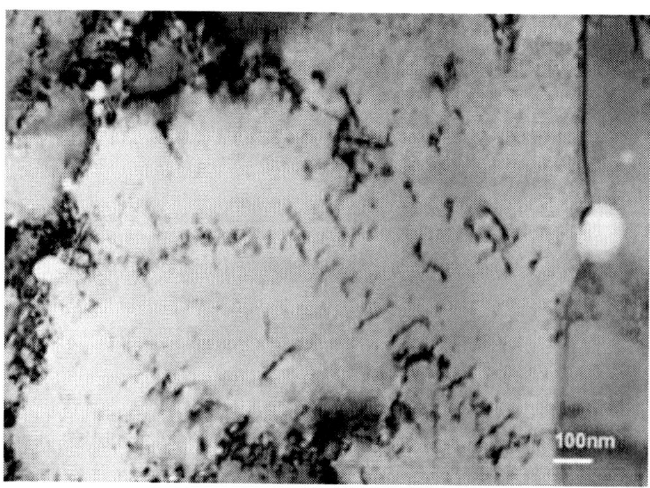

Figure 24. TEM micrograph of in-situ 15 vol%TiB_2/Cu deformed to failure at a strain amplitude of 0.4% [43].

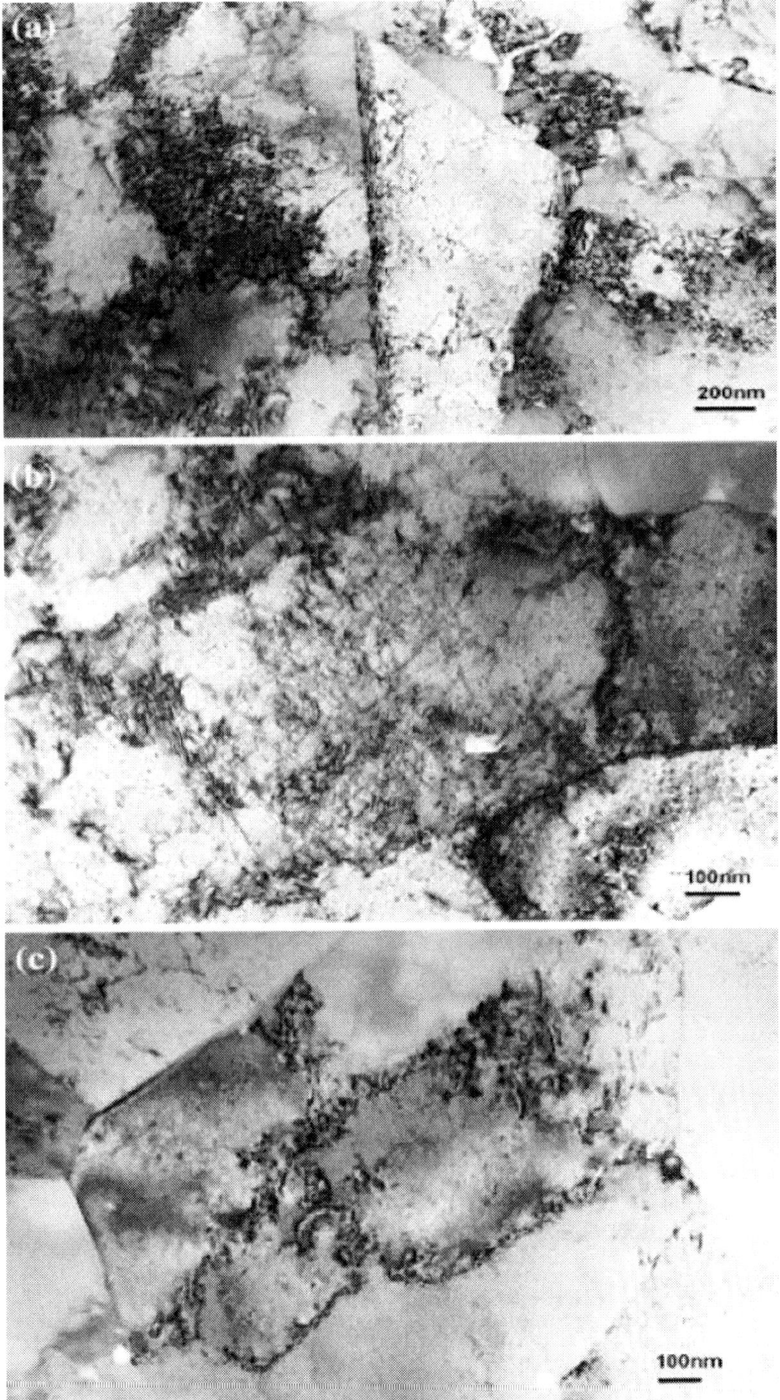

Figure 25. (a) TEM micrograph of in-situ 15 vol%TiB_2/Cu composite cycled to failure at a strain amplitude of 0.6 % showing formation of dislocation cells. Dislocation tangles and loop patches near the grain-boundaries of a central grain can be readily seen; (b) and (c) other regions of the same specimen showing formation of the dislocation networks in Cu matrix [43].

TITANIUM-BASED COMPOSITES

Titanium has higher density (4.51 g/cm^3) and melting temperature (1941 K) than aluminum. Titanium and its alloys are very attractive for applications in aerospace and automotive industries. This is attributed to their larger specific stiffness and better mechanical performance at higher temperatures when compared to aluminum alloys. The TiB_2 particulates cannot be synthesized in-situ within the titanium alloy matrix because the TiB_2 phase decomposes into an orthorhombic TiB phase with whisker morphology during the composite fabrication [44]. The TiB whisker (TiB_w) is an effective reinforcement for Ti alloys due to its chemical stability with titanium matrices, high aspect ratio, low density, high stiffness and high melting point [45,46]. The strength of TiB_w reinforced Ti-based composites is mainly determined by the ability of load transfer from the matrix to the whisker. The other strengthening mechanisms include finer matrix grain size, increased dislocation density, and increased kinetics for precipitation hardening associated with the formation of in-situ reinforcement.

Tjong and coworkers prepared the 12-15 vol.% TiB_w/Ti composites by the RHP route using four reactive systems, namely Ti -TiB_2, Ti-B, Ti-B_4C and Ti-TiN [47]. The blended powder mixtures were cold compacted and hot pressed in vacuum at 1250 °C for 0.5 h. The resulting composites were finally extruded into rods. In the process, exothermic reactions between compacted reactant powders of the Ti-B, Ti-B_4C and Ti-TiN systems can take place:

$$Ti + TiB_2 \rightarrow 2\ TiB \tag{12}$$

$$Ti + B \rightarrow TiB \tag{13}$$

$$5\ Ti + B_4C \rightarrow 4\ TiB + TiC \tag{14}$$

$$(2\text{-}x)\ Ti + (1\text{-}x)\ BN \rightarrow (1\text{-}x)\ TiB + TiN_{1-x} \tag{15}$$

Solid state diffusion between powder constituents is the main growth mechanism for the reinforcements. Figures 26(a)-(d) show typical optical micrographs of in-situ Ti-composites prepared from these systems followed by extrusion. Apparently, in-situ TiB whiskers are aligned along the extrusion direction as expected. The formation of in-situ phases in these systems is verified by the XRD method (figure 27). For the Ti-B_4C system, both TiB_w and TiC_p reinforcements are simultaneously produced during RHP according to the reaction (14). A typical TEM micrograph for in-situ phases formed in the Ti-B_4C system is depicted in figure 28. An individual TiB whisker with a large aspect ratio and equiaxed TiC_p with a diameter of ~ 100 nm can be readily seen in the micrograph. Moreover, the TiB_w exhibits a clean whisker-matrix interface. Formation of both *in-situ* TiC_p and TiB_w hybrid reinforcements in the Ti-B_4C system produces higher compressive yield strength than the composite reinforced with TiB_w alone (figure 29).

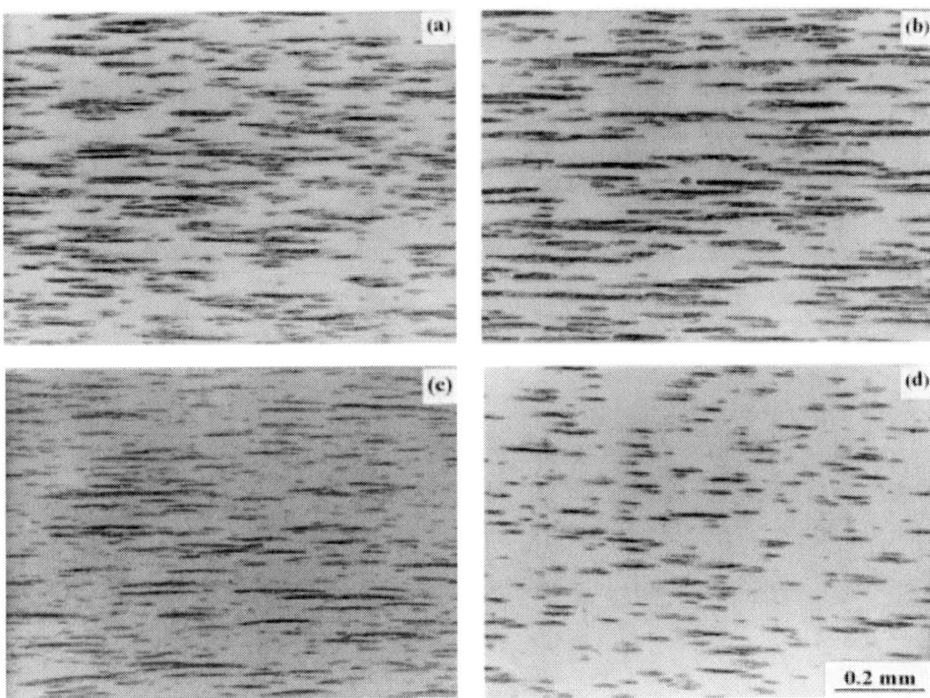

Figure 26. Optical micrographs of in-situ Ti-composites prepared from (a) Ti-B (b) Ti-TiB_2 (c) Ti-B_4C and (d)Ti-BN systems [47].

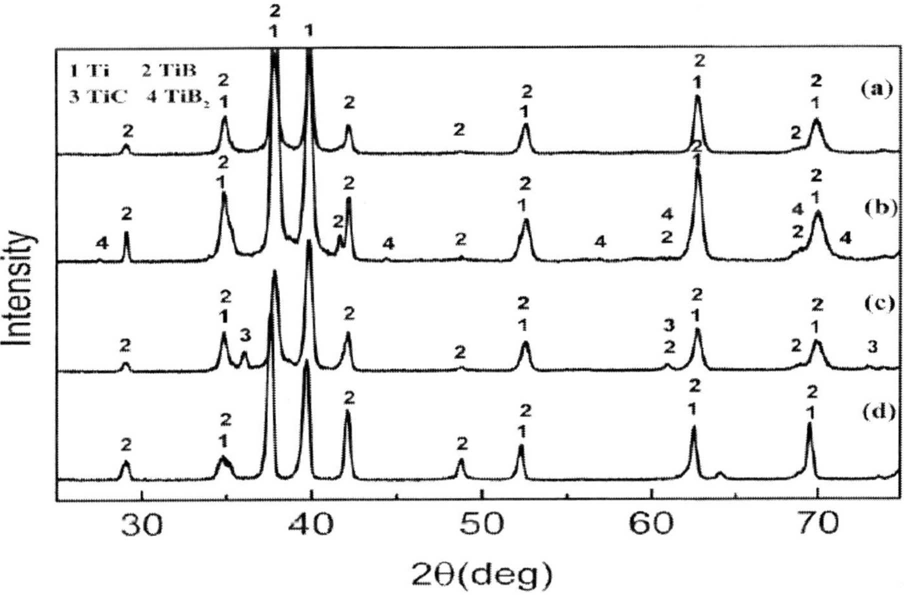

Figure 27. XRD patterns of in-situ Ti-composites prepared from (a) Ti-B (b) Ti-TiB_2 (c) Ti-B_4C and (d) Ti-BN systems [47].

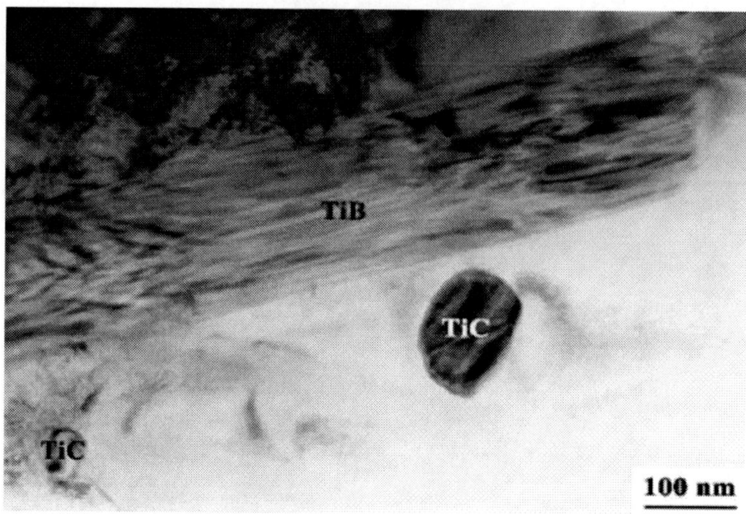

Figure 28. TEM micrograph showing a clean whisker-matrix interface for in-situ TiB and TiC reinforcements and titanium matrix in 15 vol% (TiB_w + TiC_p)/Ti composite prepared from the Ti-B_4C system [48].

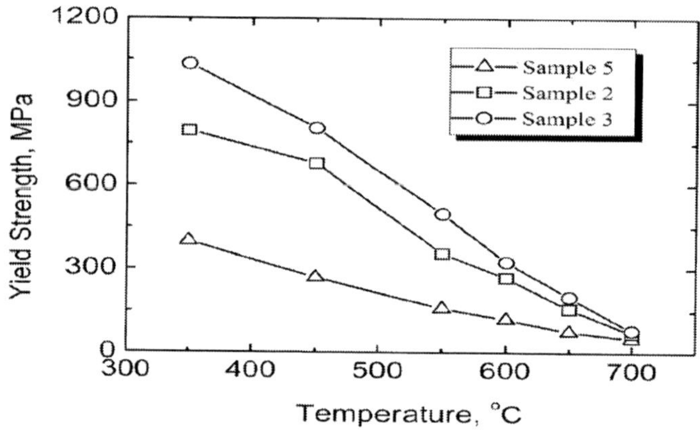

Figure 29. Variation of the compressive yield stress with temperature for pure Ti (sample 5), in-situ 15vol.%.TiB_w/Ti (sample 2) and (12.1vol.%TiB_w+2.9vol.%TiC_p)/Ti (sample 3) composites [47].

In general, there is a large scope for improvement in terms of microstructure and mechanical properties of (TiB+TiC)/Ti hybrids prepared from the Ti-B_4C system. Ni et al. reported that the size of B_4C powder had a pronounced effect on the microstructure and mechanical property of 10 vol% (TiB+TiC)/Ti hybrid [49]. Fine B_4C powder (0.5 μm) favors formation of finer TiB whiskers in the hybrid whereas coarse B_4C powder (3.5 μm) tends to yield TiB clusters. In a subsequent study, Ni et al. used RHP route to prepare 10 vol% (TiB+TiC)/Ti hybrid composites with different TiB/TiC ratios of 4:1, 2:1, 1:1, 1:2 and 1:4 from Ti (10 μm), B_4C (0.5 μm) and graphite (0.5 μm) powders [50]. They designated the resulting composites as W_8P_2, $W_{6.6}P_{3.3}$, W_5P_5, $W_{3.3}P_{6.6}$ and W_2P_8, respectively. For the Ti-B_4C system, Reaction (14) can take place, and the following reaction also occurs for the Ti-C system during RHP:

$$Ti + C \rightarrow TiC \tag{16}$$

Figure 30(a)-(c) show SEM micrographs of extruded W_8P_2, W_5P_5, and W_2P_8 composites. Both TiC_p and TiB_w disperse within Ti-matrix of the W_5P_5, and W_2P_8 hybrid composites. However, only TiB_w can be seen in the W_8P_2 hybrid due to the finer size of TiC_p formed in this sample. The W_5P_5 composite with the TiB/TiC ratio of 1:1 exhibits the highest tensile strength (1175 MPa) and elongation (1.5%) than other hybrids. The tensile strengths of representative W_8P_2 and W_2P_8 composites are 950 and 810 MPa whilst their tensile elongation values are 0.70 and 0.80%, respectively. All these hybrids definitely have higher tensile strength than that of pure titanium (800 MPa), but show much lower tensile ductility than titanium with elongation of 11.6 % (figure 31). In these hybrids, there is a synergistic effect between TiB_w and TiC_p in enhancing the tensile properties. The TiB_w undertakes applied loads but fractures easily during tensile loading due to its large aspect ratio. TiC_p cannot take up loads as efficient as TiB_w, but it can prevent crack formation and retard crack propagation more effectively than TiB_w on the basis of SEM fractographs examination. The beneficial effect of TiC_p derives from its equiaxed morphology.

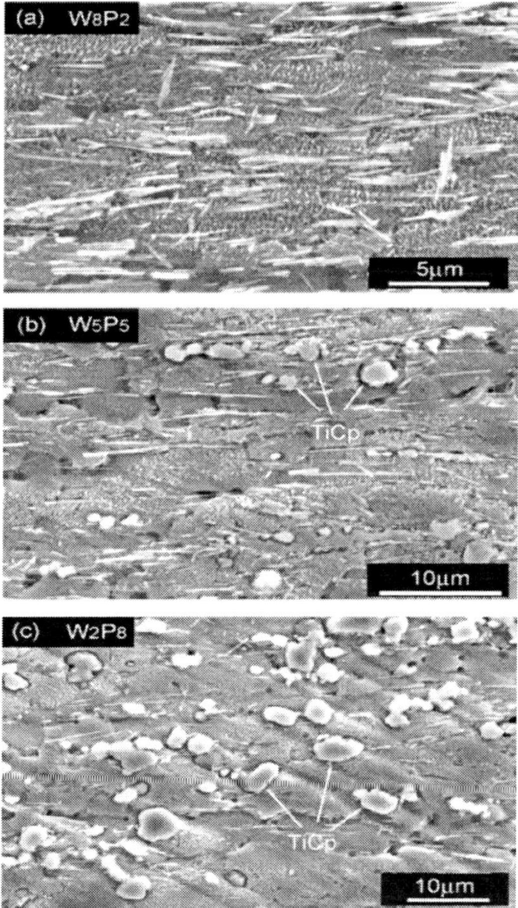

Figure 30. SEM micrographs of as-extruded composites along extrusion direction: (a) W_8P_2, (b) W_5P_5 and (c) W_2P_8. Reprinted from [50] with permission of Elsevier.

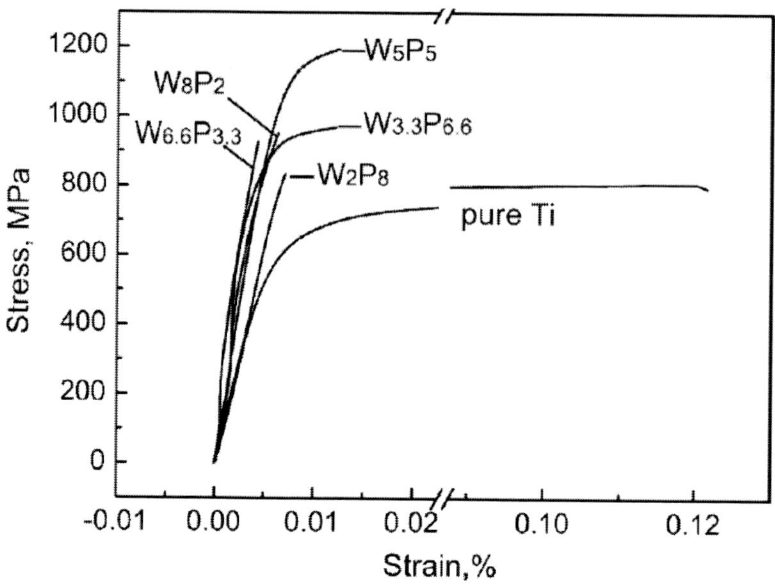

Figure 31. Tensile stress-strain curves of pure Ti and its composites prepared from Ti-B_4C-C system with different TiB/TiC ratios. Reprinted from [50] with permission of Elsevier.

Finally, the creep behavior of Ti hybrid composite is considered. Tjong and coworkers have investigated the compressive creep behavior of 15vol% (TiB_w + TiC_p)/Ti composite at 873-923 K [48]. Figure 32 shows the variation of compressive creep rate with applied stress for 15 vol% (TiB_w + TiC_p)/Ti composite. The creep data fits into straight lines with slopes equal to the value of the stress exponent, n. This implies that the creep data follow a power-law relation:

$$\dot{\varepsilon} = \frac{AGbD_o}{kT}\left(\frac{\sigma}{G}\right) exp(\left(-\frac{Q}{RT}\right) \tag{17}$$

where $\dot{\varepsilon}$ is the steady creep rate, A is a structure dependent parameter, G is the temperature-dependent shear modulus, b is the Burgers vector, D_o is the frequency factor for diffusion, k is the Boltzmann's constant, T is the absolute temperature, σ is the applied stress, Q is the activation energy, n is the stress exponent and R is the gas constant. The 15 vol% (TiB_w + TiC_p)/Ti hybrid exhibits stress exponents of 4.5, 4.6 and 4.6 at 873, 898 and 923 K, respectively. These values are very close to that for the lattice diffusion controlled climb process in α-Ti (n = 4.3) [51]. Formation of dual TiB_w and TiC_p reinforcements imparts a higher creep resistance to titanium. The creep rate of 15 vol% (TiB_w + TiC_p)/Ti hybrid is even lower than that of 15 vol% TiB_w/Ti composite, and more than one order of magnitude lower than that of unreinforced Ti (figure 33)

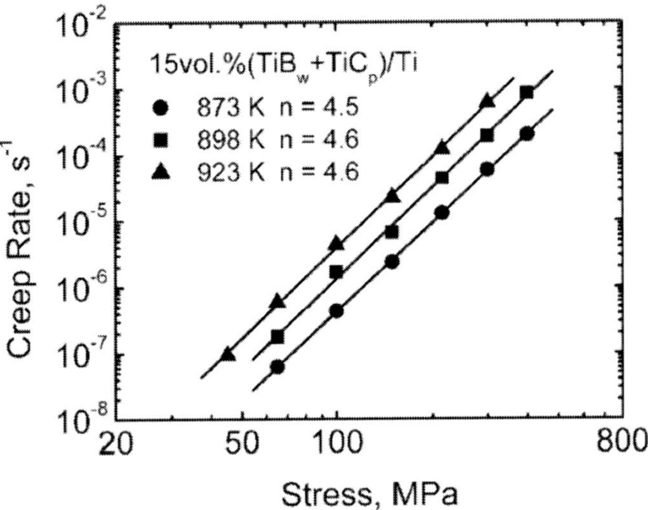

Figure 32. Variation of steady-state creep rate with applied stress for 15 vol% (TiB_w + TiC_p)/Ti composite [48].

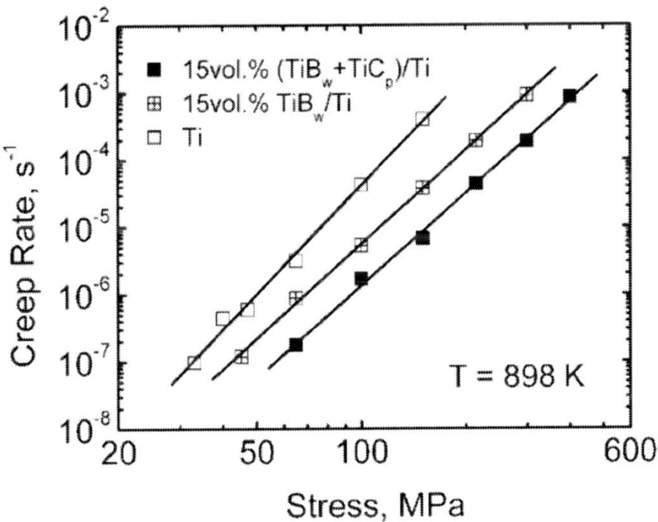

Figure 33. Comparison between compressive creep rates of pure Ti, 15vol% (TiB_w+TiC_p)/Ti hybrid and 15 vol% TiB_w/Ti composite at 898 K as a function of applied stress [48].

Conclusion

Design and development of advanced high performance composite materials and introducing them into practical use is one of the most challenging tasks of materials scientists. RHP route allows fabrication of dense Al-, Cu- and Ti-based metal matrix composites reinforced with in-situ reinforcements from elemental powders by a combination of exothermic reaction and hot pressing. Using this process, composites with novel and controlled microstructures as well as good reinforcement-matrix chemical compatibility can be created. Factors such as the selection of reinforcing materials and the optimization of mechanical properties are crucial for developing in-situ composites with high mechanical performances. Through fundamental understanding of the structure-property relationship, we can design novel in-situ MMCs with tailored properties for advanced engineering applications. Titanium carbide, titanium diboride or titanium monoboride are potential reinforcing materials for MMCs because of their outstanding properties, such as high hardness and stiffness, low density and high melting temperature. For the Al-based composites, intermetallic Al_3Ti phase forms simultaneously in-situ with TiC or TiB_2 ceramic particles during the composite fabrication. Elimination of undesirable brittle Al_3Ti phase can be accomplished by optimizing the ratio of precursor materials in the reaction system. In the case of Cu-based composites, TiB_2 particulates can be synthesized directly via reactive hot pressing of Cu, Ti and B powders. Formation of in-situ TiB_2 particulates is beneficial to improve the yield and tensile strengths of Cu considerably at the expense of tensile ductility. The TiB_2 particulates cannot be synthesized in-situ within titanium matrix because the TiB_2 phase decomposes into an TiB phase with whisker morphology during reactive hot pressing. The large aspect ratio of TiB whiskers promotes effective load transfer from the matrix to the whiskers during mechanical loadings. The titanium hybrids containing both TiB_w and TiC reinforcements exhibit higher tensile strength and creep resistance than the titanium composites reinforced with TiB_w single reinforcement.

REFERENCES

[1] Li, Y.; Langdon, T.G. *Acta Mater.* 1997, vol 45, 4797 - 4806.
[2] Yadav, S.; Chichili, D.R.; Ramesh, K.T. *Acta Metal Mater.* 1995, vol 43, 4453 -4464.
[3] Tjong, S. C.; Lau, K. C. *Compos. Sci. Technol.* 1999, vol. 59, 2005-2013.
[4] Tjong, S. C.; Ma, Z. Y. *Compos. Sci. Technol.* 1999, vol. 59, 1117-1125.
[5] Spowart, J. E.; Miracle, D. B. *Mater Sci. Eng. A.* 2003, vol 357, 111-123.
[6] Watson, I. G.; Forster, M. F.; Lee, P. D.; Dashwood, R. J.; Hamilton, R.W.; Chirazi, A. *Composites Part A.* 2005, vol 36, 1177-1187.
[7] Westwood, A. R. *Metall Mater Trans. A.* 1998, vol 19, 749 - 758.
[8] Brinkman, H. J. ; Duszczyk, J.; Katgerman, L. *J. Mater Res. Vol.* 14, 1999, 4246 -4250.
[9] Ma, Z. Y.; Tjong, S. C. *Metall Mater Trans. A.* 1997, Vol 28, 1931-1942.
[10] Tjong, S. C. *Mater Sci. Eng. R.* 2000, vol 29, 49-113.
[11] Li, Y. X. ; Hu, J. D.; Wang, H. Y.; Guo, Z. X. *Adv. Eng. Mater.* 2007, vol 9, 689 -694.
[12] Song, M. S.; Zhang, M. X.; Zhang, S. G.; Huang, B.; Li, J. G. *Mater Sci. Eng. A.* 2008, vol 473, 166 -171.
[13] Hanabe, M. R. ; Aswath, P. B. *J. Mater Res.* 1996, vol 11, 1562- 1569.
[14] Zhang, G. J.; Jin, Z. Z. *Ceram. Int.* 1996, vol 22, 143-147.
[15] Ghosh, G. ; Vaynman, S. ; Fine, M. E. *Ceram. Int.* 1999, vol 25, 649 -659.
[16] Doty, H.; Abbaschian, R. *Mater Sci. Eng. A.* 1995, vol 195, 101 -111.
[17] Inoue, M.; Suganuma, K. ; Nichara, K. *Intermetallics.* 2000, vol 8, 1035 -1042.

[18] Shackelford, J.F.; Alexander, W. *CRC Materials Science and Engineering Handbook*, 3rd ed., CRC Press, Boca Raton (Florida), 2001.

[19] Roy, D.; Ghosh, S.; Basumallick, A.; Basu, B. *Mater Sci. Eng. A.* 2006, vol 415, 202-206.

[20] Roy, D.; Ghosh, S.; Basumallick, A.; Basu, B. *J. Alloy Compd.* 2007, vol 436, 107 - 111.

[21] Roy, D. ; Basu, B.; Mallick, A. B. ; Kumar, B. V.; Ghosh, S. *Composites Part A.* 2006, vol 37, 1464 -1472.

[22] Roy, D.; Basu, B.; Mallick, A.B. *Intermetallics*. 2005, vol 13, 733 -740.

[23] Zeng, X. W.; Zhang, W.G. ; Wei, N. ; Liu, R. P. ; Ma, M. Z. *Mater Sci. Eng. A.* 2007, vol 443, 224-228.

[24] Lu, L.; Lai, M. O. ; Yeo, J. L. *Compos. Struct.* 1999, vol 47, 613-618

[25] Albiter, A. ; Conteras, A.; Bedolla, E. ; Perez, R. *Composites Part A.* 2003, vol 34, 17-24.

[26] Yi, H. ; Ma, N.; Li, X.; Zhang, Y.; Wang, H. *Mater Sci. Eng. A.* 2006, vol 419, 12-17.

[27] Kumar, S.; Subramanya Sarma, V.; Murty, B. S. *Mater Sci. Eng. A.* 2007, vol 465, 160 -164.

[28] Tjong, S. C. ; Wang, G. S. ; Mai, Y.W. *Mater Sci. Eng. A.* 2003, vol 358, 99-106.

[29] Brarin, I.; Knacke, O. Thermochemical Properties of Inorganic Substances, Springer, Berlin, 1989.

[30] Tee, K.L.; Lu, L.; Lai, M.O. *Mater Sci. Eng. A.* 2003, vol 339, 227 -231.

[31] Tjong, S. C.; Wang, G. S. *Adv. Eng. Mater.* 2004, vol 6, 964-968.

[32] Chawla, N.; Habel, U.; Shen, Y. L.; Andres, C. ; Jones, J. W.; Allison, J .E. *Metall Mater Trans. A.* 2000, vol 31, 531-539.

[33] Lorca, J.; Poza, P. *Acta Metall Mater.* 1995, vol 43, 3959-3969.

[34] Vyletel, G. M.; Allison, J. E.; van Aken, D. C. *Metall Mater Trans. A.* 1993, vol 24 2545 -2557.

[35] Han, N. L.; Yang, J. M.; Wang, Z .G *Scripta Mater*. 2000, vol 43, 801-805.

[36] Han, N. L.; Wang, Z. G.; Yang , J. M.; Zhang, G. D. *Mater Sci. Eng. A.* 2002, vol 337, 140 -145.

[37] Tjong, S. C.; Wang, G. S. *Mater Sci. Eng. A.* 2004, vol 386, 48 -53.

[38] Tjong, S.C.; Lau, K. C. *Mater Sci. Eng. A.* 2000, vol 282, 183 -186.

[39] Tjong, S.C.; Lau, K. C. *Mater Lett.* 2000,vol 43, 274-280.

[40] Ma, Z. Y.; Tjong, S. C. *Mater Sci. Eng. A.* 2000, vol 284, 70 - 76.

[41] Zarrinfar, N.; Kennedy, A. R.; Shipway, P.H. *Scripta Mater*. 2004, vol 50, 949 -952.

[42] Biselli, C.; Morris, D. G. ; Randall, N. *Scripta Metall. Mater*. 1994, vol 30, 1327-1331.

[43] Tjong, S.C.; Wang, G.S. *J. Mater Sci.* 2006, vol 41, 5263 -5268.

[44] Sahay, S.S.; Ravi Chandran, K. S.; Atri, R. *J. Mater Res.* 1999, vol 14, 4214-4223.

[45] Panda, K; Ravi Chandran, K. S. *Acta Mater.* 2006, vol 54,1641-1657.

[46] Ravi Chandran, K. S; Panda, K.B, Sahay, S. S. *JOM.* 2004, vol 56, 42-48.

[47] Ma, Z.Y; Tjong, S. C. *Scripta Mater.* 2000, vol 42, 367-373.

[48] Ma, Z.Y.; Tjong, S.C.; Meng, X.M. *J. Mater Res.* 2002, vol 17, 2307 -2313.

[49] Ni, D.R.; Geng, L.; Zhang, J.; Zheng, Z.Z. *Scripta Mater*. 2006, vol 55, 429-432.

[50] Ni, D.R.; Geng, L.; Zhang, J.; Zheng, Z.Z. *Mater Sci. Eng.* 2008, in-press.

[51] Malakondaiah, G.; Rao, P.R. *Acta Metall.* 1981, vol 29, 1263-1268. .

In: Powder Metallurgy Research Trends
Editors: L. J. Smit and J. H. Van Dijk

ISBN: 978-1-60456-852-3

Chapter 9

P/M PROCESSING OF HIGH-STRENGTH ALUMINIUM BY MECHANOSYNTHESIS AND SINTERING

J. Cintas, F.G. Cuevas and J.M. Montes
Metallurgy and Materials Engineering Group, University of Seville,
Camino de los Descubrimientos s/n, 41092 Sevilla, Spain

ABSTRACT

A simple route to produce high strength bulk nanostructured aluminium is presented in this work,. Firstly, a thermally stable nanocomposite Al powder is prepared by attrition milling of elemental Al in the presence of ammonia gas or urea. Both substances are dissociated during milling, enriching the nitrogen and carbon content of aluminium powder.

Then, the milled powder is consolidated to full density by a conventional press-and-sinter powder metallurgy (P/M) technique. Sinterability was improved by using copper powder, as additive, during milling. Shaped specimens possess a stable microstructure, with a grain size of about 150 nm, and their ultimate tensile strength (550MPa) surpass that of most commercial Al alloys, especially at high temperatures.

This new consolidation method may be of interest with a view to manufacturing large batches of small parts, a typical task for the automotive industry.

Keywords: Aluminium, mechanical alloying, mechanosynthesis, sintering, high-strength aluminium, high-temperature aluminium.

1. INTRODUCTION

Mechanical alloying (MA) is a high-energy milling process capable of producing microstructurally fine composite metal powders. In recent years it has acquired a widespread recognition as a technique whereby chemical reactions and alloying are induced in a variety of powder systems [1, 2]. Supersaturated solid solutions, even of immiscible species, non-equilibrium phases, amorphous materials, and nanostructured powders can be successfully obtained by high-energy milling processing [3-5]. It allows, through mechanosynthesis, the dispersion of fine ceramic second phases, such as oxides and carbides, in a metal matrix. The use of mechanical alloying for the mechanochemical synthesis of materials has aroused the interest of many researchers in recent years [6]. One of the most salient features of mechanical alloying is that it can trigger reactions that usually require a high temperature at near-ambient levels.

Mechanically alloyed Al powders (MA-Al) can be prepared by attrition milling. During milling, Al particles undergo severe plastic deformation and are brought into intimate contact forming cold welds. With continued plastic deformation, the hardness of the powder increases and repetitive fracture occurs. A process control agent (PCA), usually a wax powder [7], is added to prevent excessive welding of aluminium to itself, balls, impeller and attritor vessel, and to establish a dynamic balance between fracturing and welding. An additional task of the PCA is to react with aluminium to form carbides. MA Al particles are actually nanograined composite particles [8], that contain a fine and homogeneous distribution of submicroscopic ceramic (Al_2O_3, Al_4C_3) dispersoids [9].

The properties of MA Al powders are very sensitive to experimental milling conditions [10]: milling method, energy input, powder mass, number and mass of the milling balls, addition of PCA, atmosphere, etc. Many factors, hence, affect the kinetics of milling and can determine effects such as the time needed to successfully complete a milling process [11-13], provoke reactions [11], modify the reaction times [11, 12], change the degree of material deformation [11, 12], set limits to the powder size [14] or to get amorphisation of intermetallics [15].

MA-Al powders are hard and homogeneous, and possess a roughly equiaxial morphology. Their particle surfaces are covered by very stable oxide and hydroxide layers. The presence of a surface oxide film is common to Al-base powders. The oxide thickness on Al atomised powders, for instance, is around 5 nm [16]. The oxide layer of any Al-base powder is a barrier to sintering, since it inhibits material transport and the formation of necks between metal particles. Hence, the oxide film must be disrupted to ensure close enough metal-metal contacts. Some rupture of the Al oxide film occurs during pressing, and local metal contacts are created. The extent of these contacts is very much depending on the plasticity of metal particles. In this respect, MA Al powder lacks plasticity. Both characteristics (surface oxide films and hardness) make MA Al powders very difficult to sinter. Therefore, MA Al powder is typically consolidated by a complicated process, which includes a stage of hot extrusion as the main processing step (figure 1, top) [17]. Alternatively, PM consolidation can be achieved using a double cycle involving cold pressing and vacuum sintering (figure 1, middle). The second method has been developed in our laboratory at the University of Seville [18-20].

In this work, the process to obtain consolidated MA-Al specimens by using a single cycle of cold pressing and sintering is presented. These components possess a high mechanical strength at both ambient and high temperature. To this end, aluminium powder of a high hardness and sinterability was prepared by mechanical alloying. The material was strengthened by using a methane and/or ammonia atmosphere (or an additive such as urea) during milling. This not only favoured the typical dispersion of oxides, from the outer layers of Al particles, and carbides, formed by reaction with the PCA, but also strengthened the Al matrix further with dispersed carbides and/or nitrides obtained by mechanosynthesis. Sinterability was improved by using copper powder, as additive, during milling. In this way, the advantages of liquid-phase (supersolidus) sintering in the Al–Cu system are exploited to avoid the problems posed by the presence of oxide layers on the Al particles. The technique, just described, allows MA-Al powders to be consolidated in a manner similar to that employed by the PM industry, i.e. by cold pressing and sintering (see figure 1, bottom).

The new consolidation method may be of interest with a view to manufacturing large batches of small parts, a typical task for the automotive industry.

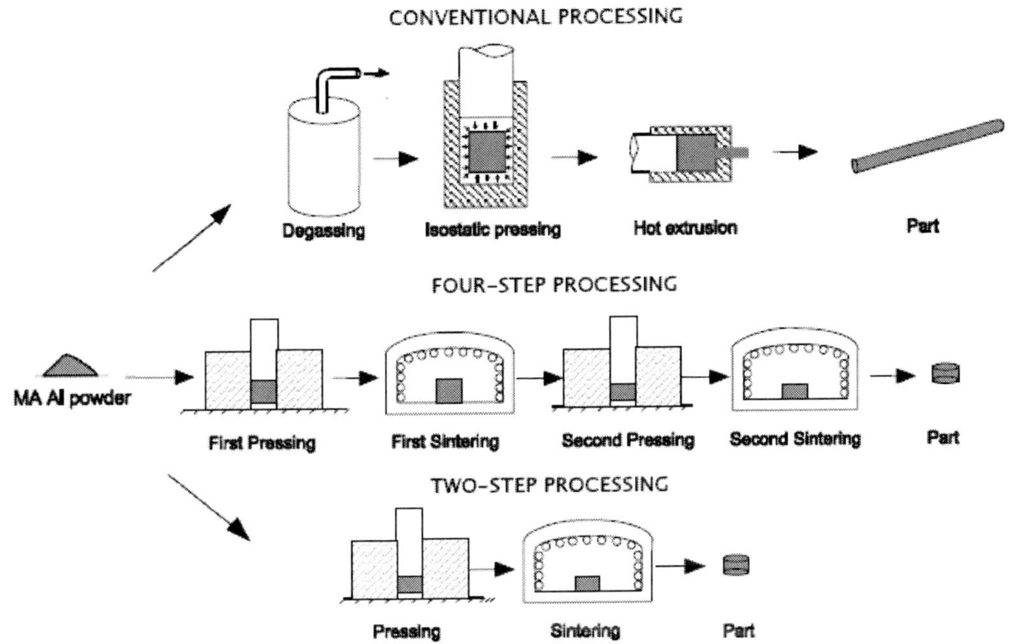

Figure 1. Conventional and alternative processing of MA-Al powder.

2. Materials and Experimental Procedure

Atomised elemental aluminium powder (ECKART), 99.7%Al, was used as starting material (AR Al). The powder was milled for 10 hours in a vertical attritor. EBS wax ($H_{35}C_{17}CONHC_2H_4NHCOC_{17}H_{35}$) was used both as PCA, during milling, and as die-wall lubricant during cold pressing. The miller was water cooled during MA and the powders were milled during 10 h at 500 rpm. Charge ratio (wt. balls / wt. powder) was 50:1. Table 1, summarizes the main milling conditions employed. A more comprehensive description can be found elsewhere [7, 21].

Powders were characterized concerning particle size, grain size, presence of second phases and powder compressibility. Milled powders were consolidated (figure 1, bottom) by using a single cycle of cold uniaxial pressing (850 MPa) and vacuum (5 Pa) sintering (650 °C, 1 h), followed by furnace cooling. The EBS wax was also employed as die-wall lubricant during pressing. Both cylindrical (12 mm diameter and mass ≈4 g), and flat "dog-bone" shaped tensile specimens (32 × 4 × 4 mm) were prepared for density, hardness, microhardness and tensile strength tests. Tensile specimens are in agreement with the corresponding MPIF [22] and ASTM [23] standards

Table 1. Milling conditions

Vessel volume	1400 cm^3
Charge ratio = balls / powder	50:1
Mass of powder	72 g
PCA percentage	3%
Rotor speed	500 rpm
Milling atmosphere	CH_4 and/or NH_3 (1.3 10^5Pa) Vacuum (5 Pa)
Refrigeration	Water at 28 °C
Milling time	10 h

Characterization and structural studies of powders and compacts were carried out by laser diffraction (Malvern Mastersizer 2000), X-ray diffraction (XRD, Bruker D8 Advance, Cu-Kα), optical microscopy (Nikon Epiphot TME 200), scanning electron microscopy (SEM, Phillips XL30), transmission electron microscopy (TEM, Phillips CM200) and an image analysis software (MIP-4). Both electronic microscopes are equipped with an EDAX energy dispersive X-ray spectrometer (EDS).Grain size values, derived from XRD patterns, were obtained by applying the Rietveld procedure [24] and Williamson-Hall [25] and Langford methods [26].

3. Main Results

3.1. Vacuum Milling

In order to show the effect of methane, ammonia and urea, some milling tests were conducted in vacuum (5 Pa), for comparison. The materials milled in vacuum were designated MA Al V.

Vacuum milling resulted in rounding and decreasing in size of the powder particles, from an average starting diameter of 44 μm (Figure 2), to 16 μm after milling (Figure 3). Due to the small size of the as-milled powder particles, it was not possible to measure its microhardness. Nevertheless, a compressibility test is an indirect method to evidence the inherent hardness of powder particles. Though the compressibility of powders depends on several powder characteristics, such as particle size, shape and distribution [27, 28], powder plasticity plays the main role. A compressibility curve is the plot of the relative green density versus the applied compaction pressure. This plot allows comparison of the hardness of different powders, especially when their size and shape characteristics are similar. The compressibility curve is also useful in choosing the correct compaction pressure, i.e. the desired green density.

The compressibility curves of the different powders are shown in figure 4. MA Al V is less compressible (hardest) than the unmilled powder. From this information, a pressure of 850 MPa was chosen to compact all the milled powders. This pressure can be employed in industrial plants, and permits production of green compacts with a relative density of around 87 %. At this level of densification, the porosity in the green compacts is of the interconnected type, and this facilitates gas evacuation during the sintering process. The soft AR Al powder reaches the same relative green density (87 %) when pressed at 150 MPa.

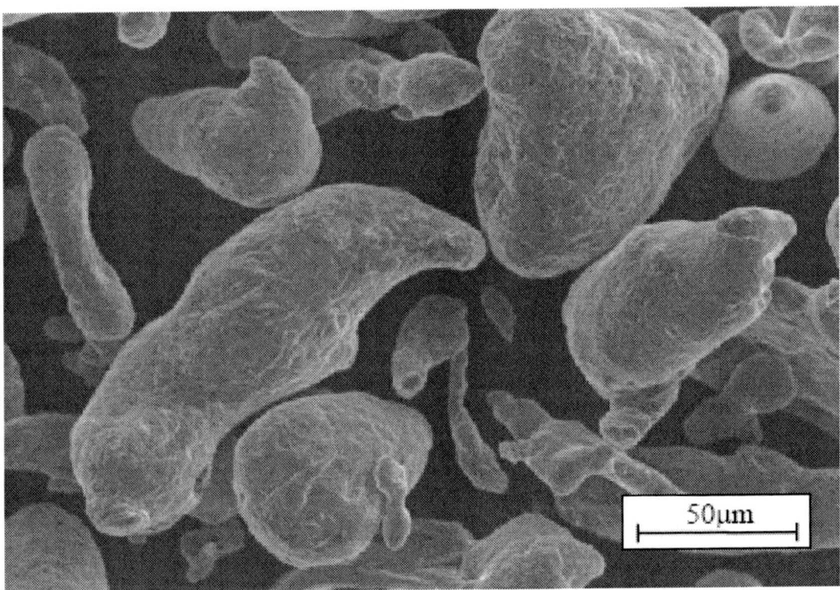

Figure 2. Shape of the AR Al powder (SEM).

Figure 3. Shape of the MA Al V powder (SEM).

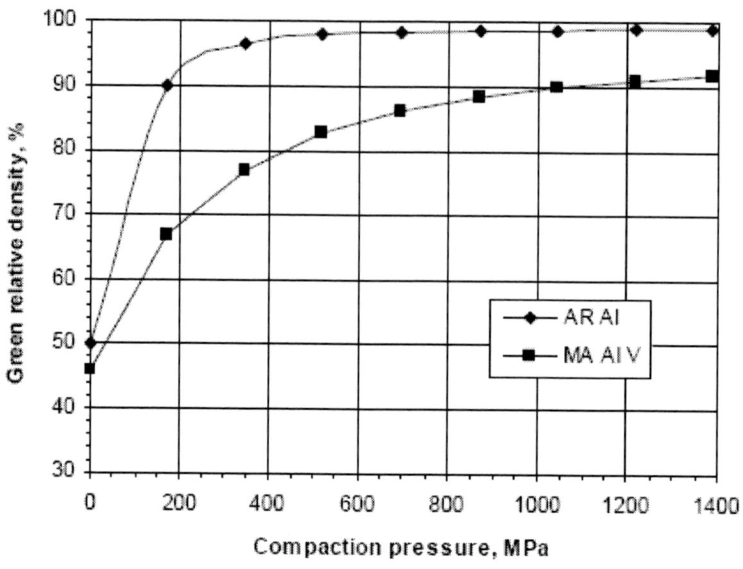

Figure 4. Compressibility curves for AR Al and MA Al V powders.

The high hardness attained by the vacuum milled powders is the result of the strengthening effect produced by plastic deformation, grain-size refinement and the presence of dispersoids or solute atoms. The chemical composition of the as-received and milled Al powders is shown in table 2. The main impurity (Fe) in the starting Al powder remains practically constant after milling. So, contamination of the milled powders can be disregarded. The increase of the percentages of oxygen, carbon, nitrogen and hydrogen during milling can be mainly attributed to reaction of Al powder with the PCA. The total oxygen content partially comes from the oxide film covering Al particles.

Table 2. Composition (wt.%) of powders

MATERIAL	Fe	O	C	N	H
AR Al	0.19	0.14	0.01	0.00	0.02
MA Al V	0.20	0.61	2.31	0.15	0.41

Figure 5 shows the XRD patterns for MA Al V in the as-milled and in the as-sintered conditions. The milled powder only exhibits the typical reflections for aluminium, whereas the sintered material also exhibits peaks associated with the formation of second phases. Specifically, there are peaks of Al_4C_3, formed from carbon present in the PCA, and small peaks of crystalline δ-Al_2O_3. The mean crystallite size, as determined by applying the procedure of Rietveld and the methods of Williamson–Hall and Langford to the XRD patterns, is *ca.* 540 nm. The consolidated MA Al V material, as observed by transmission electron microscopy (figure 6), consist of aluminium grains of roughly equiaxial morphology and Al_4C_3 particles in the form of small rods.

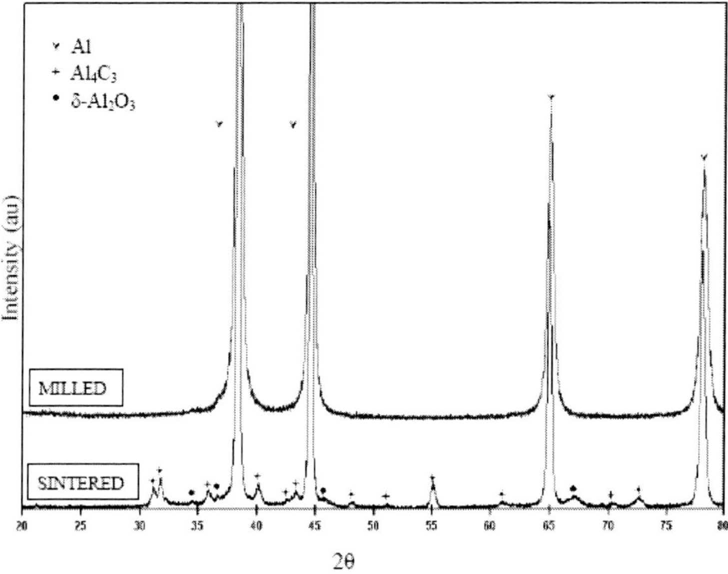

Figure 5. X ray patterns of as-milled and as-sintered MA Al V.

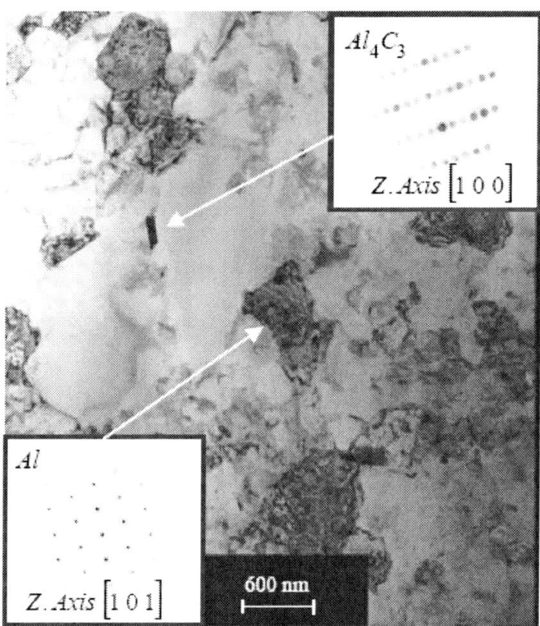

Figure 6. TEM bright-field image of sintered MA Al V. Diffraction patterns used to identify phases are also included.

3.2. Milling in a Methane Atmosphere

Methane (CH_4) is a carbon rich gas usually employed in steel carburizing. In order to increase carbon content in aluminium powder, some millings using a methane atmosphere were conducted. For aluminium to be strengthened by effect of dispersed carbides, CH_4 must be decomposed during milling. This is rather difficult as methane is a highly stable gas. However, milling involves violent collisions that produce high local temperatures; also, it alters the free energy of particles, mainly through lattice defects. Moreover, milling is a chaotic process and its outcomes are often unpredictable.

Different methods have been used in this work to try to dissociate methane during milling. Among these methods are: (i) the use of catalysts (Ni y Cu) [29] to lower the activation energy for the decomposition of methane; (ii) increasing the collision energy by using balls of a heavy material (widia); and (iii) raising the temperature inside the vessel by exploiting the exothermal nature of some reactions occurring during milling (e.g. the formation of Al_2O_3 by reaction of Al with oxygen, as residual gas in the vessel).

Any attempt in this direction failed. The methane behaved as an inert gas in all cases. However, some interesting results were obtained. Such was the case, for instance, in the investigation on nickel powder as potential catalyst for the decomposition of methane. As can be seen in table 3, both the relative density (D) and the hardness (BH) of the sintered compacts obtained from powders milled with addition of 1% Ni (designated MA Al M-1Ni) or without nickel (MA Al M), are similar. Nevertheless, their tensile properties differ markedly in some respects. Thus, the addition of Ni causes the ultimate tensile strength (UTS) to drop from 302 to 231 MPa and the elongation (E), from 1.1 to 0.2%. The poorer properties of the material milled in the presence of nickel can be ascribed to the dramatic microstructural changes induced by this metal.

Table 3. Properties of consolidated compacts, obtained from aluminium powders milled with or without nickel

MATERIAL	D, %	BH, kp/mm^2	UTS, MPa	E, %
MA Al M	97	96	302	1.1
MA Al M-1Ni	97.5	102	231	0.2
MA Al M-1Ni-S	97.3	105	327	3.2

As can be seen in figure 7, the microstructure of the final MA Al M compacts, after sintering at 650 °C for 1 h, consists of an aluminium matrix and needle-shaped particles of the Al_3Fe intermetallics (mean size 11 μm). The iron is an impurity (0.2% Fe) in the starting AR Al powder and forms this second phase during compacts sintering. On the other hand, the sintered compacts obtained from Al powder milled in the presence of 1% Ni were found to contain large layers, a few millimeters long, of an Al-Ni intermetallic compound, figure 8. This brittle intermetallics was identified by XRD as Al_3Ni. This intermetallics is the origin of the poor ultimate strength of this material, since fracture propagates across it (figure 9).

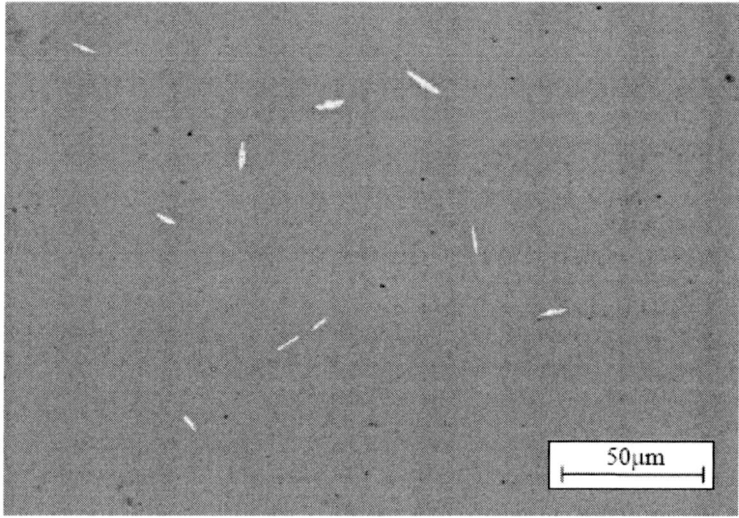

Figure 7. SEM-BSE micrograph of MA Al M sintered compacts.

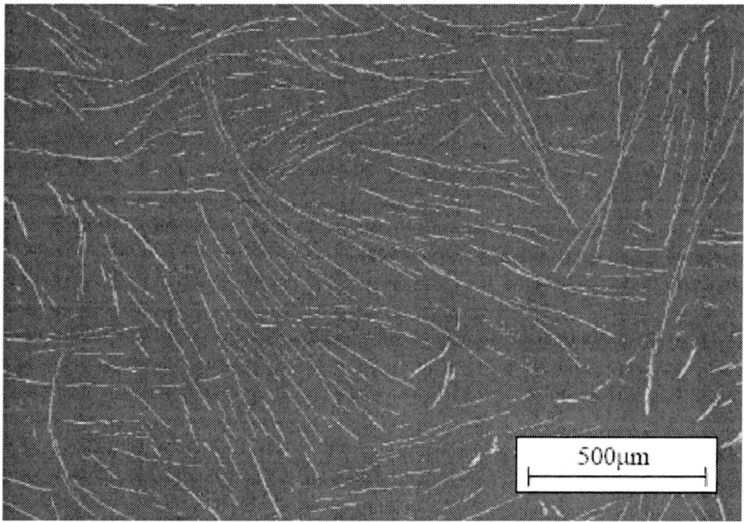

Figure 8. SEM-BSE micrograph of MA Al M-1Ni sintered compacts.

It would be very desirable to change the size and shape of the $NiAl_3$ intermetallics. The appearance of this intermetallics could be explain through the Al-Ni phase diagram (figure 10). Sintering of the MA Al M-1Ni powders, carried out at 650 °C, *i.e.*, about ten degrees above the eutectic temperature (639.9 °C) is of the supersolidus type [30]. During the cooling stage, the high diffusivity of nickel in aluminium ($D_{650°C}$ = 25.082*10^{-9} cm/s) [31] and its insolubility at subeutectic temperatures facilitate the formation of plate-like $NiAl_3$. Intermetallics formed during the heating stage, in the sintering process, disappear after heating at 650 °C for 1 h. These events can be confirmed by water-quenching the samples immediately after reaching 650 °C, and after staying at this temperature for 1 h. The first samples show particles of the intermetallic compound. In the second ones, no intermetallic particles at all are observed.

The undesirable size and shape of the intermetallic compound $NiAl_3$ can be changed using a two-step sintering instead the conventional one. In this unusual method, the first sintering step is performed at temperatures below the eutectic point (639.9 ºC) for a fairly long time (54 min). It is mainly intended to provoke the formation of the stable $NiAl_3$ phase (figure 10) as homogeneously distributed particles. The second sintering step, which is conducted at 650 ºC for a much shorter time (6 min), involves liquid-phase sintering. This improves particle binding and decreases porosity. The duration of this step must be strictly controlled to avoid complete dissolution of the intermetallic nuclei formed during the first solid-phase sintering step. In the absent of remaining intermetallic nuclei, large plate-like intermetallics appear during cooling.

Sintering in two steps facilitates the formation of Al_3Ni spheroids of small size (1.5 μm) and well distributed in the matrix (figure 11). This structure is very different from that the material obtained in one-step sintering, who shows large layers about one millimeter long, figure 8. Two-step sintering produces an improvement in tensile properties (table 3). In this way, a brittle (figure 12) and weak material (MA Al M-1Ni: UTS = 231 MPa, E = 0.2%) is converted into a ductile (figure 13) and strong material (MA Al M-1N-S: UTS = 327 MPa, E = 3.2%).

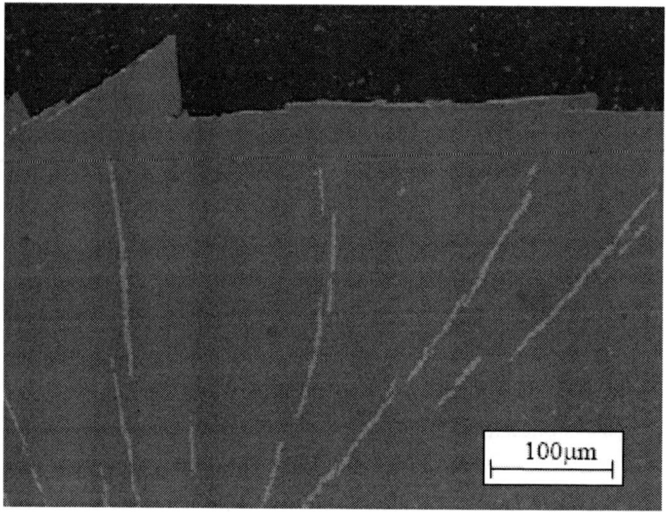

Figure 9. Fracture profile of MA Al M-1Ni sintered compacts as observed by SEM-BSE.

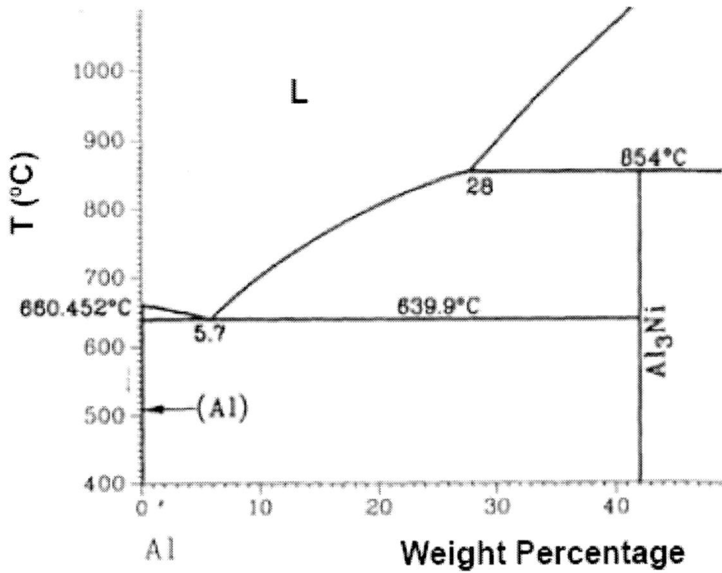

Figure 10. Al-rich portion of the Al-Ni diagram.

Figure 11. Microstructure of MA Al M-1Ni-S as observed by SEM–BSE.

The compacts treated following the two-step sintering process show a ductile fracture when observed by SEM, figure 13. The different fracture behaviour between the materials treated by the standard sintering and the materials processed by two-step sintering can easily be explained by comparison of the final microstructure obtained in each case, figures 8 and 11.

Figure 12. SEM-SE microfractograph of MA Al M-1Ni, showing cleavage surfaces.

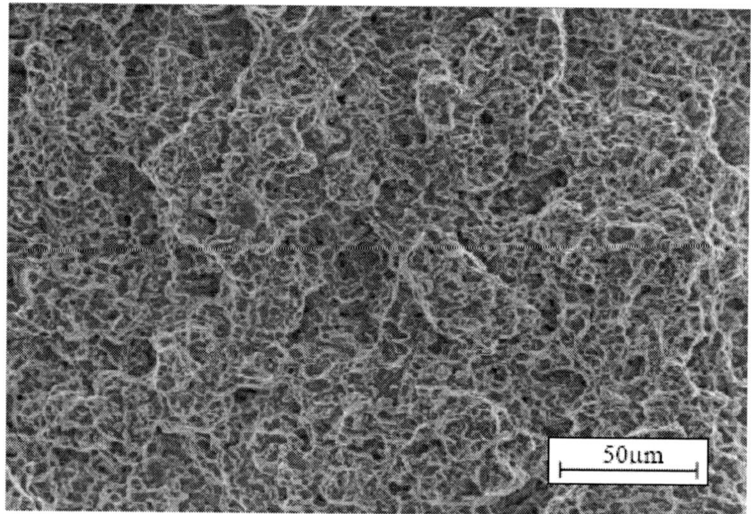

Figure 13. SEM-SE microfractograph of MA Al M-1Ni-S, showing abundant dimples.

3.3. Milling in an Ammonia Atmosphere

Gaseous ammonia is much less stable than methane and can help strengthen Al by incorporating dispersed nitrides. This was checked by milling aluminium in an NH_3 atmosphere (MA Al A).

Milled powders were analysed by XRD to determine the phases formed during milling. The diffraction patterns of both powders, milled in vacuum or ammonia gas, only exhibit the typical reflections for aluminium. That means that crystalline second phases are not present or they are bellow the limits of detection. Another possibility is that the nitrogen atoms are in solid solution. Aluminium grain size of the milled powders was measured by XRD and TEM techniques. Mean crystallite size, as determined by applying the procedure of Rietveld [24] and the methods of Williamson–Hall [25] and Langford [26] to the XRD patterns, is *ca.* 25 nm. This nanostructured nature of milled powders was supported by TEM examination after powder was embedded in Spurr's resin and ultramicrotomed.

Figure 14 shows the XRD pattern for consolidated MA Al A, as well as that for sintered specimens of vacuum milled aluminium (MA Al V), for comparison. Interestingly, the aluminium carbide (Al_4C_3), present in MA Al V, was virtually absent in MA Al A. In fact, it was replaced by aluminium oxycarbides (Al_3CON) and oxynitrides (Al_5O_6N). It should be noted that these dispersoids were effectively incorporated by milling in ammonia gas at ambient temperature. Aluminium-nitrogen compounds of the same family have also been obtained by cryomilling with N_2 [32].

As can be seen in the TEM image of figure 15, aluminium grain size is much smaller for the MA Al A sintered compacts than for those obtained from powder milled in the absence of ammonia. The pinning action of the dispersed carbides and oxides particles, in the MA Al V compacts, highly restrains grain growth during sintering. In MA Al A, the additional formation of aluminium carbonitrides and oxycarbonitrides results in an even greater effect on grain growth restriction. Thus, the equivalent mean diameter of the aluminium grains, as measured by TEM, decreases from 550 nm in MA Al V to 200 nm in MA Al A. Both PM products are ultrafine grained materials. The last one (MA Al A) is at the borderline between ultrafine-grained and nano-grained materials. This fine structure has remained despite the high processing temperature (650 ºC), very near the melting point of aluminium (660.5 ºC).

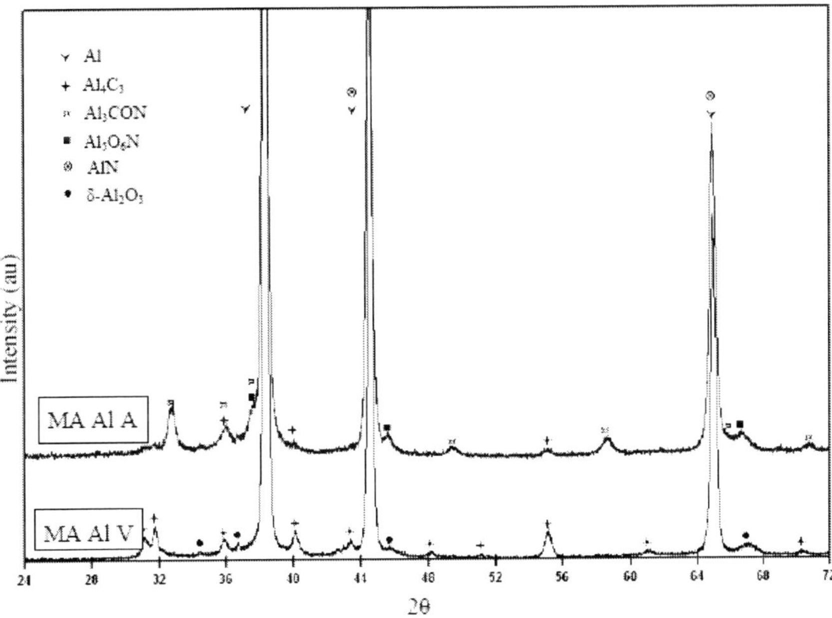

Figure 14. XRD patterns of MA Al A and MA Al V sintered specimens.

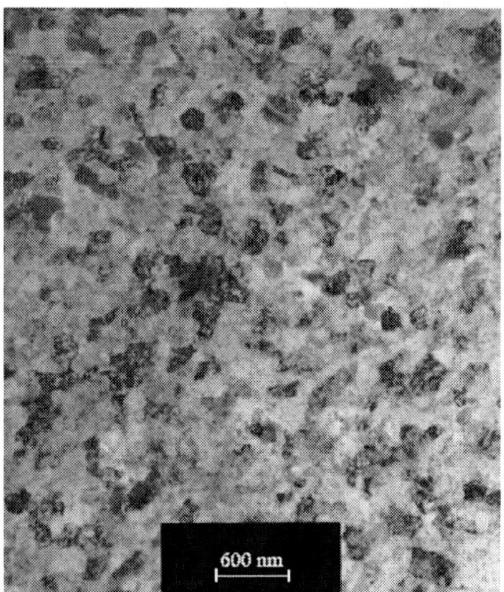

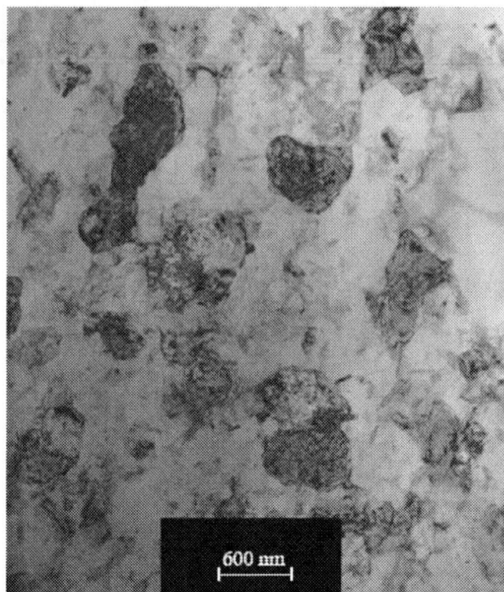

Figure 15. Aluminium grains in sintered compacts of MA Al A (left) and MA Al V (right). The different grain size of both materials is evident.

Comparing properties of the MA Al A and MA Al V specimens, table 4, the former exhibit the best values of densification, hardness and ultimate strength. Preliminary studies conducted in the authors' laboratory suggest that the good sinterability could be related to the presence of aluminium nitride or oxycarbonitrides on the surfaces of the milled powder [34]. This is supported by the observation that the surface tension of liquid aluminium decreases upon partial reaction with nitrogen gas, forming AlN [35].

The high hardness and strength of the MA Al A material results from the combined effect of a very fine grain size (200 nm), the presence of abundant nano-dispersoids and a full densified PM product. The ultimate tensile strength of the MA Al A compacts (515 MPa) surpasses that of most commercial aluminium alloys, and approaches that of the high-strength wrought Al alloys, such as the 7075-T6 [36] (UTS = 572 MPa). The outstanding UTS of the MA Al A is especially remarkable considering that the commercial Al materials are strongly alloyed and treated by ageing (precipitation hardening). However, the elongation of this PM material is low (0.6%) and it must be improved.

Table 4. Properties of MA Al V and MA Al A sintered compacts of

MATERIAL	D, %	BH, kp/mm^2	UTS, MPa	E, %
MA Al V	97	96	302	1.1
MA Al A	100	163	515	0.6

3.3. Milling in the Presence of Urea

The use of ammonia to strengthen aluminium proved highly effective. Nevertheless, its highly toxic and corrosive nature led us to explore its replacement with urea, $CO(NH_2)_2$, which is widely used as a fertilizer and more inexpensive to purchase and handle than ammonia.

Milling tests in the presence of urea were conducted in vacuum in order to avoid the side effect of the milling atmosphere. Although the amount of urea used ranged from 0.5 to 5%, only the results obtained with 2.1% urea (the amount of nitrogen equivalent to that supplied by an atmosphere consisting of 100v% NH_3) are discussed here. Such materials were designated MA Al U.

As can be seen in figure 16, the XRD pattern for the MA Al U compacts shows a slightly decreased in Al_4C_3 content in comparison with the MA Al V compacts. However, the intensity decrease is smaller than in the case of MA Al A. The MA Al U pattern includes reflections for Al_3CON; nevertheless, such reflections are weaker than in MA Al A. The reflections for Al_5O_6N are even weaker, practically imperceptible.

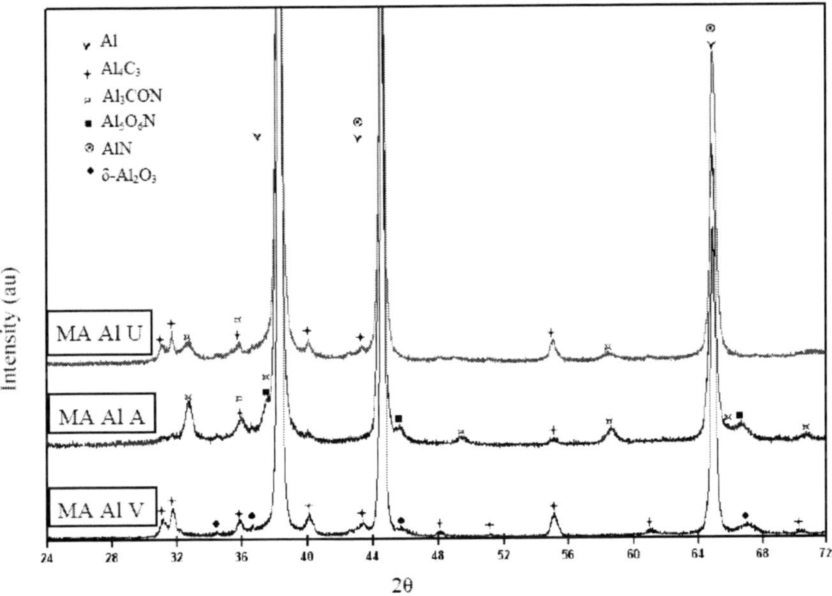

Figure 16. XRD patterns of MA Al V, MA Al A and MA Al U sintered compacts.

Both MA Al U and MA Al A were found to be structurally very similar by TEM, their grain size being about 200 nm (figures 15 and 16, left). These materials, due to the small grain size, can be considered ultrafine grained materials, or even nanostructured materials. This is a remarkable result since both Al-base materials were sintered at 650 °C for 1 hour. TEM examination at higher magnification reveals the presence of many spheroidal particles, roughly 30 nm in size (figure 16, right). In principle, such particles might correspond to the oxycarbonitrides, oxynitrides and/or nitrides detected by XRD.

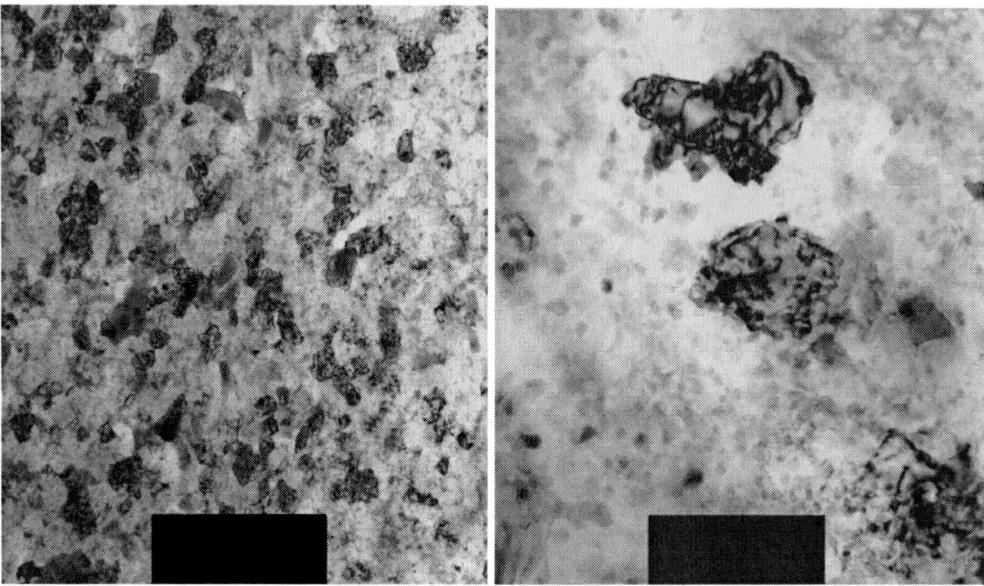

Figure 1. TEM bright-field images of MA Al U sintered compacts showing Al grains (left) and also non-fully identified second phases (right),.

A comparison of the hardness and tensile strength of the MA Al U compacts with those of the MA Al A compacts (table 5) reveals that urea constitutes an effective replacement for ammonia; in fact, it additionally resulted in a similar compact density and a higher tensile strength (550 MPa *versus* 515 MPa). These differences can be ascribed to the chemical composition of urea - $CO(NH_2)_2$ - that forms carbon and/or oxygen compounds, in addition to the nitrogen compounds. Both MA Al U (0.7%) and MA Al A (0.6%) elongations are very low, but they can be improved, as described below.

Table 5. Properties of MA Al V, MA Al A and MA Al U sintered compacts

MATERIAL	D, %	BH, kp/mm^2	UTS, MPa	E, %
MA Al A	100	163	515	0.6
MA Al U	100	185	550	0.7

3.4. Improving Ductility

In order to improve sinterability of mechanically alloyed aluminium base powders, a small amount of elemental copper powder was added during milling. The use of copper as milling additive produces a good distribution of Cu in Al. This gives rise liquid-phase sintering, of the supersolidus type, during sintering at 650 °C. The liquid thus formed can flow between the aluminium and the alumina layer and strip the latter, as it has been reported for the case of elemental Al powder [37]. In this way, metal particle binding can be enhanced.

Compacts were prepared from aluminium powder milled with additions of copper ranged from 0.5 to 1.5 % Cu. Milling was carried out in an ammonia atmosphere. For reference, Al and Cu were also milled in vacuum in some experiments. Milled powders were consolidated according to the standard procedure, i.e. cold uniaxial pressed at 850 MPa, and sintered at 650 °C during 1 h.

The liquid phase formed during sintering is highly homogeneous and produces a network that facilitates binding of the residual particles that remained solid. For example, this effect is apparent from figure 17 (right), which corresponds to a consolidated sample, coming from an Al powder milled with 1% Cu in an ammonia atmosphere (MA Al A-1Cu). All Cu-containing sintered compacts show this feature.

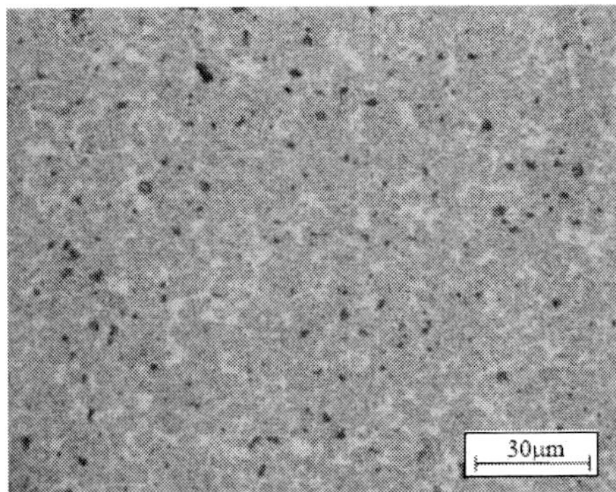

Figure 17. Optical micrographs of MA Al V (left) and MA Al A-1Cu (right) sintered compacts. A light-grey "liquid-phase" network is observed in the right micrograph.

The presence of Cu induces other microstructural changes. Thus, the MA Al V microstructure, which is similar to the MA Al M one (figure 7), shows needle-shaped Al_3Fe intermetallics, 11 μm long. In contrast, consolidated samples containing copper show two types of intermetallic compounds of smaller particle size (*ca.* 1.5 μm). They were identified, by XRD and EDX, as Al_2Cu and Al_3Fe. The Al_2Cu particles are rosette-shaped, and the Al_3Fe particles are relatively needle-shaped. However, the needles of these Al_3Fe particles are rounder, *i.e.* less sharp (figure 18), than in the MA Al V samples (figure 17, left). Under a metallurgical point of view, the microstructure of the Cu-containing sintered compacts is more desirable than that of the Cu-absent ones (figure 17, left).

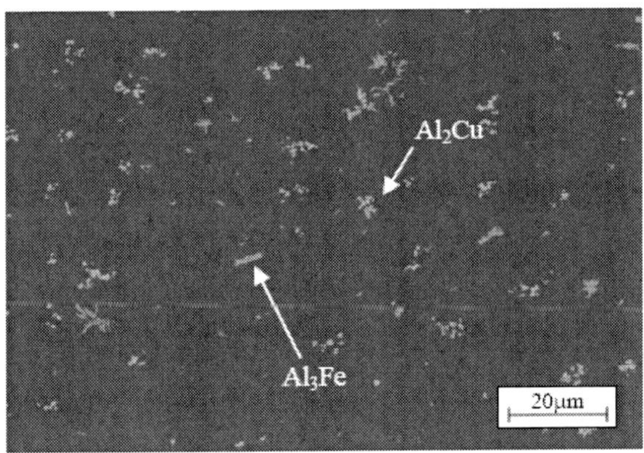

Figure 18. Microstructure of MA Al A-1Cu compacts showing two different second phases.

The enhanced cohesion of the matrix in the Cu-contaning sintered samples, due to the "liquid-phase" network, and the changes in morphology, size and nature of the intermetallic compounds, improves the tensile properties. Figure 19 shows the UTS and E values of compacts prepared from aluminium powder milled with different amounts of copper (ranging from 0 to 1.5 %Cu), in the presence of ammonia gas,. In all cases, increasing the weight percentage of Cu, decreases the strength and increases the ductility. It should be noted that only an addition of 0.5% Cu increases the elongation from 0.6% to 2%. In addition, samples containing 0.5 %Cu have a high UTS (around 540 MPa). With 1.5 % Cu, elongation increases up to 2.5%, with a tensile strength of 480 MPa.

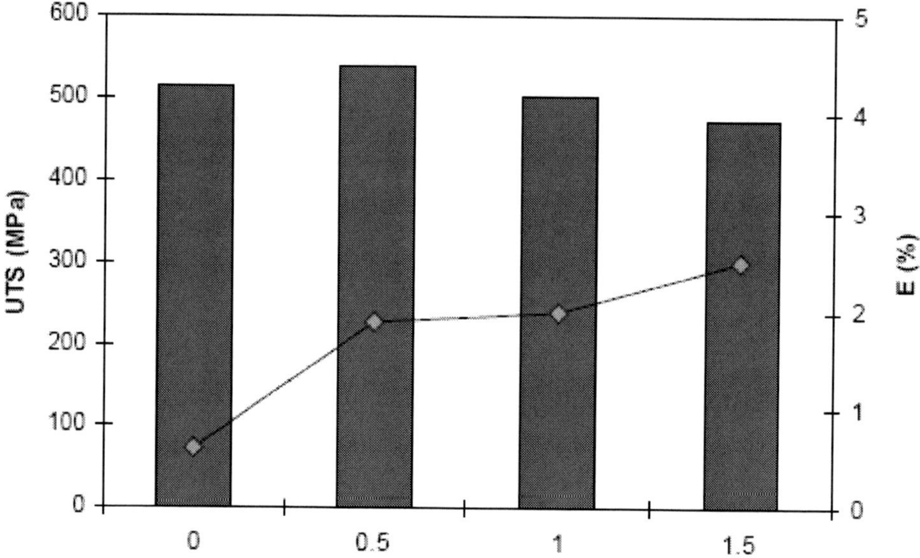

Figure 19. UTS (bars) and elongation (lines) of samples as function of copper content. Powders were milled in 100v%NH_3 atmosphere.

These properties, particularly the ultimate strength, surpass those of most commercial aluminium alloys, including the conventional Al-wrought alloys [38], as shown in table 6. The high UTS of the developed PM alloys is more remarkable, when it is considered that the commercial Al materials are strongly alloyed and treated by ageing. Nevertheless, the PM materials have, generally, lower ductility.

Table 1. Strength and elongation of commercial casting and wrought Al alloys

ALLOY	UTS, MPa	E, %
EN AC-21100-T6	300	3
EN AC-45300-T6	280	<1
EN AW-2014-T6	440	6
EN AW-6082-T6	310	6
EN AW-7075-T6	510	7

3.4. High Temperature Performance

One of the greatest weaknesses of commercial aluminium alloys is their low mechanical strength at high temperatures. The use of commercial aluminium alloys is limited to applications below around 150 °C. The reason for this is that strengthening of these alloys is based on non-stable second phases. For instance, at 260 °C, the UTS of two well-known high-strength wrought Al alloys, 7075-T6 and 2024-T6, is only 75 MPa [36], due to overaging.

A method to improve aluminium high temperature strength is reinforcing the Al matrix with submicroscopic refractory dispersoids. This is the approach followed in this work, where the amount of dispersoides of conventional MA Al materials has been increased, by milling in ammonia gas or urea. For example, figure 20 shows the UTS and E values at 25 and 250 °C for compacts prepared from Al powder milled with 0.5 %Cu in ammonia gas (MA Al A 0.5Cu). This material has UTS values higher than 325 MPa at 250 °C, being its high-temperature behavior much better than that of the conventional high-strength Al alloys. Concretely, the developed PM material is, at 250 °C, about 4.5 times stronger than high-strength commercial alloys (table 6).

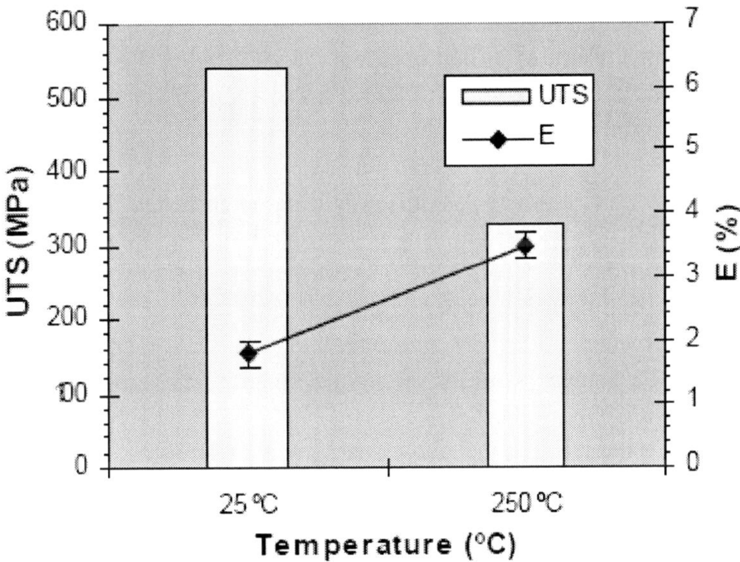

Figure 20. Tensile properties, at 25 and 250 °C, of sintered compacts of aluminium powder milled in ammonia gas with 0.5 %Cu.

These outstanding properties are more remarkable considering the simplicity of the powder consolidation method used (*viz.* cold uniaxial pressing at 850 MPa + sintering at 650 °C for 1 h). On the other hand, the new processing technique allows to obtain large numbers of specimens without subsequent machining. These advantages have led the authors to develop two patents on the described processes [39, 40].

Table 2. UTS, at 25 and 250 °C, of two commercial wrought Al alloys and one of the PM materials obtained in this work

ALLOY	UTS, MPa 25 °C	UTS, MPa 250 °C
EN AW-2014-T6	440	66
EN AW-7075-T6	510	75
MA Al A 0.5Cu	539	327

4. CONCLUSION

The purpose of this work was to explore the possibility of obtaining mechanically alloyed aluminium materials consolidated by cold pressing and sintering. MA-Al powder was prepared by attrition milling in the presence of 3% EBS, as PCA, for 10 h. Milling was carried out in an atmosphere either of methane and/or ammonia. Besides, some milling experiments were carried out in vacuum with addition of 0.5 to 5% urea, instead of ammonia gas. In some tests, a small amount of copper powder (0.5–1.5 wt.%) was also added during milling to improve sinterability of milled powders. Mechanically alloyed Al powders were, normally, consolidated by cold pressing at 850 MPa and vacuum sintering at 650 °C for 1 h. Results attained lead to the following conclusions:

- Methane behaves as an inert gas during milling, even in the presence of 1%Ni as a potential catalyst for its decomposition. Results, however, have revealed interesting features concerning the control of microstructures and properties of the sintered specimens, depending on the sintering procedure.
- Aluminium powder milled in ammonia atmosphere, or vacuum milled in the presence of 2.1% urea, exhibits very good sinterability, so that sintered samples reach almost full densification.
- The sintered materials, in the above paragraph referred to, possess an ultra-fine grained structure. Such materials can be consider nearly nano-structured materials, since there grain size is around 200 nm. This is largely the result of the presence of abundant submicroscopic dispersoids of Al_4C_3, Al_2O_3, Al_3CON and $Al_{2.81}O_{3.56}N_{0.44}$. In consequence, these materials have a high hardness (163–185 BH) and tensile strength (515–550 MPa). However, their elongations are low (*ca.* 0.6%).
- The ductility of the sintered specimens, coming from powders milled either in ammonia gas or with urea, can be improved by adding a small amount of Cu powder during milling. This induces supersolidus sintering and improves the properties. For instance, with 1.5 %Cu, the elongation increases up to 2.5%, the UTS being 480 MPa. A broad spectrum of UTS-elongation combinations can be obtained by changing the proportion of Cu powder. Thus, the addition of 0.5% Cu powder provides an UTS of 540 MPa and an elongation of 2%.
- The Cu-containing materials also exhibit a good combination of strength and elongation at 250 °C. Thus, the specimens containing 0.5%Cu have, at this

temperature, a strength of 327 MPa and an elongation of 3.5%. Such a high strength can hardly be achieved in most conventional alloys.

- On the other hand, recent work aimed at developing an ensuing industrial process, has allowed to simplify even more the proposed technique. So, (i) the milling time has been reduced to only 5 h, (ii) the cold pressure, to 700 MPa, and (iii) the sintering time, to 20 min. Finally, the substitution of nitrogen gas for the vacuum, as sintering atmosphere, is currently investigated.

ACKNOWLEDGEMENTS

The financial support of the MEC/FEDER, Madrid, through the research project ENE2006-13267-C05-05 is gratefully acknowledged.

5. REFERENCES

[1] Zakeri, M.; Allahkarami, M.; Kavei, Gh.; Khanmohammadian, A.; Rahimipour, M.R. *J. Mat. Sci.* 2008, *43 (5)*, 1638-1643

[2] Lü, L.; Lai, M.O. *Mechanical alloying*; Kluwer Academic Publishers: Boston, MA, 1998.

[3] Bakker, H.; Zhou, C.F.; Yang, H. *Prog. Mater. Sci.* 1995, *39*, 159-241.

[4] Suryanarayana, C. *Prog. Mater. Sci.* 2001, *46*, 1-184.

[5] Il Moon, K.; Sub Lee, K. *J. Alloys Comp.* 2002, *333 (1-2),* 249-259.

[6] Gutman, E.M. *Mechanochemistry of materials*, Cambridge International Science Publishing: Cambridge, UK, 1998.

[7] Cintas, J.; Montes, J. M.; Cuevas, F. G.; Gallardo, J.M. *Mat. Sci. Forum* 2005, *514-516*, 1279-1283.

[8] Herrero, J.L.; Rodríguez, J.A.; Herrera, E.J. In *Proceedings of the 1998 Powder Metallurgy World Congress*; EPMA: Bellstone, UK, 1998; Vol. 1, pp. 384-389.

[9] Singer, R.F.; Oliver, W.C.; Nix, W.D. *Metall. Trans. A* 1980, *11*, 1895-1901.

[10] Rodríguez, J.A.; Gallardo, J.M.; Herrera, E.J. *J. Mater. Sci.* 1997, *32*, 3535-3539.

[11] Schaffer, G.B; Mccormick, P.G. *Metall. Trans. A* 1992, *23*, 1285-1290.

[12] Schaffer, G.B; Forrester, J.S. *J. Mat. Sci.* 1997, *32*, 3157-3162.

[13] Zhang, H.; Liu, X. *Int. J. Refr. Met. and Hard Mat.* 2001, *19*, 203-208.

[14] Ryu, H.J.; Hong, S.H.; Baek, W.H. *J. Mat. Processing Techn.* 1997, *63,* 292-297.

[15] Park, Y.H.; Hashimoto, H.; Watanabe, R. *Mat. Sci. Forum* 1992, *88-90*, 59-66.

[16] Naranjo, M; Rodríguez, JA; Herrera, EJ. *Scripta Mater.* 2003, *49*, 65-70.

[17] Nachtrab, W.T.; Roberst, P.R. In *Advances in Powder Metallurgy and Particulate Materials*; MPIF: Princeton, NJ, 1992; Vol. 4, p. 321.

[18] Cintas, J.; Rodríguez, J.A.; Gallardo, J.M.; Herrera, E.J. *Revista de Metalurgia* 2001, *37(1)*, 370-375.

[19] Rodríguez, J.A.; Gallardo, J.M.; Herrera, E.J. *Mater. Trans., JIM* 1995, *36 (2)*, 312-316.

[20] Fuentes, J.J.; Rodríguez, J.A.; Herrera, E.J. *Mater. Sci. Forum* 2003, *426-432*, 4331-4336.

[21] Cuevas, F.G.; Cintas, J.; Montes, J.M.; Gallardo, J.M. *J. Mater. Sci.* 2006, *41*, 8339-8346.

[22] MPIF Standard 10, MPIF: Princeton, NJ, 2002.

[23] ASTM A370, ASTM: West Conshohocken, PA, 2003.

[24] Young R.A. *The Rietveld method;* Oxford Univ Press: New York, 2000.

[25] Williamson, G.K.; Hall, W.H. *Acta Metall.* 1953, *1*, 22-31.

[26] Langford, J.I. *J. Appl. Cryst.* 1978, *11*, 10-14.

[27] German RM. *Particle packing Characteristics*; MPIF: Princeton, NJ, 1989, p. 238.

[28] *ASM Handbook*; ASM International: Materials Park, OH, 2002; Vol. 7, p. 302.

[29] Meng-Sheng, L.; Quian-Er, Z. J. *Molecular Catalysis A* 1998, *136 (2)*, 185-194.

[30] German, R.M. *Liquid Phase Sintering*, Plenum Press: New York, 1985.

[31] *Smithells Metals Reference Book*, 6th ed.; Editor Brandes, E.A.; Butterworth: London, 1983; pp. 13-15.

[32] Zhou, F.; Rodriguez, R.; Lavernia, E.J. *Mat. Sci. Forum*, 2002, *386-388*, 409-414.

[33] Cintas, J.; Cuevas, F.G.; Montes, J.M.; Herrera, E.J. *Scripta Mater.* 2005, *53*, 1165-1170.

[34] Saravanan, R.A.; Molina, J.M.; Narciso, J.; García-Cordobilla, C.; Lewis, E. *Scripta Mater.* 2001, *44*, 965-970.

[35] *ASM Handbook*; ASM International: Materials Park, OH, 2002; Vol. 2, p. 114.

[36] Schaffer, G.B.; Sercombe, T.B.; Lumley, R.N. *Mater. Chem. Phys.* 2001, *67*, 85-91.

[37] UNE-EN 586-2; AENOR: Madrid, 2002.

[38] Herrera, E.J; Cintas, J.; Rodríguez, J.A. Nitruration of powders by reactive milling in the presence of some nitrogen compounds (in Spanish), Patent P2003-01963, 2003.

[39] Cintas, J.; Herrera, E.J; Rodríguez, J.A., Manufacturing of aluminium-based composite materials by mechanosynthesis and hot consolidation (in Spanish), Patent P2003-01964, 2003.

INDEX

B

C

D

E

F

G

H

I

N

O

P

Q

R

S

T

U

V

W

X

Y

Z